FOREWORD

This volume contains a portion of the over 240 ASME papers which were presented at the 32nd National Heat Transfer Conference held in Baltimore, Maryland, August 8–12, 1997. For over 40 years, the National Heat Transfer Conference has been the premiere forum for the presentation and dissemination of the latest advances in heat transfer. The work contained in these volumes range from studies of fundamental phenomena to applications in the latest heat transfer equipment.

This year's conference included over 350 papers presented in sessions sponsored by The American Society of Mechanical Engineers (ASME), The American Institute of Chemical Engineers (AIChE), The American Institute of Aeronautics and Astronautics (AIAA), and The American Nuclear Society (ANS). This represents the work of hundreds of dedicated individuals (authors, session organizers, and reviewers), to whom your colleagues extend their sincerest thanks.

Thanks are especially due to the dedicated staff of ASME, without whose tireless effort the Conference and these volumes would not be possible.

Matthew D. Kelleher
Program Chair

CONTENTS

FUNDAMENTALS OF BUBBLE AND DROPLET DYNAMICS

Introduction

Satish G. Kandlikar
Rochester Institute of Technology
Rochester, New York

Cristina H. Amon
Carnegie Mellon University
Pittsburgh, Pennsylvania

Many papers presented in this session are based on extensive numerical modeling of bubbles of drops, and in most cases, supporting experimental results verifying the numerical predictions. The emphasis in most cases is on modeling. The problems considered include film boiling, binary pool and flow boiling., bubble nucleation and growth, evaporation of water droplets, pressure drop in subcooled flow boiling, bubble shapes in the presence of electric fields, acoustic breakup of liquid jets, solidification of droplets, holdup in a jet bubble column, and interactions between two buoyant bubbles. Even though topics are quite diverse, these papers reflect the growing trend in the investigation of bubble and droplet dynamics in which modeling and numerical studies have taken an important role in providing further understanding of the complex underlying phenomena in different situations.

Son and Dhir present a further investigation on numerical modeling of film boiling at high heat fluxes. Their earlier model was able to correctly predict the nodes and antinodes of a Taylor wave from where bubbles are released. At higher heat fluxes, their current work indicates that vapor jets are formed on both nodes and antinodes, and bubbles are released from the top of vapor columns.

Kandlikar presents a two-part paper on modeling binary heat transfer. In Part I, the pool boiling heat transfer is modeled by considering the effect of diffusion on bubble growth rates. The model provides a predictive method without incorporating any empirical constants. The model is able to predict the pool boiling data for a number of diverse binary systems.

In Part II, Kandlikar extended the pool boiling model developed in Part I to predict the flow boiling heat transfer for binary mixtures. Different levels of heat transfer suppression were observed, and the convective boiling was seen to dominate in the severely suppressed regions with further reductions in the nucleate boiling contributions. The results compared well with the experimental data available in the literature.

Kandlikar et al. present the results of an experimental study on bubble nucleation and growth characteristics of subcooled flow boiling conditions. The effect of flow around the nucleating bubble is numerically investigated. Bubble growth rates are seen to be influenced considerably due to the imposed flow.

The effect of the transient radial conduction on CHF is numerically investigated by Watwe and Bar-Cohen. The presence of a substrate is shown to improve the CHF in thin heaters. Their study also highlights the effect of the wettability of the surface on CHF.

The evaporation rates of a small water droplet with potassium acetate and sodium iodide additives are investigated by King et al. The water droplets are placed on a heated stainless steel surface. A simple model is proposed to model the evaporation rates.

Fleischer et al. developed an analytical model to predict the influence of bubble formation on pressure drop in uniformly heated vertical channels. The model also predicts a threshold heat flux below which flow instability may not occur during flowrate reductions.

Hader et al. present the result of a numerical investigation conducted to study the effect of the bubble shape on the flow field and heat transfer for bubbles of one dielectric fluid suspended in another dielectric fluid. Their results indicate that the bubble shape is an important parameter affecting the flow field and heat transfer.

Mitra-Majumdar et al. studied the turbulent air-water flow through a jet bubble column and obtained numerical and experimental results for gas holdup. In addition, their results show the inadequacy of using a single bubble diameter in modeling the flow.

Esmaeeli and Arpaci have modeled the interactions of two buoyant, deformable and interacting bubbles in a superheated liquid. The numerical solutions provide an insight into the bubble shapes as a function of their separation distance and density ratio.

The impact of a molten metal droplet on a solid surface and its solidification have been investigated numerically by Delplanque. His results show that internal circulation within the droplet has negligible effects while the droplet deformation should be considered in the solidification process.

Breakup of the liquid sheet emerging from a conical spray as a result of acoustic modulation is investigated by Chung et al. They presented the results of the breakup length as a function of the driving frequency and input power for liquids with different viscosities.

Yang and Seng numerically analyzed the effect of Marangoni convection around a vapor bubble under micro and earth-gravity conditions. Their results indicate that Marangoni convection is enhanced by the interface evaporation, and the isotherms are strongly modified around a bubble at high values of Marangoni number.

We wish to thank the authors for their contributions in improving our understanding of the fundamentals underlying the behavior of the bubbles and droplets. We also would like to thank our reviewers for their efforts in improving the quality of the papers. We hope the papers presented under these sessions will encourage the researchers to address the unresolved issues in this area.

HTD-Vol. 342, National Heat Transfer Conference
Volume 4
ASME 1997

NUMERICAL ANALYSIS OF FILM BOILING NEAR CRITICAL PRESSURES WITH A LEVEL SET METHOD

Gihun Son and Vijay K. Dhir

Mechanical and Aerospace Engineering Department
University of California, Los Angeles
Los Angeles, California 90095

ABSTRACT

Recently, attempts have been made to numerically simulate film boiling on a horizontal surface. It has been observed from experiments and numerical simulations that during film boiling the bubbles are released alternatively at the nodes and antinodes of a Taylor wave. Near the critical state, however, hydrodynamic transition in bubble release pattern has been reported in the literature. The purpose of this work is to understand the mechanism of the transition in bubble release pattern through complete numerical simulation of the evolution of the vapor-liquid interface. The interface is captured by a level set method which is modified to include the liquid-vapor phase change effect. It is found from the numerical simulation that at low wall superheat the interface moves upwards, bubbles break off and the interface drops down alternatively at the nodes and antinodes. However, with increase in wall superheat stable vapor jets are formed on both the nodes and antinodes and bubbles are released from the top of the vapor columns. The numerical results are compared with the experimental data and visual observations reported in the literature and found to be in good agreement with the data.

NOMENCLATURE

c_p = specific heat at constant pressure
$\vec{g}$ = gravity vector
Gr = Grashof number, $(\rho_v^2 g l_o^3/\mu_v^2)(\rho_l/\rho_v - 1)$
H = step function
h = grid spacing
h_{fg} = latent heat of evaporation
k = thermal conductivity
l_o = characteristic length, $\sqrt{\sigma/g(\rho_l - \rho_v)}$
$\vec{m}$ = mass flux vector defined in Eq. (11)
Nu = Nusselt number, $l_o q_w/k_v \Delta T$
p = pressure
Pr = Prandlt number, $c_{pv}\mu_v/k_v$
q = heat flux
R = radius of each circular region
r = radial coordinate, $R - |x|$
Re = Reynolds number, $\rho_v u_o l_o/\mu_v$
$\vec{S}_u$ = source term vector
T = temperature
ΔT = temperature difference, $T_w - T_{sat}$
u = x-directional velocity
$\vec{u}$ = velocity vector, (u, v)
$\vec{u}_i$ = interfacial velocity vector
u_o = characteristic velocity, $\sqrt{g l_o}$
v = y-directional velocity
v_{fg} = $\rho_v^{-1} - \rho_l^{-1}$
x = horizontal coordinate
y = vertical coordinate
β = the ratio of sensible heat to latent heat, $c_{pv}\Delta T/h_{fg}$
β' = $\beta/(1 + 0.5\beta)$
δ = vapor film thickness
κ = interfacial curvature
λ_{d2} = 2-dimensional "most dangerous" wavelength
μ = viscosity
ρ = density
σ = surface tension
ϕ = level set function

Subscripts

l, v	= liquid, vapor
sat, w	= saturation, wall
t	= partial differentiation with respect to t
$x,\ y$	= partial differentiation with respect to $x,\ y$

INTRODUCTION

In recent years significant efforts have been made to understand bubble dynamics associated with phase change processes, such as film boiling, condensation on the underside of a surface and melting or sublimation of a substrate placed beneath a pool of heavier liquid.

Berenson(1961) developed a semi-empirical model of saturated film boiling on a horizontal surface. He assumed that vapor bubbles were placed on a square grid with a spacing equal to two-dimensional "most dangerous" Taylor wavelength, $\lambda_{d2}(= 2\pi\sqrt{3\sigma/g(\rho_l - \rho_v)})$, and a thin vapor film of uniform thickness connected the neighboring bubbles. By further assuming that the mean bubble height and bubble diameter were proportional to the bubble spacing and two bubbles were supported per λ_{d2}^2 area of the heater, he predicted Nusselt number for film boiling as

$$Nu_B = 0.425(Gr\ Pr/\beta)^{1/4} \tag{1}$$

The predictions from Eq. (1) were found to compare well with Berenson's data. However, Berenson's model did not account for the time variation of either the film thickness and the bubble height or bubble diameter and the flow field in the liquid.

Dhir et al.(1977) studied pseudo film boiling during sublimation of a slab of dry ice placed beneath a pool of warm liquid. The heat transfer rate was determined by noting a change in the enthalpy of the overlying liquid pool. Data were found to compare favorably with prediction from an equation similar to Eq. (1) when the lead constant was reduced to 0.36. It was argued that the reduction of about 15% in the lead constant resulted from the fact that during sublimation only one bubble was supported per λ_{d2}^2 area instead of $\lambda_{d2}^2/2$ as assumed by Berenson. This was based on their experimental observation that the bubbles were released from the same location rather than alternately from the nodes and antinodes as observed during film boiling on flat plates or during early period of sublimation.

Film boiling on a horizontal platinum wire in water near its critical state was investigated by Reimann and Grigull(1975). They observed that the bubbles were released in different ways depending on the magnitude of heat flux from the wire. At low heat fluxes, discrete bubbles were released from the heater, which is similar to the observations during film boiling at one atmosphere pressure. However, with increase in heat flux, vapor left the heater in the form of jets. Subsequently, similar transitions were observed by Dhir and Taghavi-Tafreshi(1981) during dripping of a liquid from underside of a horizontal tube.

Klimenko(1981) carried out a somewhat generalized analysis of film boiling on horizontal flat plates. Employing a basic formulation similar to that of Berenson, Klimenko developed a correlation that included data near critical pressures. According to his correlation, Nusselt number for film boiling on an upward facing horizontal surface was expressed as

$$Nu_K = 1.90 \times 10^{-1} Gr^{1/3} Pr^{1/3} f_1; \text{ for } Gr < 4.03 \times 10^5 \tag{2}$$

$$Nu_K = 2.16 \times 10^{-2} Gr^{1/2} Pr^{1/3} f_2; \text{ for } Gr > 4.03 \times 10^5 \tag{3}$$

where

$$\begin{aligned} f_1 &= 1 && \text{for } \beta > 0.71 \\ &= 0.89\beta^{-1/3} && \text{for } \beta < 0.71 \\ f_2 &= 1 && \text{for } \beta > 0.50 \\ &= 0.71\beta^{-1/2} && \text{for } \beta < 0.50 \end{aligned}$$

Equations (2) and (3) suggest that for small values of $\beta(= c_{pv}\Delta T/h_{fg})$, the heat transfer coefficient should vary as $\Delta T^{-1/3}$ or $\Delta T^{-1/2}$ depending on the magnitude of Grashof number, Gr. For high values of β, the heat transfer coefficient is predicted to be independent of wall superheat.

The above described studies have provided us with data and semi-empirical models for description of film boiling on flat plates. However, many of the assumptions made in the models (e.g. constant film thickness between bubbles as assumed in Berenson's model for the heat transfer coefficient in film boiling) remain not validated. Very recently, Son and Dhir(1996) have carried out a complete numerical simulation of the evolution of the vapor-liquid interface during saturated film boiling on a horizontal surface. For an axi-symmetric case they have used a coordinate transformation technique supplemented by a numerical grid generation method. In the method the matching conditions at the interface can be imposed accurately as long as computational grids can be constructed numerically. From the numerical simulation, the film thickness and in turn the heat transfer coefficient are found to vary both spatially and temporally. Another numerical study of film boiling, independent of Son and Dhir's work, has been reported by Juric and Tryggvason(1996). They used a two-dimensional front-tracking method which could handle breaking of the interface. In their method, the interface was described as a transition region of finite thickness rather than as a surface separating two fluids. However, the heat transfer rates predicted from both numerical studies were lower than those predicted by Berenson's correlation. This under-prediction could probably be caused by 3-dimensional effects of the bubble release pattern. Generally, the front-tracking method when the interface is described explicitly is very hard to use for 3-dimensional problems.

In this paper, a level set approach is adopted as a method that can not only handle breaking and merging of the interface but can also be extended easily to three-dimensional problems. This method is appropriate for investigation of a complicated flow such as that occurs during saturated film boiling near critical pressures. The numerical algorithm developed by Sussman et al.(1994) for incompressible two-phase flow is modified to include the effect of liquid-vapor phase change.

NUMERICAL ANALYSIS

In this study, the computations are performed for two-dimensional incompressible flow which is described in axi-symmetric coordinates. The computation domain is chosen as the circular regions around the nodes and antinodes of the Taylor wave as shown in Fig. 1. Each circle has an area $\lambda_{d2}^2/2$ and its origin is located on the center of a node and an antinode where the bubbles are released. By using this geometry it is possible not only to describe vapor bubbles as spherical rather than cylindrical but also to compute the bubbles released alternatively. A horizontal coordinate, x, is defined in the interval of $-R \leq x \leq R$. Here the radius of each circle, R, is evaluated as

$$R = \lambda_{d2}/\sqrt{2\pi} = \sqrt{6\pi\sigma/g(\rho_l - \rho_v)} \quad (4)$$

At the boundary of each circular region, $x = 0$, fluid velocity and temperature are assumed to be continuous. Then, the equations describing the flow in both circular regions can be obtained if r and $\partial/\partial r$ included in the equations describing purely axi-symmetrical flows are replaced by $r = R - |x|$ and $\partial/\partial x$.

The interface separating two phases is captured by a level set function, ϕ, which is defined as a signed distance from the interface: the negative sign is chosen for the vapor phase and the positive sign for the liquid phase.

The equations governing the momentum conservation including gravity and surface tension are written as:

$$\rho(\vec{u}_t + \vec{u}\cdot\nabla\vec{u}) = -\nabla p + \rho\vec{g} - \sigma\kappa\nabla H + \nabla\cdot\mu\nabla\vec{u} + \vec{S}_u \quad (5)$$

where $\vec{u} = (u, v)$ and $\vec{S}_u = (S_u, S_v)$. S_u and S_v are defined as:

$$S_u = r^{-1}(r\mu u_x)_x + (\mu v_x)_y - \mu r^{-2}u$$
$$S_v = r^{-1}(r\mu u_y)_x + (\mu v_y)_y$$

In Eq. (5), σ is the surface tension, H is a step function($H = 0$ for $\phi < 0$ and $H = 1$ for $\phi > 0$) and κ is the interfacial curvature expressed as

$$\kappa = \nabla\cdot\frac{\nabla\phi}{|\nabla\phi|} \quad (6)$$

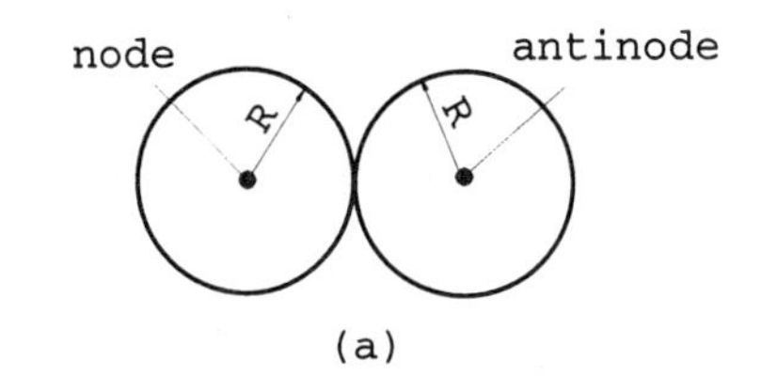

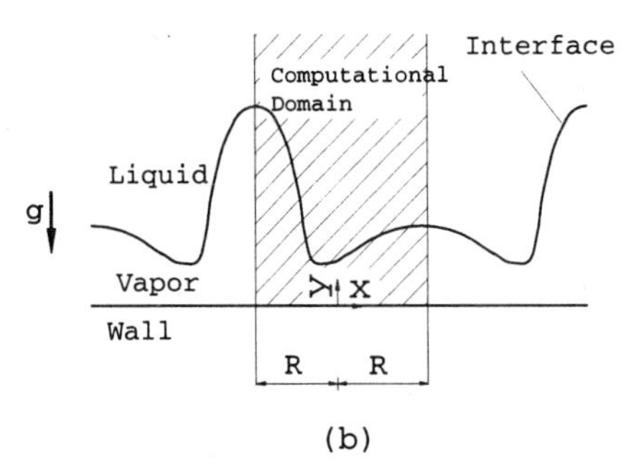

Fig. 1 Two-circular regions used in numerical simulation: (a) top view and (b) side view

The surface tension force, $-\sigma\kappa\nabla H$, is implemented in the volume form to avoid the need for explicitly describing the interface (Brackbill et al., 1992). Also, density and viscosity fields are described as:

$$\rho = \rho_v + (\rho_l - \rho_v)H \quad (7)$$
$$\mu^{-1} = \mu_v^{-1} + (\mu_l^{-1} - \mu_v^{-1})H \quad (8)$$

Patankar(1980) and Alexiades and Solomon(1993) have shown that the viscosity formulation given by Eq. (8) can be more effective in evaluating such transport properties as viscosity and conductivity than that is possible by the form given by Eq. (7).

The energy equation for film boiling is derived to satisfy such a condition that the vapor-liquid interface is maintained at the saturation temperature:

$$\rho_v c_{pv}(T_t + \vec{u}\cdot\nabla T) = \nabla\cdot k\nabla T \quad \text{for } H < 1$$
$$T = T_{sat} \quad \text{for } H = 1 \quad (9)$$

where the effective thermal conductivity, setting $k_l^{-1} = 0$, is evaluated:

$$k^{-1} = k_v^{-1}(1 - H) \quad (10)$$

The equation governing the mass conservation should include the effect of volume expansion due to liquid-vapor phase change. It is derived from the conditions of the mass continuity and energy balance at the interface:

$$\vec{m} = \rho_v(\vec{u_i} - \vec{u_v}) = \rho_l(\vec{u_i} - \vec{u_l}) \quad (11)$$
$$\vec{m} = -k\nabla T/h_{fg} \quad (12)$$

Then,

$$\vec{u_l} - \vec{u_v} = v_{fg}\, \vec{m} \qquad (13)$$

where $v_{fg} = \rho_v^{-1} - \rho_l^{-1}$. Using Eq. (13) in a manner similar to that used for implementation of the surface tension into the momentum equation, the equation governing mass conservation can be formulated as:

$$\nabla \cdot \vec{u} \;=\; v_{fg}\, \vec{m} \cdot \nabla H \qquad (14)$$

The boundary conditions used in this study are as follows:
at the wall($y = 0$),

$$\vec{u} = 0; \qquad T = T_w \qquad (15)$$

at the planes of axi-symmetry($x = \pm R$),

$$u = v_x = T_x = 0 \qquad (16)$$

at the top of computational domain,

$$u_y = v_y = T_y = 0 \qquad (17)$$

In the level set formulation, the vapor-liquid interface is described as $\phi = 0$. The zero level set of ϕ is advanced by the interfacial velocity while solving the following equation:

$$\phi_t = -\vec{u}_i \cdot \nabla\phi \qquad (18)$$

where

$$\vec{u}_i = \vec{u} + \rho^{-1}\, \vec{m} \qquad (19)$$

It is noted by Sussman et al.(1994) that the level set function should be maintained as a distance function for all time to prevent excessive numerical errors. Therefore, at each time step the level set function is reinitialized as a distance function, $|\nabla\phi| = 1$, from the interface by obtaining a steady state solution of the equation:

$$\phi_t = \frac{\phi_o}{\sqrt{\phi_o^2 + h^2}}(1 - |\nabla\phi|) \qquad (20)$$

where ϕ_o is a solution of Eq. (18) and h is a grid spacing. To prevent numerical instability due to discontinuous material properties, a step function is smoothed as:

$$\begin{aligned} H &= 1 && if\ \phi \geq +1.5h \\ &= 0 && if\ \phi \leq -1.5h \\ &= 0.5 + \phi/(3h) + \sin[2\pi\phi/(3h)]/(2\pi) && if\ |\phi| \leq 1.5h \end{aligned}$$

This equation implies that the interface separating two phases is replaced by a transition region of finite thickness. The interface thickness can be maintained as constant for all the time while keeping the level set function as a distance function.

When discretizing the governing equations temporally, the diffusion terms are treated by a fully implicit scheme and the convection and source terms by a first-order explicit method. Then, the discretized governing equations are expressed as

$$\begin{aligned} (\phi^{n+1} - \phi^n)/\Delta t &= -\vec{u}_i^n \cdot \nabla\phi^n && (21) \\ |\nabla\phi^{n+1}| &= 1 \qquad \text{for } \phi \neq 0 && (22) \\ (T^{n+1} - T^n)/\Delta t &= -\vec{u}^n \cdot \nabla T^n \\ &\quad + \nabla \cdot k\nabla T^{n+1}/\rho_v c_{pv} && (23) \\ \rho(\vec{u}^{n+1} - \vec{u}^n)/\Delta t &= -\rho\vec{u}^n \cdot \nabla\vec{u}^n - \nabla p^{n+1} - \sigma\kappa\nabla H \\ &\quad + \rho\vec{g} + \nabla \cdot \mu\nabla\vec{u}^{n+1} + \vec{S}_u^n && (24) \\ \nabla \cdot \vec{u}^{n+1} &= v_{fg}\, \vec{m}^{n+1} \cdot \nabla H && (25) \end{aligned}$$

where superscript n and $n+1$ represent n and n+1 time steps respectively. Fluid properties and a step function are evaluated from ϕ^{n+1}.

In order to obtain the governing equation for pressure which achieves mass conservation, the fractional-step method, or projection method, is used (refer to Bell et al., 1988, Son and Dhir, 1996). In this study, the projection method is formulated in a staggered grid system in which the locations for velocity components are displaced from those for pressure and temperature. This is done to avoid the difficulty caused by pressure boundary conditions. While discretizing the differenential equations spatially, a second-order central difference method is used for the diffusion terms and the interfacial curvature, κ. However, to prevent numerical oscillations a second-order ENO method described by Chang et al.(1995) is adopted for the convection terms and the distance function, $|\nabla\phi| = 1$.

The discretized equations are solved iteratively by a line-by-line TDMA(Tridiagonal-Matrix Algorithm) supplemented by Gauss-Seidel method which was suggested by Patankar(1980). To enhance the rate of iteration convergence, a relaxation factor is obtained from an orthogonal-residual method. Also, the iterative procedure to solve the pressure equation is combined with a sawtooth type of multigrid method (Sonneveld and Wesseling, 1985).

RESULTS AND DISCUSSION

In carrying out numerical simulation, characteristic length, l_o and characteristic velocity, u_o, are defined as

$$l_o = \sqrt{\sigma/g(\rho_l - \rho_v)}; \qquad u_o = \sqrt{g l_o} \qquad (26)$$

Also, the following parameters are defined:

$$Re = \frac{\rho_v u_o l_o}{\mu_v}; \quad Pr = \frac{c_{pv}\mu_v}{k_v}; \quad \beta = \frac{c_{pv}\Delta T}{h_{fg}}; \quad Nu = \frac{l_o q_w}{k_v \Delta T}$$

where $\Delta T = T_w - T_{sat}$ and q_w is a wall heat flux.

First, the computations are made to validate the volume source term, $v_{fg}\, \vec{m} \cdot \nabla H$, that has been included in Eq.

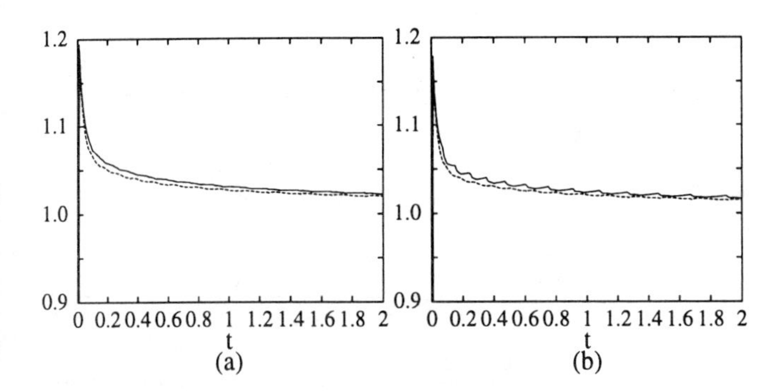

Fig. 2 Ratios of the numerical solutions to the exact solutions for an unsteady one-dimensional phase change problem with 30 grid points: (a) for $\rho_l = 1000\rho_v$ and (b) for $\rho_l = 2\rho_v$. The solid lines and the dashed lines represent the interfacial velocities and the liquid velocities respectively.

(25) to account for liquid-vapor phase change. An unsteady one-dimensional problem with phase change is chosen as a test problem, which can be solved analytically (Alexiades and Solomon, 1993):

$$\rho_v c_{pv} T_t = k T_{yy} \quad \text{for } 0 < y < \delta \qquad (27)$$

$$T = T_{sat} \quad \text{for } y \geq \delta \qquad (28)$$

$$T = T_w \quad \text{at } y = 0 \qquad (29)$$

$$\rho_v \delta_t h_{fg} = -k T_y \quad \text{at } y = \delta \qquad (30)$$

Here, δ is the vapor film thickness and its time-derivative can be obtained analytically:

$$\delta\delta_t = \frac{k_v \Delta T}{\rho_v h_{fg}} \frac{2z}{\sqrt{\pi} erf(z)} \qquad (31)$$

where z is the root of the equation:

$$z e^{z^2} erf(z) = \beta/\sqrt{\pi} \qquad (32)$$

The liquid velocity caused by liquid-vapor phase change is expressed as:

$$v_l = \delta_t \left(1 - \rho_v/\rho_l\right) \qquad (33)$$

For $\beta = 1, Re = 1, Pr_v = 1, \mu_v = \mu_l$ and $\rho_l = 2\rho_v$ or $\rho_l = 1000\rho_v$, the test problem is computed numerically using the algorithm developed in this study. Initially, the dimensionless interface height, δ, is taken to be 0.5 and the temperature profile is assumed to be linear. The numerical results are compared with the exact solutions as shown in Fig. 2. The curves in Fig. 2 represent the ratios of the interfacial velocities and the liquid velocities obtained from the computation to those obtained analytically. The numerical errors are less than 2% at $t = 2$, when using the 30 mesh points, regardless of the density ratios. As the mesh points are increased by a factor of two the errors are reduced to less than 1%.

Table 1 Properties of water near critical pressures

T_{sat} (^oC)	ρ_v/ρ_l	l_o (mm)	μ_v/μ_l	Gr	Pr	c_{pv}/h_{fg} ($1/^oC$)
373.3	0.66	0.20	0.90	2110	147.	4.034
373.8	0.77	0.11	0.96	216	420.	17.60

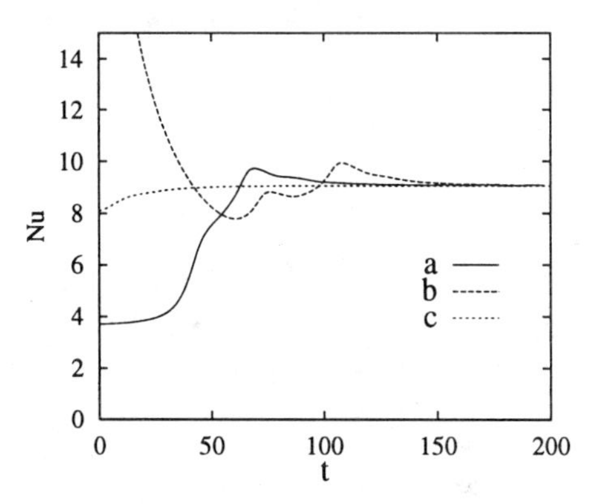

Fig. 3 Dependence of Nusselt number averaged over a cell area on initial conditions for $T_{sat} = 373.3^oC$ and $\Delta T = 30^oC$

During numerical simulations of film boiling near critical pressures, the properties of water listed in Table 1 are used. Initially the vapor-liquid interface is disturbed asymmetrically as:

$$\delta = 0.3 + 0.2 J_o(3.832 r/R) \quad \text{for x} < 0$$

$$= 0.3 + 0.1 J_o(3.832 r/R) + 0.1 J_o(3.832) \quad \text{for x} > 0$$

where J_o is Bessel's function of order 0 and a value of 3.832 is the first root satisfying $dJ_o/dr = 0$. The initial vapor temperature profile is taken to be linear and fluid velocity is set to zero. Since in reality film boiling is a cyclic process, the computations should be carried out over several cycles until the effect of initially specified conditions disappears. For $T_{sat} = 373.3^oC$ and $\Delta T = 30^oC$, the computations are performed with grid points of 128×256 using three different initial conditions for the temperature profile: (a) it is linear, (b) it is uniform at the saturation temperature and (c) it is extrapolated from a nearly steady state solution obtained on coarse grids. Figure 3 shows the dependence of the Nusselt numbers averaged over a cell area on the initial conditions. It is seen that the curves for the three cases attain the same values at dimensionless time of 150. At this time the effect of initial conditions appears to have completely vanished.

To select an appropriate mesh size, convergence for grid resolutions is tested with mesh points of 32×64, 64×128, 128×256 and 256×512. Table 2 lists the Nusselt numbers at a nearly steady state averaged over a cell area for the

Table 2 Dependence of Nusselt number on grid resolutions for $T_{sat} = 373.3^oC$ and $\Delta T = 30^oC$

No. of Grids	Nu
32 x 64	6.09
64 x 128	8.14
128 x 256	9.06
256 x 512	9.10

different mesh sizes. As the mesh points increase the relative difference of Nusselt numbers between successive mesh sizes becomes small. For 128×256 and 256×512 meshes, the difference is less than 1%. Therefore, most of computations in this study are done on 128×256 grid points to save the computing time without losing the accuracy of numerical results. During the computations, time steps are chosen to satisfy the CFL condition, $\Delta t \leq (|u| + |v|)/h$, due to the explicit treatment of the convection terms and the condition that the numerical results should not change if the time steps are halved. An appropriate time step is approximately a dimensionless time of 0.01.

Figure 4 shows the evolution of the vapor-liquid interface for different wall superheats at $T_{sat} = 373.3^oC$. The corresponding saturation pressure is 21.90 MPa, which is 0.99 times the critical pressure for water. It is found from Fig. 4 (a) that at the low wall superheat, $\Delta T = 10^oC$, discrete vapor bubbles are released alternatively at the nodes and antinodes like film boiling at low pressures. After the vapor bubble pinches off, the vapor stem also breaks off as shown at $t = 105$ and $t = 113$. Such a breakoff leads to formation of small secondary bubbles. After bubble breakoff, the interface in the peak region drops down rapidly because of the restoring force of surface tension. When the wall superheat is increased to $\Delta T = 22^oC$, the bubble release pattern changes as shown in Fig. 4 (b). A stable vapor jet starts to be formed on the node ($x = -R$) while the pattern of discrete bubble release still exists on the antinode ($x = R$). At high superheat, $\Delta T = 30^oC$, stable jets are formed on both the node and antinode. Fig. 4 (c) shows that the interface is nearly stationary except the fact that bubbles are released from the upper end of the vapor columns. Also, the interface is found to be almost symmetric with respect to $x = 0$ at the nearly steady state. These hydrodynamic transitions in bubble release pattern are consistent with those observed by Reimann and Grigull(1975). They observed that during film boiling of water on 0.1mm diameter platinum wire at $T_{sat} = 373.3^oC$, vapor left in several different modes as shown in Fig 5. With increase in heat flux, the modes of vapor removal shift from (a) to (d).

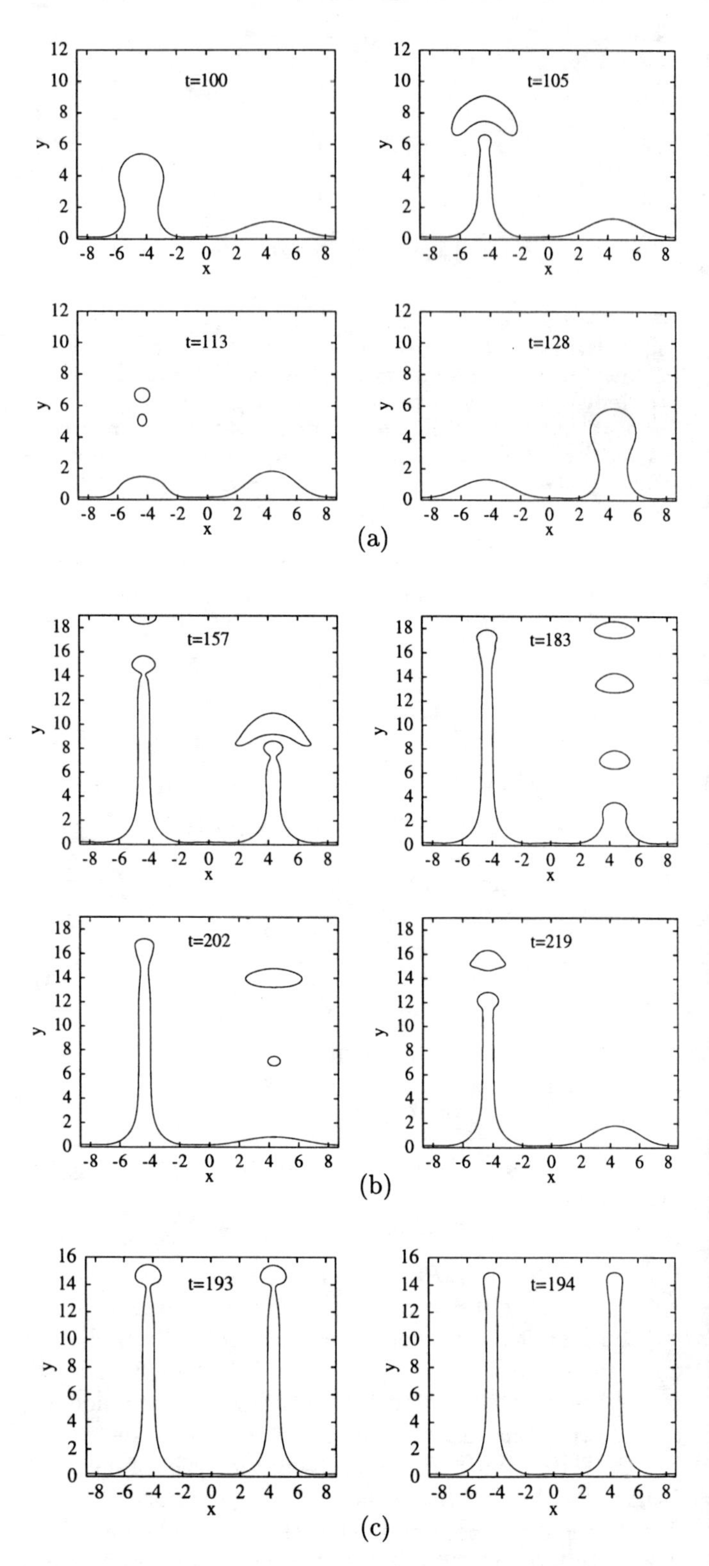

Fig. 4 Evolution of the interface for different wall superheats at $T_{sat} = 373.3^oC$: (a) $\Delta T = 10^oC$, (b) $\Delta T = 22^oC$ and (c) $\Delta T = 30^oC$

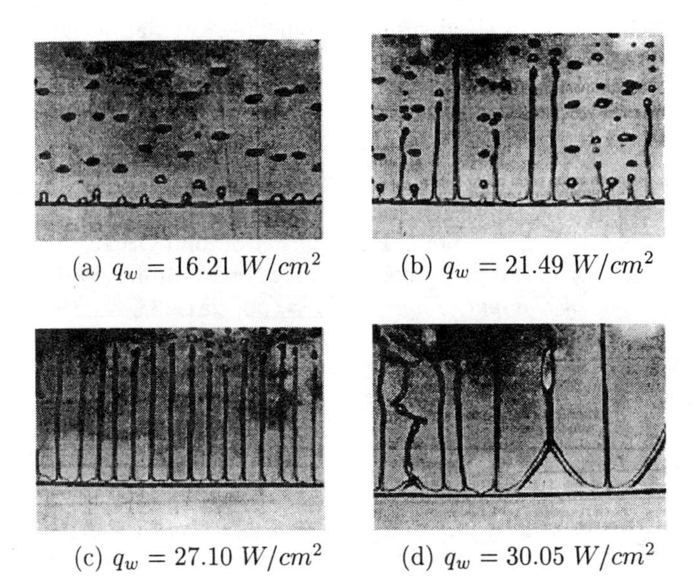

Fig. 5 Hydrodynamic transition in bubble release pattern observed by Reimann and Grigull(1975), which was presented in the paper of Dhir and Taghavi-Tafreshi(1981).

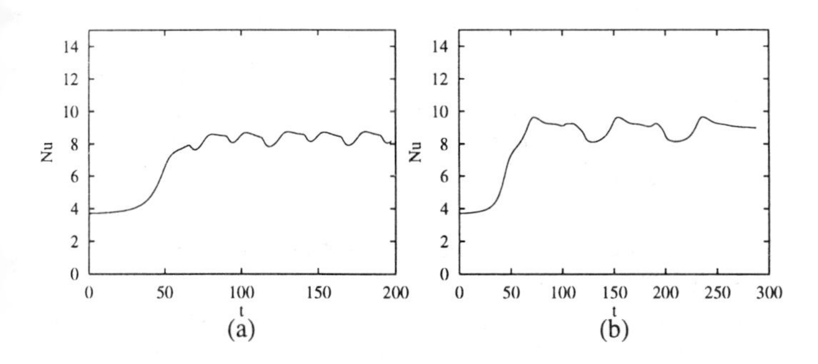

Fig. 6 Nusselt number averaged over a cell area for $T_{sat} = 373.3^oC$: (a) $\Delta T = 10^oC$ and (b) $\Delta T = 22^oC$

The modes (a), (b) and (c) are similar to the bubble release pattern simulated numerically in the present study. It is interesting to note that the heat fluxes for Fig. 5 (b) and (c) are comparable to the heat fluxes, $q_w = 21.6(W/cm^2)$ and $q_w = 29.9(W/cm^2)$ in Table 3, obtained numerically for different bubble release modes though the geometry of boiling surface is not the same.

Nusselt numbers averaged over the cell area for $\Delta T =$ 10 and 22^oC are plotted in Fig. 6 for several cycles. It is seen that the temporal variation of Nusselt numbers is nearly cyclic after the effect of the initial condition disappears. Just after the bubble pinch off, the heat transfer rate decreases. This can be explained by noting that the surface tension acting as a restoring force pushes down the interface at the peak and in turn the interface in the valley region moves upward to conserve vapor volume. Thereafter, the Nusselt number increases again as the peak moves upward and the interface in the valley region moves downward. The Nusselt numbers averaged over one cycle are listed in Table 3. In this table the values of Nu are compared with the predictions from Berenson's model (Eq. 1), Nu_B, and the correlation of Klimenko (Eq. 2), Nu_K. In Eq. 1, $\beta'(=\beta/(1+0.5\beta))$ rather than β is used to account for sensible heat transfer. If we had simply used β instead of β' as originally proposed by Berenson, correlation equation (1) will predict Nusselt numbers much less than those predicted from the present work. This difference is due to the fact that at low pressures sensible heat correction is small in comparison to latent heat of vaporization. But this is not true near the critical pressure. Near the critical pressures, while the Nusselt numbers predicted from the correlations remain almost constant at large β, those predicted numerically increase slightly with ΔT. This increase results from the fact that the formation of stable vapor jets at high wall superheat increases the pressure difference between the peak and the valley region of the interface, which provides more efficient flow passages for vapor removal. Such hydrodynamic transition in bubble release pattern was not considered in Berenson's and Klimenko's models, where the ratio of bubble height to the bubble spacing was assumed to be constant regardless of the wall superheat. It is interesting to note that the Nusselt numbers obtained from the present analysis are bounded by those predicted from Berenson's and Klimenko's correlations. Table 3 shows that the numerical Nusselt numbers are closer to the values obtained from the modified Berenson's model than those predicted by the Klimenko's correlations, which are about 31% $\sim$ 19% higher.

Table 3 Comparison of Nusselt number

T_{sat}	ΔT	β	Nu_B	Nu_K	Nu	q_w	N_j
373.3	10	40.3	8.53	12.9	8.44	9.28	0
373.3	20	80.7	8.49	12.9	8.85	19.5	0
373.3	22	88.7	8.48	12.9	8.91	21.6	1
373.3	30	121.	8.47	12.9	9.06	29.9	2
373.8	4	70.4	6.25	8.54	6.57	5.52	0
373.8	5	88.0	6.24	8.54	6.90	7.25	1
373.8	6	106.	6.23	8.54	6.90	8.69	1
373.8	7	123.	6.23	8.54	6.97	10.2	2
373.8	8	141.	6.23	8.54	7.02	11.8	2

(T_{sat}, ΔT are in ^oC and q_w in W/cm^2. N_j is the number of stable vapor jets. For Nu_B, β in Eq. (1) is replaced by β'.)

The Nusselt numbers averaged over a cell area are plot-

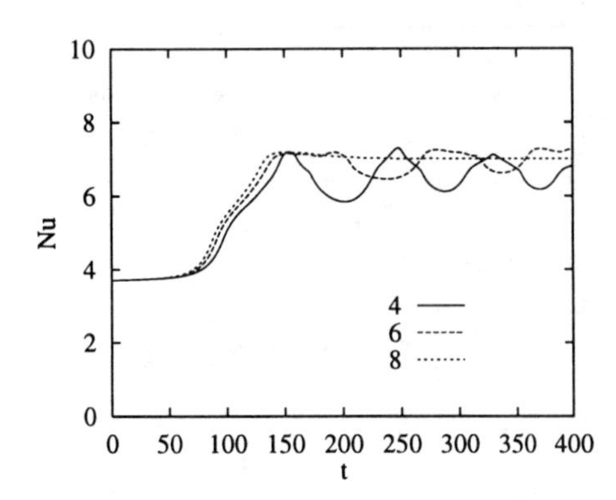

Fig. 7 Nusselt number averaged over a cell area for $T_{sat} = 373.8^oC$. The numbers in the figure represent values of ΔT.

ted in Fig. 7 for a pressure of 22.03 MPa and $T_{sat} = 373.8^oC$. It is observed that at low wall superheat, the Nusselt numbers vary cyclically. However, at the highest superheat an asymptotic value is obtained. Interestingly, the upper values obtained during cyclic variations are bounded by the Nusselt number for the highest wall superheat(8^oC).

Table 3 lists the number of stable jets as well as the average Nusselt numbers for the different system pressures and wall superheats. It is found that a critical value of $\beta(= c_{pv}\Delta T/h_{fg})$ exists at which the bubble release pattern changes. Approximately, for $\beta \geq 120$ stable vapor jets are formed on both the nodes and antinodes. For $85 \leq \beta \leq 120$, vapor release occurs in the forms of stable vapor jets and bubbles. However, for $\beta \leq 80$ no stable vapor jets exist.

CONCLUSIONS

1. The numerical algorithm that can handle breaking and merging of the interface and account for the effect of liquid-vapor phase change has been developed successfully.
2. A numerical simulation of the evolution of the vapor-liquid interface during saturated film boiling near the critical pressures has been carried out.
3. From the numerical simulation, it is shown that at low wall superheat discrete bubbles are released alternatively at the nodes and antinodes. However, with increase in wall superheat vapor jets are formed on both the nodes and antinodes. This is consistent with visual observations reported in the literature.
4. The Nusselt numbers obtained from the present analysis are bounded by those predicted from Berenson's and Klimenko's correlations. However, it is noted that Nusselt number increases very weakly with wall superheat.

ACKNOWLEDGEMENT

This work received support from the National Science Foundation.

REFERENCES

Alexiades, V., and Solomon, A.D., 1993, *Mathematical Modeling of Melting and Freezing Processes*, Hemisphere, Washington, D.C., pp. 34-37, pp. 215-216.

Bell, J.B., and Colella, P., 1989, "A Second-Order Projection Method for the Incompressible Navier-Stokes Equations", *J. of Comput. Phys.*, Vol. 85, pp. 257-283.

Berenson, P.J., 1961, "Film Boiling Heat Transfer From a Horizontal Surface", *J. Heat Transfer*, Vol. 83, pp. 351-362.

Brackbill, J.U., Kothe, D.B., and Zemach, C., 1992, "A Continuum Method for Modeling Surface Tension", *J. of Comput. Phys.*, Vol. 100, pp. 335-354.

Chang, Y.C., Hou, T.Y., Merriman, B., and Osher, S., 1996, "A Level Set Formulation of Eulerian Interface Capturing Methods for Incompressible Fluid Flows", *J. of Comput. Phys.*, Vol. 124, pp. 449-464.

Dhir, V.K., Castle, J.N., and Catton, I., 1977, "Role of Taylor Instability on Sublimation of a Horizontal Slab of Dry Ice", *J. Heat Transfer*, Vol. 99, pp. 411-418.

Dhir, V.K., and Taghavi-Tafreshi, K., 1981, "Hydrodynamic Transitions During Dripping of a Liquid from Underside of a Horizontal Tube", *ASME Paper*, 81-WA/HT-12.

Juric, D., and Tryggvason, G., 1996, "Computations of Film Boiling", presented at the ASME FED summer meeting, San Diego, July 7-11.

Klimenko, V.V., 1981, "Film Boiling on a Horizontal Plate-New Correlation", *Int. J. Heat Mass Transfer*, Vol. 24, pp. 69-79.

Patankar, S.V., 1980, *Numerical Heat Transfer and Fluid Flow*, Hemisphere, Washington, D.C., pp 44-47, pp 61-66.

Reimann, M., and Grigull, U., 1975, "Warmeubergang bei freier Konvektion und Filmsieden im kritischen Gebiet von Wasser und Kohlendioxid", *Warme- und Stoffubertragung*, Vol. 8, pp 229-239.

Son, G., and Dhir, V.K., 1996, "Nonlinear Taylor Instability with Application to Film Boiling and Melting", *Proceeding of Japan-U.S. Seminar on Two Phase Flow Dynamics*, pp. 301-309.

Sonneveld, P., and Wesseling, P., 1985, "Multigrid and Conjugate Gradient Methods as Convergence Acceleration Techniques", *Multigrid Methods for Integral and Differential Equations*, ed. Paddon, D.J. and Holstein, H., Oxford, Clarendon Press.

Sussman, M., Smereka, P., and Osher, S., 1994, "A Level Set Approach for Computing Solutions to Incompressible Two-Phase Flow", *J. of Comput. Phys.*, Vol. 114, pp. 146-159.

HTD-Vol. 342, National Heat Transfer Conference
Volume 4
ASME 1997

BUBBLE NUCLEATION AND GROWTH CHARACTERISTICS IN SUBCOOLED FLOW BOILING OF WATER

Satish G. Kandlikar

Viktor Mizo[1] Michael Cartwright[2] Emeka Ikenze[3]

Department of Mechanical Engineering
Rochester Institute of Technology
Rochester, NY
USA
Ph: (716) 475-6728; e-mail: sgkeme@rit.edu

ABSTRACT

Experimental results are presented for the onset of nucleation, bubble growth, and heat transfer in subcooled flow boiling of water near atmospheric pressure. The theoretical analysis for nucleation criterion is extended by including the results of a numerical study to locate the stagnation streamline in the liquid flowing around the bubble. The results obtained from the analysis show a very good agreement with the experimental data and further validation with other fluids is suggested. The model by Bergles and Rohsenow (1964) provides the lower limit of wall superheat required for bubble nucleation. The bubble growth rates are also measured and are found to be different than those under pool boiling conditions.

NOMENCLATURE

d_C - cavity diameter, m
D_h - hydraulic diameter of the flow channel, $= 4WH/(W+H)$, m
H - height of the flow channel, (=0.003 m), m
h - distance of the bubble center above the wall, m
h_{lv} - latent heat, J/kg
p - pressure, Pa
Re - Reynolds number, $\rho u_m D_h/\mu_f$
Re_b - bubble Reynolds number, $2\rho u_m r_b/\mu_f$
r_b - bubble radius, m
r_c - cavity radius, m
T_{WALL} - wall temperature, °C
T_{BULK} - bulk temperature, °C
u_m - mean velocity in flow channel, m/s
W - width of the flow channel, (=0.040 m), m
x - coordinate axis, normal to the flow direction along the width of the channel
y - coordinate axis, normal to the heated wall
z - coordinate axis, opposite to the flow direction in the channel

Current Affiliations of the Graduate Students:
[1] General Electric Company, Erie, PA
[2] Delphi Harrison Thermal Systems, Lockport, NY
[3] Hewlett Packard Company, Vancouver, WA

Greek Letters

β - contact angle between vapor and liquid at the wall, radians
θ_w - wall superheat, $T_{WALL}-T_{SAT}$, K
θ_{sub} - liquid subcooling, $T_{sat}-T_{BULK}$, K
μ - viscosity, kg/m s
ρ - density, kg/m^3
σ - surface tension, N/m

Subscripts

SAT - saturation state
c - cavity
f, L - liquid
v - vapor

INTRODUCTION

Bubble Nucleation

The early work of Hsu (1962), Bergles and Rohsenow (1964), Sato and Matsumura (1964), and Davis and Anderson (1966) provide the basis for analyzing nucleation characteristics of cavities in flow boiling. The nucleation criterion for a cavity was developed by Hsu (1962) by considering a truncated spherical bubble sitting at the mouth of the cavity as shown in Fig. 1a. The pressure inside the bubble is higher than the outside liquid due to the curvature of the liquid-vapor interface, and the difference is given by the Young-Laplace equation as $2\sigma/r_b$ for a spherical bubble of radius r_b. The corresponding saturation temperature inside the bubble is obtained from Clausius-Clapeyron equation. In order for this bubble to grow, Hsu postulated that the minimum temperature surrounding the bubble (which occurs at the top of the bubble, being farthest from the heater surface) should be at least equal to or greater than the saturation temperature of the vapor inside the bubble. The temperature in the liquid near the bubble top was estimated by assuming a linear temperature profile in the thermal boundary layer of thickness, δ_t=k/h. Hsu assumed the bubble top to be at a distance $1.6r_b$ away from the wall, which translates into a contact angle of 53.1 degrees. The resulting quadratic equation provides the following solution for the range of active cavities under a given wall superheat condition.

$$\{r_{c,min}, r_{c,max}\} = \frac{\delta_t}{4}\left(\frac{\theta_W}{\theta_W+\theta_{SUB}}\right) \times \left[1 \mp \sqrt{1 - \frac{12.8\,\sigma T_{SAT}(p_l)(\theta_W+\theta_{SUB})}{\rho_v h_{lv} \delta_t \theta_W^2}}\right] \quad (1)$$

Bergles and Rohsenow (1964) considered a hemispherical bubble as shown in Fig. 1b, and proposed a tangency criterion which results in the same expression as derived by Sato and Matsumura (1964) with liquid saturation temperature at the top of the hemispherical bubble surface.

$$\{r_{c,min}, r_{c,max}\} = \frac{\delta_t}{2}\left(\frac{\theta_W}{\theta_W+\theta_{SUB}}\right) \times \left[1 \mp \sqrt{1 - \frac{8\,\sigma T_{SAT}(p_l)(\theta_W+\theta_{SUB})}{\rho_v h_{lv} \delta_t \theta_W^2}}\right] \quad (2)$$

Davis and Anderson (1966) considered the truncated bubble shape similar to Hsu's model, but introduced the contact angle β as a variable. The resulting equation for the range of active cavities is obtained following a similar analysis.

$$\{r_{c,min}, r_{c,max}\} = \frac{\delta_t \sin\beta}{2(1+\cos\beta)}\left(\frac{\theta_W}{\theta_W+\theta_{SUB}}\right) \times \left[1 \mp \sqrt{1 - \frac{8\,\sigma T_{SAT}(p_l)(\theta_W+\theta_{SUB})}{\rho_v h_{lv} \delta_t \theta_W^2}}\right] \quad (3)$$

Kenning and Cooper (1965) evaluated the temperature in the liquid adjacent to the bubble top by considering the location of the stagnation point on the front face of a hemispherical bubble as shown in Fig. 1c. Large air bubbles were introduced in the liquid flow and the height of the stagnation point was observed visually. The height y_s of the stagnation point above the surface was obtained as a function of the bubble Reynolds number and the cavity radius.

$$\frac{y_s}{r_c} = 0.54\left[1 - \exp\left(-\frac{Re_b}{45}\right)\right] \quad (4)$$

Kenning and Cooper postulated that the temperature of the liquid adjacent to the top of the bubble is the same as that at the stagnation height in the thermal boundary layer since the streamline just above the stagnation location would sweep over the bubble. The resulting equations for the active cavity range could not be reduced to a closed form, but a simple numerical scheme could be employed to solve them.

Bubble Growth Rates

Nucleate boiling heat transfer involves nucleation of bubbles, their growth and subsequent departure from existing cavities into the adjacent liquid flowing over a heated surface. The heat transfer during nucleation in pool boiling has been extensively studied and a number of heat transfer mechanisms (microlayer evaporation, microconvection, vapor-liquid exchange, etc.) have been proposed in the literature. In the case of flow boiling, the bubble growth rates have not received the same level of attention as in the pool boiling case.

A qualitative description of the bubble growth mechanism can be given as follows. The bubble growth rate depends on the rate of heat transfer to the interface. The heat transfer at the microlayer near the base of the bubble, described under the evaporation microlayer model, and the heat transfer from the superheated liquid layer surrounding the bubble described under the relaxation microlayer model, are both jointly responsible for the bubble growth. In the initial phase of the growth, the bubble is small and is surrounded by a highly superheated liquid layer near the wall. This results in a rapid bubble growth in which the inertia forces are dominant. Later as the bubble grows, it comes in contact with the liquid farther away from the wall with a lower superheat. The heat transfer is then through the transient conduction in the thin liquid film surrounding the bubble (relaxation microlayer). This region is the thermally controlled growth region.

Mikic et al. (1970) established a general relationship for bubble growth rates in uniformly superheated liquid in both inertia and thermally controlled regions. Their analysis showed that the bubble growth is proportional to t (time) for the inertia controlled region while it is proportional to $t^{1/2}$ in the thermally controlled region. Van Stralen (1968) proposed a different form of the bubble growth curve applicable in both regions using an exponential function, and Van Stralen, et al., (1975) found good agreement between their experimental data and predictions for pure and binary mixtures.

The above studies focus on the bubble growth in pool boiling. In flow boiling, the dependence of the bubble shape and the front and the rear contact angles on flow velocity was investigated by Kandlikar (1995). Their model considered the deflection of the bubble in the flow direction, and included the upstream and downstream contact angles in the analysis. The effective bubble removal is therefore initiated either at the upstream or the downstream edge of the bubble. Four possible mechanisms for bubble departure were identified: (i) sweep removal at the front edge, (ii) lift removal at the front edge, (iii) sweep removal at the rear edge, and (iv) lift removal at the rear edge. Experimental results were presented for the effect of flow on the contact angles and the departure bubble diameters.

OBJECTIVES OF THE PRESENT WORK

The main objective of the present work is to gain further understanding of the nucleation (onset and suppression) and bubble growth characteristics under flow conditions. To accomplish this, the nucleation characteristics of a polished aluminum heater surface under subcooled flow of water are studied using a microscope and a high speed camera. The effects of flow rate, surface temperature, and subcooling are investigated to establish the nucleating cavity sizes and to determine the range of active cavities under a given set of conditions. Heat transfer data is also obtained to identify the presence of nucleate boiling at high flow rates. The experimental results for boiling incipience are compared with the present work and the available models.

The bubble growth rates are measured over different cavities for different values of subcooling, flow rate and wall temperature. The results are compared with the pool boiling models available in the literature and need for future work is identified.

EXPERIMENTAL SETUP

The experimental setup is the same as the one described by Kandlikar and Stumm (1995). A brief description of the setup is given here. It consists of

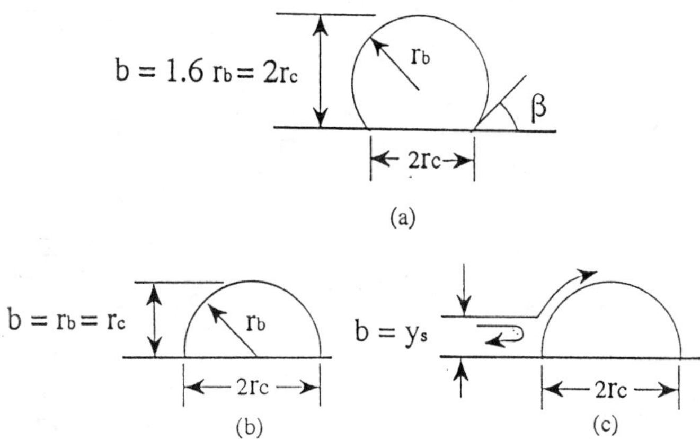

Figure 1. Bubble shapes employed in nucleation analysis, (a) truncated sphere, Hsu (1962), (b) hemispherical, Bergles and Rohsenow (1964), (c) hemispherical, Kenning and Cooper (1965)

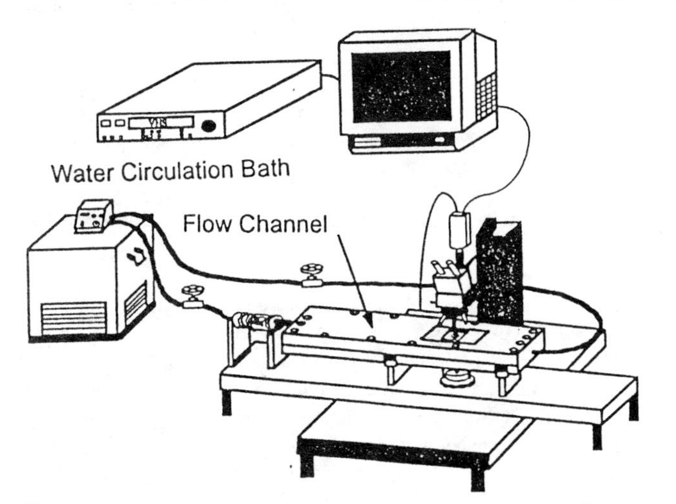

Figure 2. Schematic of the experimental loop

a horizontal, rectangular flow channel 3 mm x 40 mm cross-section with a heated aluminum rod of 10 mm diameter placed in the center of the lower (40 mm wide) wall. Water from the constant temperature bath near atmospheric pressure flows through the channel. The accuracy of flow and temperature measurements are ±3 percent and ±0.1°C respectively. The bubble nucleation and its growth are observed through a microscope. Top-views of the bubbles are obtained by looking down directly on the bubble through a viewing window. Side views of the bubbles are obtained through a front-surface silicon mirror placed at 45 degrees adjacent to the heated surface, parallel to the flow. A microscope and a video camera are employed to provide an effective magnification of up to 1350X on the video monitor. The video camera is able to capture images at frame rates of up to 6000 frames per second (fps) using a Kodak Ektapro camera with image intensifier. A schematic of the experimental setup is shown in Fig. 2.

The surface of the aluminum heater is polished on a cloth covered metallographic polishing wheel using 1 micron particle size alumina in water suspension. The polished aluminum surface is placed flush with the bottom wall of the flow channel. Figure 3 shows a schematic view of the heater assembly. Four thermocouples are placed at locations 7-10 in the 10 mm diameter section at equal distances along the aluminum rod. A circular

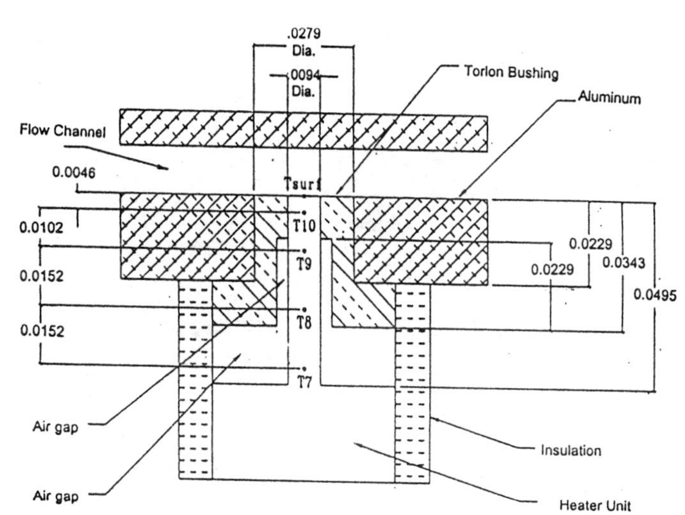

Figure 3. Schematic of the heater assembly used in thermal network analysis

heater is wrapped around the 25 mm diameter lower section of the aluminum rod. A special low thermal conductivity, high temperature plastic bushing made of Torlon is used near the channel wall in which the aluminum rod is press-fitted.

EXPERIMENTAL PROCEDURE

Experiments were conducted to obtain the heat transfer data and to observe the bubble nucleation and growth over cavities for different conditions of flow and heater power settings. The constant temperature bath was filled with distilled water. Prior to any experiments, water was heated to 90°C and continuously circulated through the test section for several hours to allow the removal of any dissolved gases. The procedure was repeated over several days before conducting actual test runs.

The water bath temperature and the flow rate were set at desired values. The heater was then powered with a DC power source. After allowing a sufficient time to attain steady-state conditions (about 20 minutes), bubble activity was observed through the microscope using appropriate magnification and frame rate for the high-speed camera. The presence of bubble activity in many cases could be observed only by recording at high speed (500 to 6000 fps) and playing it back at slow speed (30 fps). The heater input was increased in small increments and the bubble activity over specific sites was recorded. Simultaneous measurements of water and heater temperatures were made and recorded. Each run was identified with a specific run identifier displayed by the camera. A log was also maintained to clearly identify the flow and temperature data at different video tape counter locations.

HEATER SURFACE TEMPERATURE ESTIMATION

The heater surface temperature and the surface heat flux are obtained from the measured temperatures through a thermal network model of the heater assembly shown in Fig. 3. From this data, heat transfer coefficient on the heater surface is calculated. It is estimated that the error in the surface temperature calculation is within ±0.2 C and in the heat transfer coefficient is within ±5%.

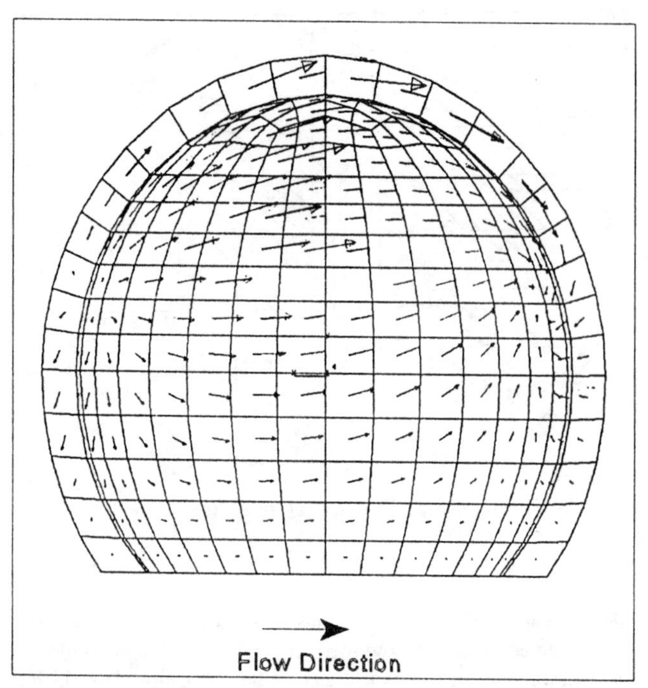

Figure 4. Flow distribution over a truncated bubble surface

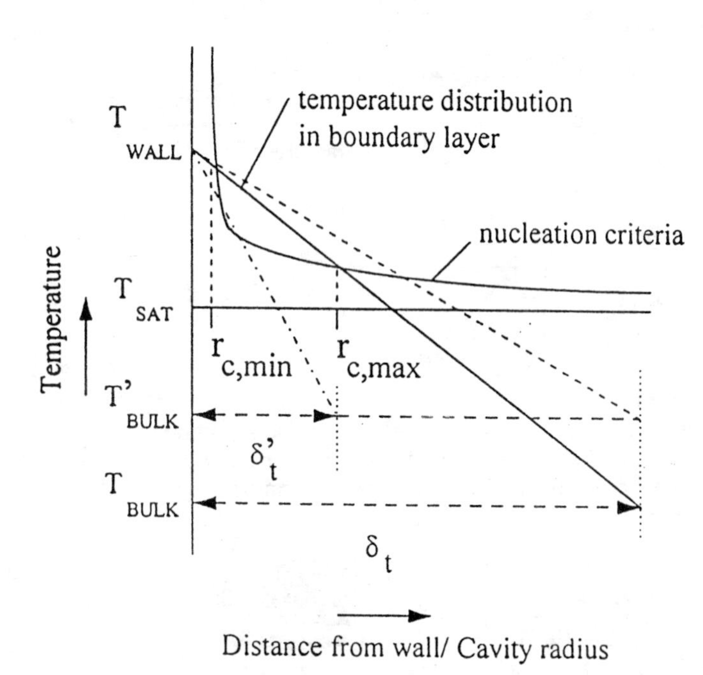

Figure 5. Representation of nucleation criteria in subcooled flow boiling

THEORETICAL ANALYSIS

The bubble shape and the temperatures of the wall and the liquid adjacent to the bubble are responsible for growth or collapse of a bubble nucleus. Side views of bubbles obtained in earlier investigations (see Kandlikar, 1995) show that the bubbles are of a truncated spherical shape. The lowest temperature in the liquid around the bubble occurs at the top of the bubble. The liquid temperature at this location has been estimated by earlier investigators (Hsu, 1962, Bergles and Rohsenow, 1964, Davis and Anderson, 1966) from a linear temperature profile in the thermal boundary layer. However, Kenning and Cooper (1965) conducted an experimental study to find the location of the stagnation point on the front surface of the bubble facing the flow using a dye injection technique as discussed earlier. The bubbles in Kenning and Cooper's experiments were much larger (above 5 mm) and were generated using air injection.

The bubbles of interest to us in determining the range of active cavities are of the same order of magnitude as the cavities, which are typically smaller than about 30-40 micrometers in diameter. For these conditions, it was decided in the present investigation to model the flow around a bubble attached to the wall using a CFD software, STAR-CD. The CFD mesh is created using P3/PATRAN preprocessor. The mesh is transferred to STAR-CD version 2.2.1. The programs are run on a model 750 Hewlett-Packard workstation with 128 MB RAM and a UNIX operating system.

The mesh consists entirely of 8-noded hexahedra. It utilizes a fine mesh in the immediate vicinity of the bubble, gradually transforming into a coarser mesh farther away from the bubble. The liquid-vapor interface on the spherical surface is modeled as a no-slip condition, representing a no shear surface. This assumption is reasonable since the vapor shear stress inside the bubble is negligible compared to the shear stress in the liquid flow, and no surface contaminants are assumed to be present. Further details are given by Ikenze (1996).

The location of the stagnation point is determined as a function of bubble diameter, contact angle, and flow velocity. A contact angle of 54° was employed based on the visual observations made in this investigation for the aluminum-water combination. Figure 4 shows a typical result of the flow field around the bubble. The length of a vector shown represents the magnitude of the velocity in that direction. The location of the stagnation point is identified from a similar pressure distribution plot around the bubble. Results are generated for different flow velocities and a range of contact angles from 30° to 60°. It is found that the stagnation point is located at an angle of 20° to 24° above the horizontal plane (parallel to the heater surface) passing through the bubble center, and is not significantly affected by the flow velocity and the contact angle in the range covered in this investigation. The results could be represented within 4% by the following simple form.

$$\frac{y_s}{r_b} = 1.10 \qquad (5)$$

where y_s is the distance of the stagnation point above the heater surface, and r_b is the bubble radius.

Using the temperature in the liquid boundary layer at a distance y_s (see Fig. 1c) as the temperature at the top surface of the bubble of radius r_b, the nucleation criteria is rederived. If this temperature is below the saturation temperature inside the bubble, the bubble will collapse, otherwise the bubble will grow and the cavity is activated. The final equation for the range of nucleation cavity radii is obtained as follows.

$$\left\{r_{c,\min}\ ,\ r_{c,\max}\right\} = \frac{\delta_t \sin\beta}{2.2}\left(\frac{\theta_W}{\theta_W+\theta_{SUB}}\right) \times \left[1 \mp \sqrt{1 - \frac{9.2\,\sigma\, T_{SAT}(p_l)(\theta_W+\theta_{SUB})}{\rho_v h_{lv} \delta_t \theta_W^2}}\right] \qquad (6)$$

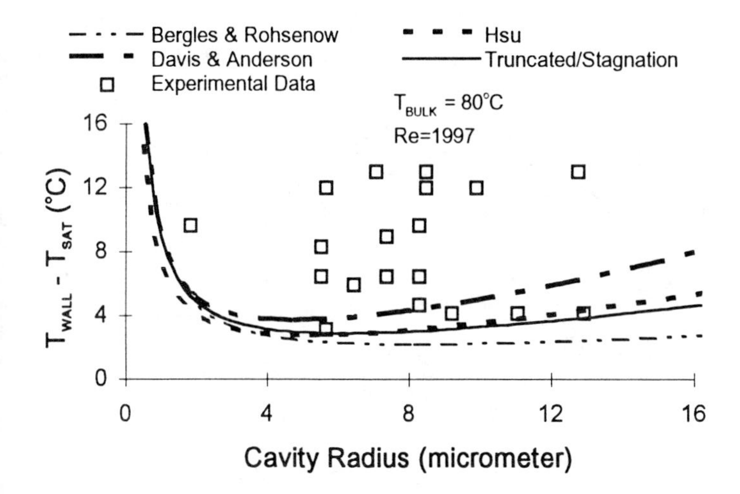

Figure 6. Nucleation criteria in subcooled flow boiling, T_{BULK}=80°C and Re=1997.

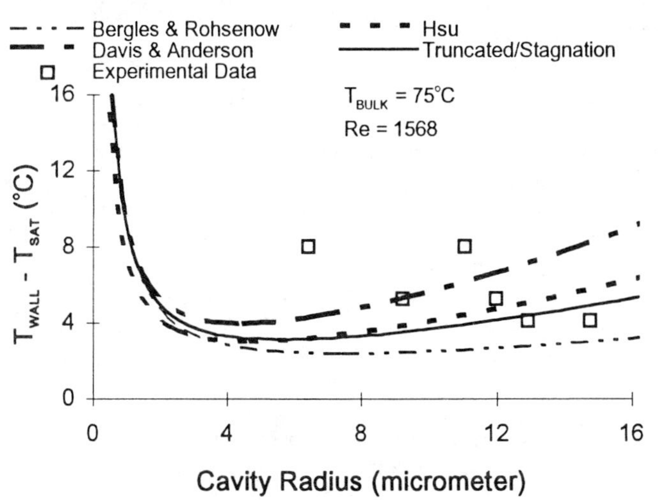

Figure 7. Nucleation criteria in subcooled flow boiling, T_{BULK}=75°C and Re=1568.

The nucleation criterion in subcooled flow boiling is represented on a plot of liquid superheat vs. the distance from the wall as shown in Fig. 5. The bulk liquid is represented by a point at its subcooled temperature T_b at a distance of δ_t from the wall. The line joining the wall temperature point at T_{WALL} with the bulk fluid point represents the linear temperature profile in the boundary layer. The intersection of this line with the nucleation criterion gives the minimum and the maximum cavity radii as given by equation (6). This figure is used in discussing the effects of subcooling on nucleation characteristics.

RESULTS AND DISCUSSION

Nucleation Criterion

Nucleation in flow boiling is primarily dependent on the wall superheat, liquid subcooling, and flow rate. The effects of these parameters are examined with the help of Fig. 5. As the subcooling is reduced to T'_{BULK}

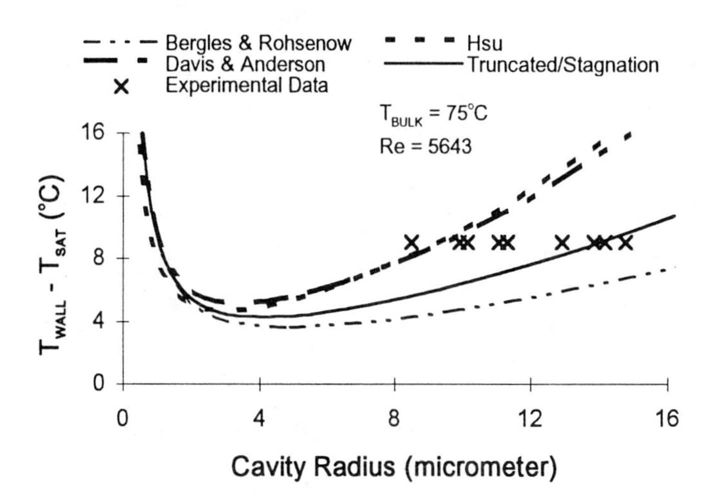

Figure 8. Nucleation criteria in subcooled flow boiling, T_{BULK}=75°C and Re=5643.

while keeping the flow rate and the wall temperature constant, the minimum cavity radius decreases slightly, but the maximum cavity radius and the range of active cavities increase considerably. On the other hand, as the flow rate is increased, the thermal boundary layer thickness decreases to δ'_t causing the minimum and maximum cavity radii and the range of active cavities to decrease. At very high flow rates, the boundary layer thickness may become so small that no cavities are activated unless substantially higher wall temperatures are applied. For saturated water at atmospheric pressure and a wall superheat of 12 °C, a heat transfer coefficient of approximately 100,000 $W/m^2{}^\circ C$ is needed to suppress all cavities. The effect of an increase in the wall temperature (not shown in Fig. 5) can also be predicted with the help of Fig. 5 by raising T_{WALL}. It will cause the minimum cavity radius to decrease slightly and the maximum cavity radius to increase.

The results of the nucleation experiments are plotted in Fig. 6 as wall superheat versus radii of active cavities as the heater power setting is raised between the two successive runs. The ranges of active cavity radii predicted by different models are also shown. The intersection of a constant temperature line corresponding to the wall superheat with a curve yields the range of cavities. A cavity of a given radius will nucleate if the wall temperature is above the line indicated by the individual curves. The curves indicate the minimum superheat needed to activate a given cavity. The experimental data shown correspond to a bulk temperature of 80 °C and Re=1997. The Davis and Anderson's (1966) and Hsu's (1962) models overpredicted the wall temperature for many cavities. The present model is able to predict nearly all the data well. The Bergles and Rohsenow (1964) model represents the lower limit of the wall superheat which is somewhat lower than the present model.

Figures 7 and 8 show a similar plot of the wall superheat plotted against the active cavity radius for a water temperature of 75 °C at two flow rates. The agreement between the data and different models is similar to that observed in Fig. 6.

It can be seen from Figs. 6-8 that the Bergles and Rohsenow's model covers the broadest range. At higher Reynolds number, both Hsu's model and Davis and Anderson's model are too restrictive, and the experimental data often cross the r_{max} limit. The lower limits for the active cavity radii from different models are quite close to each other except for Hsu's model.

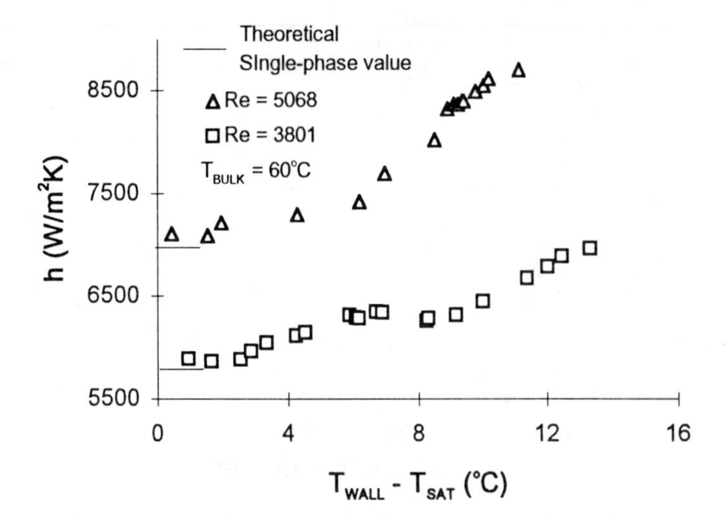

Figure 9. Experimentally determined heat transfer coefficient in subcooled flow boiling, T_{BULK}=60°C and Re=5068 and 3801.

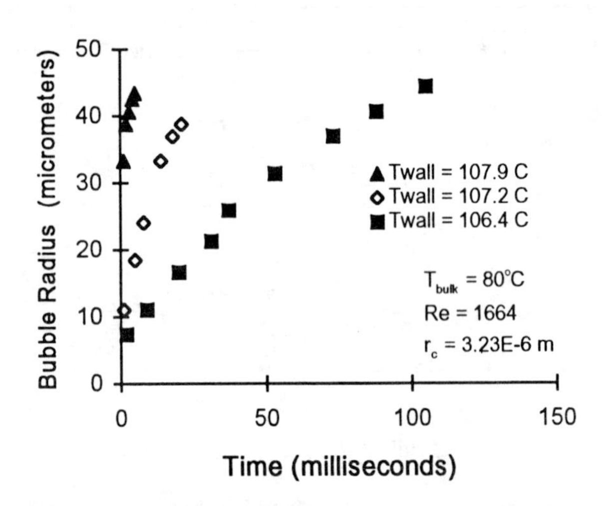

Figure 11. Effect of wall temperature on the bubble growth for T_{BULK} = 80 °C, Re = 1710, and cavity diameter = 6.5 μm.

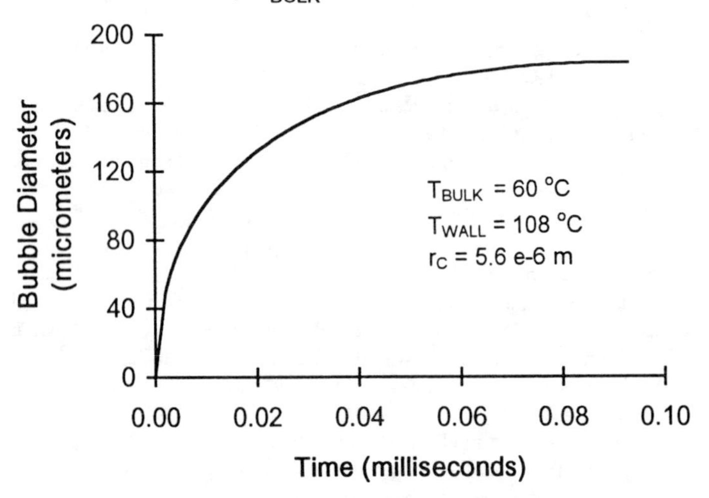

Figure 10. Pool boiling bubble growth curve from Mikic and Rohsenow's (1969) model

No bubble activity was identified near the r_{min} range, primarily because of the insufficient magnification available to visualize feature sizes smaller than 2 micrometers, and the limited camera speed (6000 fps).

The presence of nucleation was also identified by observing the variation in heat transfer coefficient with wall superheat as shown in Fig. 9 for two flow rates. It can be seen that initially the heat transfer is by single phase mode to liquid. As the superheat increases above a certain value dependent on the individual test conditions, heat transfer coefficient starts to rise with the wall superheat as more cavities are activated. An interesting fact was observed in many cases that no bubbles were visually detected with the video camera but the heat transfer coefficient plot indicated a significant contribution due to nucleate boiling (higher h than the single phase value). To explore this fact further, a single nucleating cavity was observed under a low heat flux condition with a 30 fps video camera. As the heat flux was increased, no nucleation could be observed. The 30 fps camera was then replaced with the high speed camera set at 500 fps. Bubble activity was once again observed this time at a faster rate. Further increase in heat flux caused the bubbles to disappear, but they could be observed again at higher camera speeds. After further increases in the heat flux, the bubble activity could no longer be seen even at the highest available speed of 6000 fps. The bubble activity was perhaps still going on over this cavity, but could not be detected due to the limited camera speed. Complete suppression of nucleation can therefore be ascertained in the subcooled flow only through the heat transfer coefficient measurements in the absence of a higher speed camera. So long as there is a wide range of cavities present on the surface, and the nucleation criterion is satisfied, it is reasonable to assume that nucleation will be present on the surface.

Bubble Growth Rates

Before reviewing the experimental data on bubble growth rates in flow boiling, it would be useful to see the theoretically predicted growth curve for pool boiling using Mikic and Rohsenow's (1969) model. Figure 10 shows one such plot drawn for T_{BULK} =60 °C, T_{WALL}=108 °C, and a cavity diameter of 11.1 μm. These conditions are the same as those employed in one of the flow boiling data point obtained in the present experiments.

The growth curve shown in Fig. 10 indicates that for the stated conditions under pool boiling, a bubble growing on this cavity grows to a size of 180 μm in about 100 μsec. The bubble growth is extremely rapid and in order to visually observe say 10 frames during the bubble growth, a frame rate of 140,000 fps is needed.

Figure 11 shows the bubble growth curves obtained in the present investigation for r_c=3.23 μm, T_{BULK}=80 °C, and Re=1664 for three different wall temperatures of 106.4, 107.2 and 107.9 °C. A change of 0.6 to 0.7 °C in wall temperature causes significant changes in the bubble growth rates. The growth time reduces from 130 msec to 22 msec, and further down to 5 msec with increasing wall temperatures. The departure bubble diameters are around 80 to 90 μm. This data was obtained using high speed camera at 1000 fps.

With further increase in wall temperature, the growth rate becomes even faster. Figure 12 corresponds to the same conditions as for Fig. 11, except for the higher wall temperatures of 109.6 and 110.4 °C. The bubble growth time reduced drastically to 1.7 and 1.5 msec, while the departure bubble diameter decreased to about 68 μm. The data in Fig. 12 was obtained using the high-speed camera at 3000 and 4000 fps respectively for

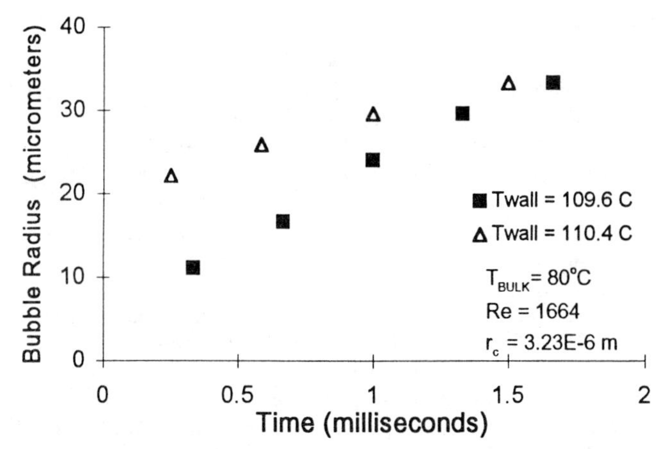

Figure 12. Effect of wall temperature on the bubble growth for TBULK = 80 °C, Re = 1710, and cavity diameter = 6.5 μm (at 3000 and 4000 fps).

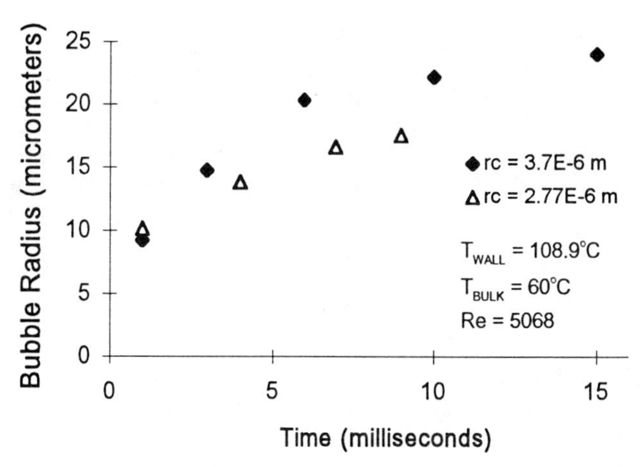

Figure 14 Effect of cavity diameter on the bubble growth for T_{BULK} = 60 °C, Re = 5068, and cavity diameters of 5.5 and 7.4 μm (at 500 fps).

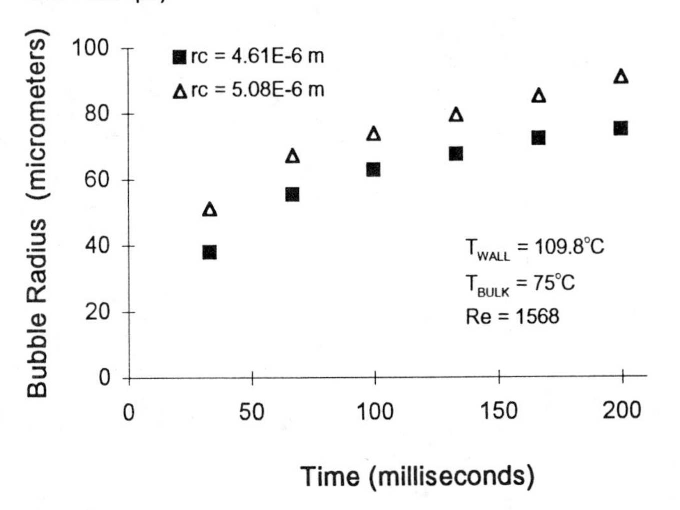

Figure 13. Effect of cavity diameter on the bubble growth for T_{BULK} = 75 °C, Re = 1568, and cavity diameters of 9.2 and 10.2 μm (at 30 fps).

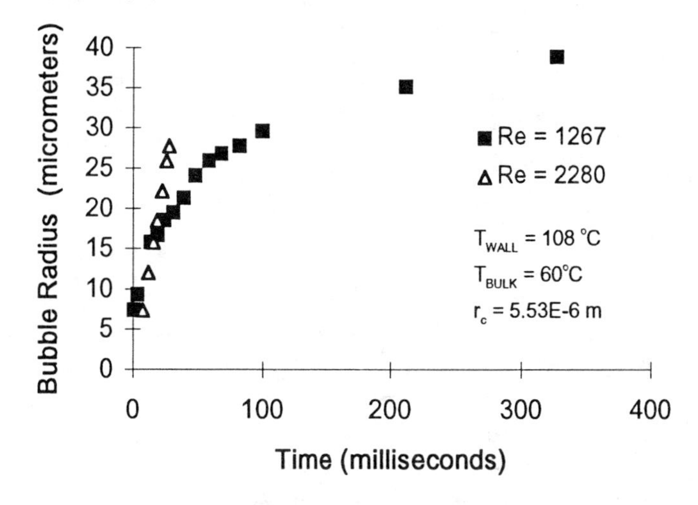

Figure 15. Effect of Reynolds number on the bubble growth for T_{BULK} = 60 °C, cavity diameter of 11.1 μm, (at 1000 fps).

the two sets. The increasing growth rate with increasing wall temperature is still evident at the higher wall superheat values as well (note the first data points in the two sets).

Figure 13 shows the effect of cavity diameter on the bubble growth. For T_{WALL}=109.8 °C, T_{BULK}=75 °C, and Re=1568, the growth rates for two cavities 9.2 and 10.2 μm diameter are plotted. The two curves are almost parallel and the smaller cavity yields a smaller bubble indicating the strong influence of the wall superheat on bubble growth.

Bubble frequency from the larger diameter cavities increased with flow rate until eventually bubbles could not be observed from these sites even at the highest camera speed available. Figure 14 shows the growth curves at Re=5068 for cavities of 5.5 and 7.4 μm diameter. At higher flow rates, smaller cavities are activated and their growth rate is faster in spite of a reduction in wall temperature to 108.9 °C from 109.8 °C as shown in Fig. 13.

The effect of flow on the bubble growth rates is further examined in Fig. 15. For Re=1267, the bubble grows to 78 μm diameter, while at Re=2280, the bubble grows only to a diameter of 56 μm. The growth time for the bubble is also shortened at higher flow rates, and the bubble frequency is also observed to increase with flow rate. At higher flow rates, the active cavities become smaller, and the growth becomes even more rapid as seen by comparing Figs. 13-15. This could be explained with the help of Fig. 5. As the flow rate increases, the boundary layer thickness decreases and the liquid temperature line intersects the nucleation criteria curve at smaller cavity diameters. At some point, as the boundary layer thickness reduces further at higher flow rates, the tangency condition is reached and complete suppression will be achieved with further increase in flow rate.

The effect of subcooling can be seen by comparing Fig. 12 with T_{BULK}=80 °C, and Fig. 13 with T_{BULK}=75 °C. The cavity diameters are 6.5 and 9.2 μm respectively. Although the cavity size also influences the growth curve, the difference in the growth time in these two figures is of two orders of magnitudes, and is largely attributable to the 5 degree change in the bulk temperature. At higher bulk temperatures above 80 °C, the bubble activity could not be seen even with the camera speed of 6000 fps.

Comparison with Pool Boiling Bubble Growth Rates

The data presented in Fig. 15 can be compared with the pool boiling growth curve presented in Fig. 10 for the same conditions of T_{WALL}, T_{BULK} and r_c From this comparison, following observations can be made. (i) The bubble growth rate curves are significantly slowed down (by several orders of magnitudes) for a cavity of a given size and a given wall superheat under subcooled flow boiling. (ii) Departure bubble diameter becomes smaller with increasing flow velocity. This is due to the effects caused by the changes in the temperature profile in the liquid boundary layer due to the flow.

CONCLUSIONS

An experimental investigation is conducted on the nucleation and bubble growth characteristics of cavities under subcooled flow boiling of water on an aluminum heater surface. The effects of subcooling, flow rate and wall temperature on these characteristics have been studied. Based on the present work, the following conclusions are made.

The nucleation criteria of Bergles and Rohsenow utilizing a hemispherical bubble shape provides the lower limit of wall superheat required to nucleate a given cavity, and is recommended as the nucleation criterion. Although the current model with a more detailed description of the bubble shape and the temperature field near the interface provides a slightly better agreement, the analysis is applicable only to the conditions tested in the current experiments. Further validation is needed with other fluids.

Effect of Flow Rate

Increasing flow rate deactivates larger cavities. The temperature profile in the liquid boundary layer becomes steeper and the active cavity range shifts toward smaller radii. The departure diameter decreases from the pool boiling conditions as the flow rate increases. The growth rate also reduces, and further analysis is warranted on bubble growth rates in flow boiling.

Effect of Subcooling

The effect of subcooling is quite complex on nucleating cavity radii and is strongly coupled with the flow rate as expressed by eq. (2). At higher subcooling, the range of active cavities shifts toward smaller radii, and a higher wall superheat is required to initiate nucleation on a cavity of a given radius.

The effect of subcooling on bubble growth is quite interesting. At high subcooling, the bubble growth is slow. With higher bulk temperatures, the bubble growth is rapid and the departure bubble diameters are smaller. The bubble growth is so strongly dependent on the bulk temperature that above 80 °C, no bubble activity could be captured in spite of careful monitoring of the surface temperatures in small steps, although the heat transfer data indicated the presence of nucleation. It is possible that small bubbles (smaller than 5-10 μm diameter) are ejected at high speeds (in excess of 6000 bubbles/second) and go undetected under the present imaging capabilities. Visual observations therefore cannot be relied upon to establish complete suppression of nucleate boiling.

Effect of Wall Superheat

At higher wall superheats, the range of active cavities becomes wider and the heat transfer coefficient increases due to the increased nucleation. The wall superheat has a major influence on the bubble growth rate. The growth rate starts to increase very rapidly near a certain value of wall superheat, dependent on the operating conditions. With as little as a 1 °C change in the wall superheat, the bubble growth time decreases by an order of magnitude.

REFERENCES

Bergles, A. E., and Rohsenow, W. M., 1964, "The Determination of Forced-Convection Surface-boiling Heat Transfer," Journal of Heat Transfer, Trans. ASME, Series C, Vol. 86, No. 3, pp. 365-372.

Davis, E. J., and Anderson, G. H., 1966, "The Incipience of Nucleate Boiling in Forced Convection Flow," AIChE Journal, Vol. 12, No. 4, pp. 774-780.

Hsu, Y.Y., 1962, "On the Size Range of Active Nucleation Cavities on a Heating Surface," Journal of Heat Transfer, ASME, Series C, Vol. 84, pp. 207-216.

Hsu, Y. Y., and Graham, R. W., 1961, "An Analytical and Experimental Study of the Thermal Boundary Layer and Ebullition Cycle in Nucleate Boiling," NASA TN-D-594.

Ikenze, C. M., 1996, "A CFD Study of Fluid Flow Characteristics in the vicinity of a Truncated Sphere in a Long, Narrow Channel," M. S. Thesis, Mech. Eng. Dept., Rochester Institute of Technology, Rochester, NY.

Kandlikar, S. G., 1992, "Bubble Behavior and Departure Bubble Diameter of Bubbles Generated Over Nucleating Cavities in Flow Boiling," Pool and External Flow Boiling, Proceedings of The Engineering Foundation Conference on Pool and External Flow Boiling, ASME, March 22-27, Santa Barbara, CA, pp. 447-452.

Kandlikar, S. G., and Stumm B. J., 1995, "A Control Volume Approach for Investigating Forces on a Departing Bubble under Subcooled Flow Boiling," Journal of Heat Transfer, Vol. 117, pp. 990-997.

Kandlikar, S. G., 1994, "Measurement of Departure Bubble Diameter and Advancing and Receding Contact Angles in Subcooled Flow Boiling of Water," Paper presented at the joint ASME and ISHMT Conference, Jan. 5-8, BARC, Bombay, India.

Kenning, D. B. R., and Cooper, M. G., 1965, "Flow Patterns Near Nuclei and the Initiation of Boiling During Forced Convection Heat Transfer," Paper presented at the Symposium on Boiling heat Transfer in Steam Generating Units and Heat Exchangers held in Manchester, IMechE (London) Sept. 15-16.

Mikic, B. B., Rohsenow, W. M., and Griffith, P., 1970, "On Bubble Growth Rates," International Journal of Heat Mass Transfer, Vol. 13, pp. 657-666.

Nukiyama, S., 1934, "The Maximum and Minimum Values of Heat q Transmitted from Metal Surface to Boiling Water under Atmospheric Pressure," Journal of Society of Mechanical Engineers (Japan), Vol. 37, pp. 367-374, 553-554.

Sato, T., and Matsumura, H., 1964, "On the Conditions of Incipient Subcooled Boiling with Forced Convection," Bulletin of JSME, Vol. 7, No. 26, pp. 392-398.

Van Stralen, S. D. J., 1968, "The Growth Rate of Vapor Bubbles in Superheated Pure Liquids and Binary Mixtures," International Journal of heat Mass Transfer, Vol. 11, pp. 1467-1489.

Van Stralen, S. D. J., Sohal, M. S., Cole, R., and Sluyter, W. M., 1975, "Bubble Growth Rates in Pure and Binary Systems: Combined Effect of Relaxation and Evaporation Microlayers," International Journal of heat Mass Transfer, Vol. 18, pp. 453-467.

HTD-Vol. 342, National Heat Transfer Conference
Volume 4
ASME 1997

BOILING HEAT TRANSFER WITH BINARY MIXTURES PART I- A THEORETICAL MODEL FOR POOL BOILING

Satish G. Kandlikar
Mechanical Engineering Department
Rochester Institute of Technology
Rochester, NY 14623
USA
Ph: (716)475-6728; e-mail: sgkeme@rit.edu

ABSTRACT

Experimental evidence available in the literature indicates that the pool boiling heat transfer with binary mixtures is lower than the respective mole- or mass-fraction averaged value. In the present work, a theoretical analysis is presented to estimate the mixture effects in binary pool boiling. A new pseudo-single component heat transfer coefficient replaces the commonly used reciprocal mole-fraction averaged ideal heat transfer coefficient. The liquid composition and the interface temperature at the interface of a growing bubble are predicted analytically and their effect on the heat transfer is estimated. The present model is compared with the theoretical model of Calus and Leonidopoulos (1974), and two empirical models, Calus and Rice (1972) and Fujita et al. (1996). The present model is able to predict the heat transfer coefficients and their trends in azeotrope forming mixtures as well as mixtures with widely varying boiling points.

NOMENCLATURE

A_{12} - thermodynamic factor
c_p - specific heat, J/kg K
D_{12} - Diffusion coefficient of 1 in mixture of 1 and 2
D_{12}^0 and D_{12}^0 - Diffusion coefficient of 1 present in infinitely low concentration of liquid mixture
F_D - diffusion factor, $=\alpha/\alpha_{id}$
g - $=(x_1-x_{1,s})/(y_{1,s}-x_{1,s})$, defined by equation (13)
Δh_{LG} - latent heat of vaporization, J/kg
Ja_0 - modified Jakob number, defined by equation (12)
$\dot{m}$ - mass flux, kg/m^{2}s
$\dot{q}$ - heat flux, W/m^2
R - bubble radius, m
T - temperature, K
t - time from bubble inception, sec
ΔT_{bp} - boiling point range, difference between the dew point and bubble point temperatures, K
ΔT_{id} - wall superheat for ideal mixture as defined by earlier investigators, K
ΔT_s - $= (T_s - T_{sat})$, K
$\Delta T_1, \Delta T_2$ - wall superheats for 1 and 2 in pool boiling
$v_{m,1}, v_{m,2}$ - liquid molar specific volumes, m^3/kg-mol
V_1 - volatility parameter, defined by equation (23)
x_1, x_2 - mass fraction of 1 and 2 in liquid phase
$\tilde{x}_1, \tilde{x}_2$ - mole fraction of 1 and 2 in liquid phase
y_1, y_2 - mass fraction of 1 and 2 in vapor phase
$\tilde{y}_1, \tilde{y}_2$ - mole fraction of 1 and 2 in vapor phase
z - distance from bubble interface in liquid phase, m

Greek Letters

α - heat transfer coefficient, W/m^2 K
β_0 - contact angle, radians
δ_m - thickness of mass diffusion boundary layer, m
η - viscosity, kg/m s
κ - thermal diffusivity, m^2/s
λ - thermal conductivity, W/m K
ρ - density, kg/m^3
σ - surface tension, N/m
ϕ - association parameter

Subscripts

1, 2 - components of a binary system
1-more volatile component
avg - average
B - binary
BL - boundary layer
D - mass diffusion
G - vapor
id - ideal
L - liquid
m - mixture
PB - pool boiling
psc - pseudo-single component
s - liquid-vapor interface of a bubble
sat - saturation

Table 1 Summary of Some Important Methods for Predicting Binary Pool Boiling Heat Transfer Coefficients

Author and Year	Correlation Scheme	Comments
Stephan and Körner (1969)	$\alpha = \frac{\alpha_{id}}{1 + A\lvert y_1 - x_1 \rvert (0.88 + 0.12p)}$ α_{id} - reciprocal mole-fraction average	The composition difference was recognized as an important factor. Empirical constant A is specific to a mixture (range (0.43-0.56), p is in bar.
Calus and Rice (1972). Binary mixtures of isopropanol with water and acetone.	$\left[\frac{Nu}{K_p^{0.7}}\right]\left[\frac{T_{sat}}{T_{sat,water}}\right]^4 = E\left[\frac{Pe}{1+\lvert y-x \rvert (\kappa / D_{12})^{0.5}}\right]^{0.7}$ $K_p = p / [g\sigma(\rho_L - \rho_G)]^{0.5}, E - empirical\,constant'$ $Pe = \frac{q}{\Delta h_v}\frac{\rho_L}{\rho_G}\frac{c_L}{\lambda_L}\left(\frac{\sigma}{g(\rho_L - \rho_G)}\right)^{0.5}$	The diffusion resistance term by Van Stralen (1966) is applied to modify the pool boiling correlation. $(c_L/\Delta h_v)(dT/dx)$ term is neglected and an empirical constant is introduced.
Calus and Leonidopoulos (1974). n-propanol/water.	$\Delta T = (\Delta T_1 x_1 + \Delta T_2 x_2)\cdot \left[1 + (y-x)\left(\frac{\kappa}{D_{12}}\right)^{0.5}\left(\frac{c_L}{\Delta h_v}\right)\left(\frac{dT}{dx}\right)\right]$	This scheme introduces the averaging equation from pure-fluid temperature differences. Original form of diffusion resistance term by Van Stralen (1966) is retained without any empirical constants. α_{id}-reciprocal mass-fraction average value
Schlünder (1982)	$\alpha = \alpha_{id}\frac{1}{1 + \frac{\Delta T_s}{\Delta T_{id}}\lvert y_1 - x_1 \rvert \left[1 - \exp\left(-\frac{B_o \dot{q}}{\beta_L \rho_L h_{LG}}\right)\right]};$ $B_o = 1, \beta_L = 2 \times 10^{-4} m/s$	Introduced the boiling range of the respective pure components at the same pressure, and a term representing the mass transfer effects. B_o represents the evaporation component, assumed to be 1, and β_L set as a constant α_{id} - reciprocal mole-fraction average
Thome and Shakir (1987) ethanol/(water, benzene), acetone/ water, nitrogen/(argon, oxygen, methane)	$\alpha = \alpha_{id}\frac{1}{1 + \frac{\Delta T_{bp}}{\Delta T_{id}}\left[1 - \exp\left(-\frac{B_o \dot{q}}{\beta_L \rho_L h_{LG}}\right)\right]};$ $B_o = 1, \beta_L = 3 \times 10^{-4} m/s$	The boiling range and the mass diffusion effects from Schlünder's model were combined. New value of β_L was determined empirically from data. α_{id} - reciprocal mole-fraction average
Wenzel et al. (1995). Binary mixtures of acetone, isopropanol, and water	$\alpha = \frac{\alpha_{id}}{1 + (\alpha_{id}/\dot{q})(T_s - T_{sat})};$ $\frac{y_{1,s} - x_1}{y_{1,s} - x_{1,s}} = \exp\left(-\frac{\dot{q}}{B_0 \beta_L \rho_L \Delta h_v}\right);$ $B_0 = 1, and\, \beta_L = 10^{-4} m/s$	Bubble interface temperature replaces saturation temperature in calculating α. Empirically determined values of the constant, B_0, and mass transfer coefficient, β_L, employed. α_{id} - reciprocal mole-fraction average
Fujita et al. (1996). methanol/(water, ethanol, and benzene), ethanol/(water and n-butanol), benzene/heptane and water/ethylene glycol.	$\alpha = \frac{\alpha_{id}}{1 + K_s(\Delta T_{bp} / \Delta T_{id})};$ $K_s = [1 - \exp(-2.8\,\Delta T_{id} / \Delta T_{sat.1-2})];$ $\Delta T_{id} = \tilde{x}_1 \Delta T_1 + \tilde{x}_2 \Delta T_2, \Delta T_{sat.1-2} = T_{sat,1}\big\rvert_p - T_{sat,2}\big\rvert_p$	It is an extension of earlier models by Stephan and Korner (1969), and Jungnickel et al. (1980). The mass diffusion effect expressed as an empirical function of boiling temperature range and difference in saturation temperatures of pure components at the same pressure. α_{id} - reciprocal mole-fraction average

INTRODUCTION

Boiling of binary and multi-component mixtures constitutes an important process in chemical, process, air-separation, refrigeration, and many other industrial applications. Reboilers feeding the vapors to distillation columns and flooded evaporators generally employ pool boiling, while the in-tube evaporation involves flow boiling. Although the multi-component boiling is of greater practical interest, a fundamental understanding of the mechanism can be obtained first with binary mixtures.

The present work is directed toward reviewing the existing theories on pool boiling heat transfer with binary mixtures, and developing a theoretical model to predict the effect of mass diffusion on the pool boiling heat transfer.

REVIEW OF PREVIOUS WORK

Binary Pool Boiling Models and Correlations

Table 1 provides a summary of some of the important models and correlations available in the literature. Stephan and Körner (1969) recognized the importance of the term $|y_1 - x_1|$ in the reduction of binary heat transfer. An empirical constant specific to the mixture was introduced along with a pressure correction. Several later investigators modified this equation, and provided values of the empirical constant. Jungnickel (1979) modified this correlation by including a heat flux multiplier.

Calus and Rice (1972) were among the first investigators to develop an empirical model based on the bubble growth theories by Scriven (1959) and Van Stralen (1966). The term representing the reduction in bubble growth in binary systems was used in predicting heat transfer as well. Since their method yields the pool boiling coefficients directly, it cannot utilize more accurate pure component data if available. Their model was unable to predict the severe suppression seen in their own data.

Calus and Leonidopoulos (1974) model is based on theoretical considerations. Although their model could not represent the effect of composition on the heat transfer well, it provided a lower mean error than Stephan and Körner (1969) and Calus and Rice (1972) correlations.

Schlünder (1982) introduced the difference between the saturation temperatures of the pure components at the same pressure as a parameter in his correlating scheme. Also, a correction factor incorporating the mass transfer coefficient was introduced to modify the Stephan and Körner's (1969) correlation. The mass transfer coefficient was treated as an empirical constant. Their scheme formed the basis for a number of later modifications.

Thome (1983) recognized the need to account for the rise in the saturation temperature at the bubble interface. He introduced the boiling range as a parameter. Later, Thome and Shakir (1987) introduced the mass transfer correction factor proposed by Schlünder (1982).

Wenzel et al. (1995) followed a similar approach to Schlünder (1982), but set out to obtain the actual value of the interface concentration by applying the mass transfer equations. The mass transfer coefficient was empirically determined to be 10^{-4} m/s. The interface concentration was then used in determining the interface temperature.

Fujita and Tsutsui (1994) modified Thome and Shakir's (1982) correlation by replacing the mass transfer term with a heat flux dependent term. The empirical constant in their correlation was evaluated from the experimental data. Fujita et al. (1996) modified the Fujita and Tsutsui correlation by introducing a term containing ΔT_{id}.

Ideal Heat Transfer Coefficient for Mixtures

The ideal pool boiling heat transfer coefficient for mixtures used extensively in the literature is defined on the basis of a mole fraction average of the wall superheat for the pure fluids, which results in the reciprocal mole-fraction average equation for α:

$$\alpha_{PB,B,id} = \left[\tilde{x}_1 / \alpha_{PB,1}\Big|_{p\ or\ T} + \tilde{x}_2 / \alpha_{PB,2}\Big|_{p\ or\ T} \right]^{-1} \tag{1}$$

The ideal temperature difference ΔT_{id} is obtained from a linear mole fraction average of ΔT_1 and ΔT_2. Calus and Rice (1972) and Calus and Leonidopoulos (1974) used mass fractions instead of mole fractions.

PSEUDO-SINGLE COMPONENT HEAT TRANSFER COEFFICIENT, $\alpha_{PB,B,psc}$

The ideal heat transfer coefficient given by eq. (1) was intended to represent the heat transfer coefficient of the mixture in the absence of any mass diffusion effects. Since the pool boiling heat transfer is a highly non-linear phenomenon there is little justification in using any of these averaging techniques to obtain the "ideal" value for the mixture. There is also no clear indication, or justification, as to the state at which the respective pure component values should be determined.

In the present work, a pseudo-single component heat transfer coefficient, $\alpha_{PB,B,psc}$, is introduced. $\alpha_{PB,B,psc}$ is evaluated by incorporating the effects of the relevant mixture properties on the pool boiling heat transfer. To determine these effects, Stephan and Abdelsalem (1980) correlation for pure fluids is employed. Using this correlation, the effect of individual properties in the region away from the critical point can be expressed as:

$$\alpha_{PB} = (C_1)\, T_{sat}^{-0.674}\, \Delta h_{LG}^{0.371}\, \rho_G^{0.297}\, \sigma^{-0.317}\, \lambda_L^{0.284}\, \beta_0^{0.066}\, \rho_L^{0.062}\, c_{p,L}^{0.042} \tag{2}$$

The average mixture heat transfer coefficient, $\alpha_{PB,B,avg}$ is defined as the heat transfer coefficient obtained with linear mass fraction averaged equations for T_{sat} and other properties in eq. (2). Alternatively, the following equation is able to predict $\alpha_{PB,B,avg}$ directly from α_1 and α_2 for pure fluids within less than 5 percent even when the individual properties were varied by a factor of 2-3.

$$\alpha_{PB,B,avg} = 0.5\left[\left(x_1\alpha_1 + x_2\alpha_2 \right) + \left(x_1/\alpha_1 + x_2/\alpha_2 \right)^{-1} \right] \tag{3}$$

The actual mixture properties are however different than the linear averaged values. $\alpha_{PB,B,psc}$ for the mixture is based on the actual mixture properties. The effect of mixture properties is incorporated through eq. (2) by neglecting the terms β_0, σ and $c_{p,L}$ which have small exponents.

$$\alpha_{PB,B,psc} = \alpha_{PB,B,avg} \left(\frac{T_{sat,m}}{T_{sat,avg}} \right)^{-0.674} \left(\frac{\Delta h_{LG,m}}{\Delta h_{LG,avg}} \right)^{0.371} \left(\frac{\rho_{G,m}}{\rho_{G,avg}} \right)^{0.297} \left(\frac{\sigma_m}{\sigma_{avg}} \right)^{-0.317} \left(\frac{\lambda_{L,m}}{\lambda_{L,avg}} \right)^{0.284} \tag{4}$$

The subscript "m" refers to the actual mixture properties, and "avg" refers to the mass-fraction averaged properties. Near the critical state, the approximation $(\rho_L - \rho_G) \approx \rho_L$ is no longer valid, and eq. (4) could be modified to include $\rho_L/\rho_{L,avg}$, with an appropriate exponent.

$\alpha_{PB,B,psc}$ offers several advantages over the commonly used $\alpha_{PB,B,id}$ since the actual mixture properties are employed. Mixtures with a high degree of non-ideality are also represented accurately. The second advantage is that α_1 and α_2 can be calculated from any appropriate correlation, such as Stefan and Abdelsalam (1982), Gorenflo (1984), Rohsenow (1950), etc. Alternatively, experimental values of α_1 and α_2, if available, could be directly employed in equation (3). Further, α_1 and α_2 may be obtained at the same total pressure, the same reduced pressure, or the same system temperature as the mixture.

BINARY POOL BOILING HEAT TRANSFER COEFFICIENT

The equilibrium interface concentrations are first calculated as a bubble grows asymptotically under the diffusion controlled growth. A one-dimensional transient heat and mass transfer analysis is then conducted to estimate the effect of mass diffusion on the heat transfer.

Liquid Concentration at the Interface

Consider a bubble growing on a heated wall in a binary system. The concentration of component 1 varies from $x_{1,s}$ at the interface to x_1 in the bulk liquid across the concentration boundary layer thickness δ_m. According to Van Stralen (1979), the interface concentration approaches an asymptotic value. Consider an instant when the bubble radius is R and the concentration gradient extends from R to R+δ_m. The evaporated vapor leaves the interface at the equilibrium vapor phase concentration of $y_{1,s}$.

The excess pressure inside a bubble due to the interface curvature is negligible in the latter stages of the bubble growth. The system pressure is therefore used in determining properties at the interface. The concentration gradient in the vapor inside the bubble is neglected. The average concentration in the boundary layer is represented by $x_{1,BL,avg}$ which is calculated from a mass balance of component 1 in the boundary layer and inside the bubble.

$$x_{1,BL,avg} = x_1 - \frac{1}{3}\frac{R}{\delta_m}\frac{\rho_G}{\rho_L}(y_{1,s} - x_1) \quad (5)$$

Following the analysis by Mikic (1969) for transient heat transfer under 1-D assumption with a planar interface, the liquid concentration at a distance z from the interface at any instant t from the bubble inception is obtained as:

$$(x_{1,z} - x_{1,s})/(x_1 - x_{1,s}) = erf\left(z/(2\sqrt{D_{12}t})\right) \quad (6)$$

The concentration boundary layer thickness is given by:

$$\delta_m = (\pi D_{12} t)^{1/2} \quad (7)$$

Combining equations (6) and (7),

$$(x_{1,z} - x_{1,s})/(x_1 - x_{1,s}) = erf\left[\left(\sqrt{\pi}/2\right)\left(z/\delta_m\right)\right] \quad (8)$$

Integrating equation (8), the average concentration in the boundary layer is obtained:

$$(x_{1,BL,avg} - x_1)/(x_{1,s} - x_1) = 0.313 \quad (9)$$

Combining equations (5) and (9), $x_{1,s}$ is obtained as:

$$x_{1,s} = x_1 - 1.06(R/\delta_m)(\rho_G/\rho_L)(y_{1,s} - x_1) \quad (10)$$

The ratio R/δ_m is obtained by applying the bubble growth equations developed by Van Stralen (1975) for the 1-D planar approximation.

$$R = \left(2/\pi^{1/2}\right) Ja_0 \, (\kappa t)^{1/2} \quad (11)$$

where $\Delta T_s = T_{sat,s} - T_{sat}$, and the modified Jakob number, Ja_0, and g are given by:

$$Ja_0 = \frac{(T_W - T_{L,sat})}{(\rho_V/\rho_L)[\Delta h_{LG}/c_{p,L} + (\kappa/D_{12})^{1/2}(\Delta T_S/g)]} \quad (12)$$

$$g = (x_1 - x_{1,s})/(y_{1,s} - x_{1,s}) \quad (13)$$

Dividing eq. (11) by eq. (7), R/δ_m is obtained as:

$$R/\delta_m = (2/\pi) Ja_0 \, (\kappa/D_{12})^{1/2} \quad (14)$$

and the interface concentration is obtained by combining equations (10) and (14):

$$x_{1,s} = x_1 - (2.13/\pi) Ja_0 (\kappa/D_{12})^{1/2} (\rho_G/\rho_L)(y_{1,s} - x_1) \quad (15)$$

Effect of Mass Diffusion Resistance on Heat Transfer

The mass diffusion is the controlling mechanism for heat transfer, and limits the availability of the more volatile component at the interface. The binary pool boiling heat transfer coefficient, $\alpha_{B,PB}$ is expressed in terms of the pseudo-single component heat transfer coefficient and a diffusion correction factor, F_D.

$$\alpha_{PB,B} = \alpha_{PB,B,psc} F_D \quad (16)$$

The heat transfer coefficient is assumed to be proportional to the evaporation rate at the bubble interface. F_D is then obtained by comparing the mass transfer rates with and without the diffusion resistance. In the absence of the mass diffusion effects, the interface temperature corresponds to the bulk liquid saturation temperature, and the heat flux due to transient conduction at the interface under 1-D approximation is given by:

$$\dot{q}_s = \lambda_L (T_W - T_{sat})/(\pi \kappa t)^{1/2} \quad (17)$$

The resulting evaporation mass flux is given by:

$$\dot{m}_s = \dot{q}_s/\Delta h_{LG} = (1/\Delta h_{LG})\lambda_L (T_W - T_{sat})/(\pi \kappa t)^{1/2} \quad (18)$$

A similar analysis could be conducted for the mass diffusion in the boundary layer. For relatively low total mass flux, the transient mass diffusion equation similar to heat diffusion equation can be applied. The resulting mass flux is given by:

$$\dot{m}_{1,s} = \rho_L D_{12} (x_1 - x_{1,s})/(\pi D_{12} t)^{1/2} \quad (19)$$

The total evaporation rate in the binary system is obtained by dividing equation (19) with ($y_{1,s}$-x_1).

$$\dot{m}_{s,D} = \frac{\dot{m}_{1,s}}{(y_{1,s} - x_1)} = \frac{\rho_L D_{12}}{(\pi D_{12} t)^{1/2}} \frac{x_1 - x_{1,s}}{y_{1,s} - x_1} \quad (20)$$

The reduction in the heat transfer coefficient, represented by the factor F_D is obtained by comparing the two evaporation rates given by eqs. (18) and (20).

$$F_D = \frac{\dot{m}_s}{\dot{m}_{s,D}} = (\frac{D_{12}}{\kappa})^{1/2} \frac{x_1 - x_{1,s}}{y_{1,s} - x_1} \frac{\Delta h_{LG}}{c_p (T_W - T_{sat})} \quad (21)$$

Equation (21) can be simplified with eqs. (15) and (12).

$$F_D = \frac{2.13}{\pi}\left[1 + (c_{p,L} / \Delta h_{LG})(\kappa / D_{12})^{1/2} (\Delta T_s / g)\right]^{-1} \quad (22)$$

At small values of the parameter ($\Delta T_s/g$), such as near the azeotropic compositions, or in the vicinity of pure fluids, eq. (20) will not be applicable as the total mass flux is much higher than the diffusion flux. Also, the diffusion factor F_D should reduce to 1 for pure fluid case. This transition is seen to occur at $\Delta T_s/g \approx 0.3$.

Further simplification is obtained by using the slope of the bubble point curve, and using y_1 in stead of $y_{1,s}$ and introducing a new parameter V_1, called as the volatility parameter.

$$V_1 = (c_{p,L} / \Delta h_{LG})(\kappa / D_{12})^{0.5} \left|(dT / dx_1)(y_1 - x_1)\right| \quad (23)$$

At V_1=0.005, the bracketed term in eq. (21) is close to 1, and a linear interpolation is used in the region $0 \le V_1 \le 0.005$, with F_D=1 at V_1=0, and F_D=0.678 at V_1=0.005. The final expression for F_D is given below.

Final Expression For F_D

For V_1>0.005

$$F_D = 0.678\left[1 + (c_{p,L} / \Delta h_{LG})(\kappa / D_{12})^{1/2} (\Delta T_s / g)\right]^{-1} \quad (24)$$

and for $0 \le V_1 \le 0.005$,

$$F_D = 1 - 64.0 V_1 \quad (25)$$

Although the expression for F_D looks similar to that by Calus and Leonidopoulos (1974) except for the leading constant, there is one major difference. The expression for g incorporates the vapor phase composition, $y_{1,s}$ at the interface condition, and not y , corresponding to the bulk condition. This requires the calculation of the interface compositions using eq. (15). An iterative scheme is therefore needed to obtain $y_{1,s}$. Using y_1 in stead of $y_{1,s}$ introduces errors which are significantly higher for mixtures with large V_1, such as water/ethylene glycol.

COMPARISON WITH EXPERIMENTAL DATA

The model developed in this work is compared with the experimental data reported by Fujita et al. (1996) and by Jungnickel et al. (1979). The mixtures represent diverse combinations; ethylene glycol/water system has a large difference in the volatility and the boiling points, and methanol/benzene, R-23/R-13 and R-22/R-12 form azeotropes. The properties of the mixtures were obtained from NIST (1995) and HYSIM

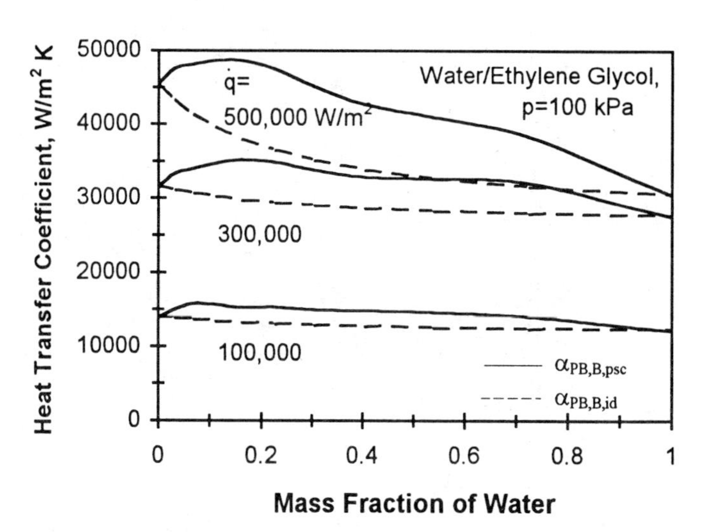

Fig. 1 Variation of pseudo-single component and ideal heat transfer coefficients with composition, water/ethylene glycol, p=100 kPa

(1996) programs. The diffusion coefficients were calculated by the Vignes correlation (1971) given below.

$$D_{12} = (D_{12}^0)^{x_2} (D_{21}^0)^{x_1} A_{12} \quad (26)$$

This correlation is very suitable for non-ideal mixtures. D^0_{12} and D^0_{21} are the self diffusion coefficients given by the Wilke-Change (1955) correlation,

$$D_{12}^0 = 1.1782 \times 10^{-16} (\phi M_2)^{1/2} T / (\eta_{L,2} v_{m,1}) \quad (27)$$

and A_{12} is the thermodynamic factor to account for the non-ideality of the mixture. As discussed by Kandlikar et al. (1975), A_{12} is assumed to be 1.0 since the error introduced is quite small. ϕ is the association factor for the solvent (2.26 for water, 1.9 for methanol, 1.5 for ethanol, and 1.0 for unassociated solvents, Taylor and Krishna, 1993), and $v_{m,1}$ is the molar specific volume.

Figure 1 shows a comparison of the ideal heat transfer coefficient, $\alpha_{PB,B,id}$, used by previous investigators, and the new pseudo-single component heat transfer coefficient, $\alpha_{PB,B,psc}$ for ethylene glycol mixture at 100 kPa. It can be seen that the two are quite different. The main reason for this is the abrupt change in the vapor density with the slight addition of water to pure ethylene glycol. The vapor phase has a high concentration of water. Pool boiling heat transfer is expected to be affected by this behavior. $\alpha_{PB,B,id}$ does not reflect this property effect, and therefore shows an almost monotonous variation with the composition. For azeotropic compositions, $\alpha_{PB,B,psc}$ represents the heat transfer coefficient for the mixture without the diffusion effects, and is expected to be more accurate than $\alpha_{PB,B,id}$.

Figure 2 shows a comparison of the Fujita et al.'s (1996) data for ethylene glycol/water mixtures with the theoretical model of Calus and Leonidopoulos (1974) and the present model. The Calus and Leonidopoulos model predicts well for low concentrations of water, but overpredicts in the rest of the range. The present model shows an excellent agreement over the entire range for all the three heat fluxes with an average error of less than 12 percent.

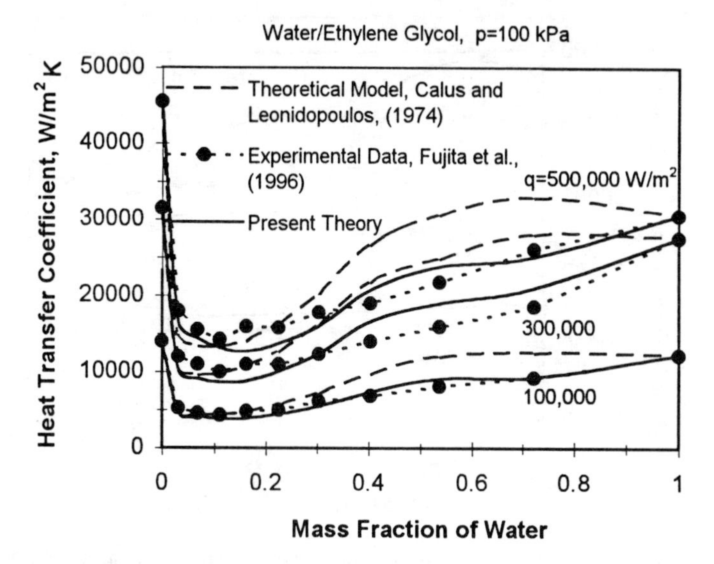

Fig. 2 Comparison of present model and Calus and Leonidopoulos (1974) model with Fujita et al.'s (1996) data, water/ethylene glycol, p=100 kPa

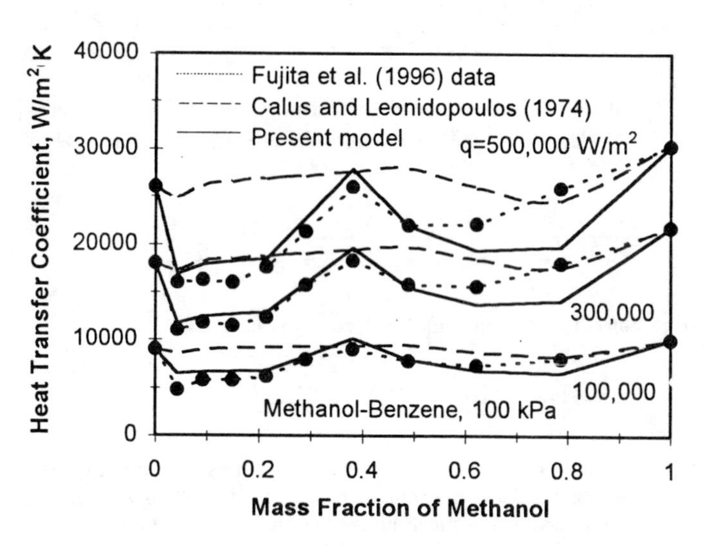

Fig. 4 Comparison of present model and Calus and Leonidoppoulos (1974) model with Fujita et al. (1996) data, methanol/benzene, p=100 kPa

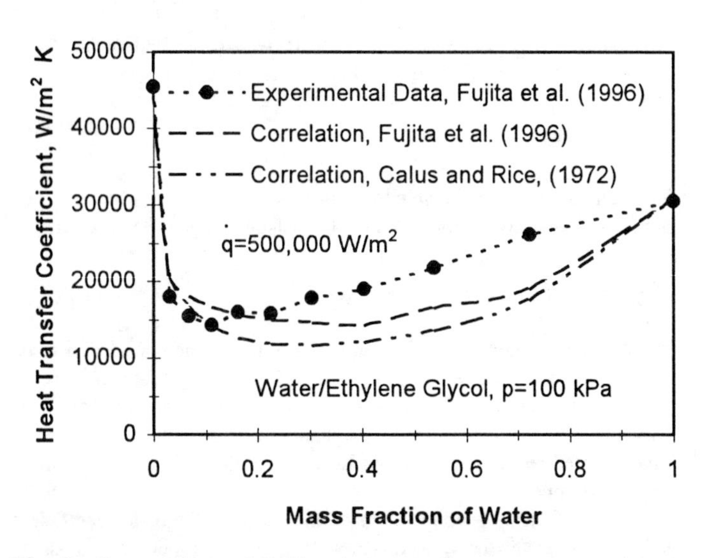

Fig. 3 Comparison of Fujita et al. (1996) and Calus and Rice (1972) correlations with Fujita et al. (1996) data, water/ethylene glycol, p=100 kPa

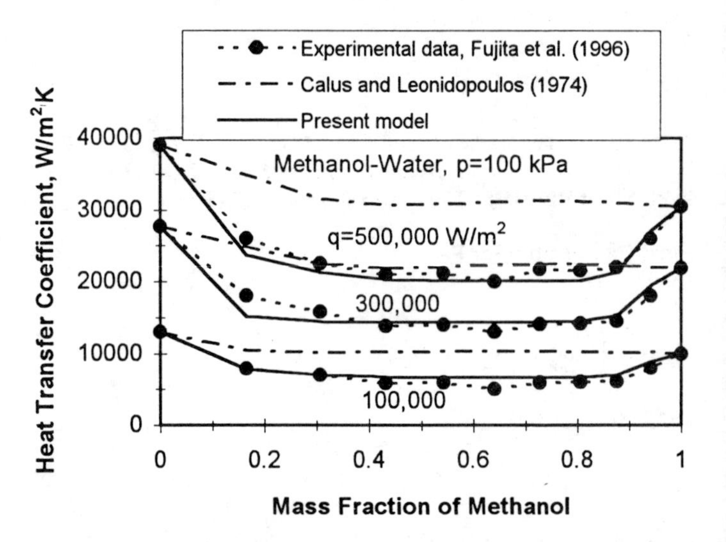

Fig. 5 Comparison of present model and Calus and Leonidopoulos (1974) model with Fujita et al. (1996) data, methanol/water, p=100 kPa

Figure 3 shows a comparison of the same data of Fig. 2 with the two empirical correlations, Calus and Rice (1972) and Fujita et al. (1996). It can be seen that both correlations perform well at lower water concentrations, while Fujita et al.'s correlation is slightly better in the higher concentration range. It may be noted that the data set under discussion was used by Fujita et al. (1996) in their correlation development.

Figure 4 shows a similar comparison as in Figure 2 with Fujita et al.'s methanol/benzene data at 100 kPa. The present model shows the same behavior as the data over the entire range including the azeotropic composition. It underpredicts the heat transfer coefficient at higher heat fluxes and at higher concentrations of methanol. However, the mean error is less than 12 percent for each heat flux data set. The Calus and Leonidopoulos (1974) model is not able to predict the suppression. The main reason is that the ethylene glycol has a large dT/dx_1 slope, and the concentration difference between the two phases at interface is quite high. For methanol/benzene, the slope dT/dx_1 is not as steep, and the concentration difference between the two phases is also smaller.

Figure 5 shows the comparison with Fujita et al.'s data for methanol/water at 100 kPa and three values of heat fluxes. The agreement between the present theoretical model and the data is excellent in the entire range with an absolute deviation between 5.6 and 9.3 percent. The model overpredicts slightly at higher heat fluxes, still within less than 10 percent mean error. The Calus and Leonidopoulos model shows a behavior similar to that for the methanol/benzene mixture shown in Fig. 4.

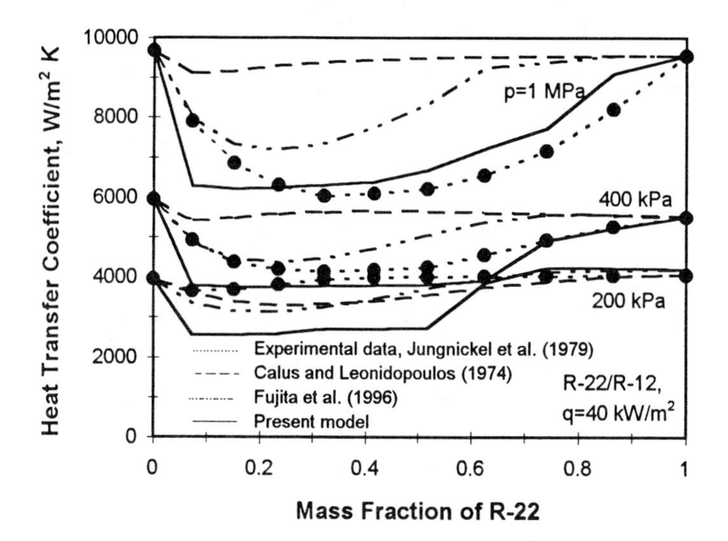

Fig. 6 Comparison of present model and Calus and Leonidopoulos (1974) model with Jungnickel et al.'s (1979) data, R-22/R-12, 40 kW/m^2

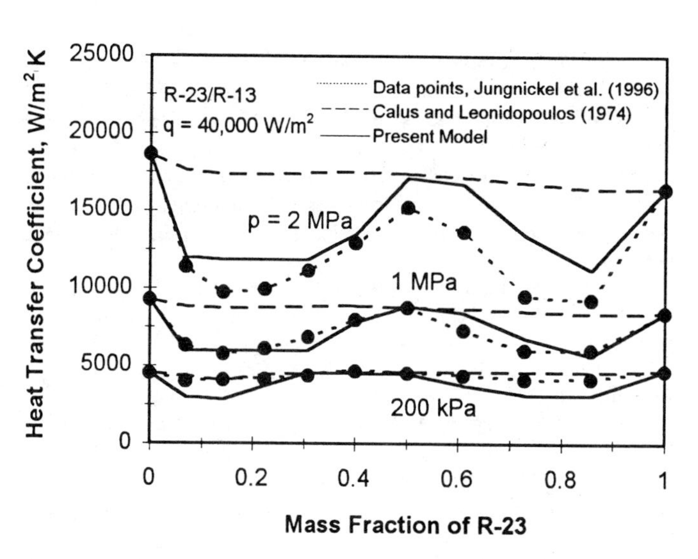

Fig. 7 Comparison of present model and Calus and Leonidopoulos (1974) model with Jungnickel et al.'s (1979) data, R-23/R-13, 40 kW/m^2

Figures 6 shows the comparison with Jungnickel et al.'s (1979) data for R-22/R-12 mixtures for three different pressures. The agreement between the proposed method and data is very good at higher pressures. At low pressure however, the proposed method predicts a lower value of heat transfer coefficient. Calus and Leonidopoulos (1974) model consistently overpredicts over the entire range. To further investigate the pressure effect, Fujita et al.'s (1996) correlation is also shown in Figure 6. Their correlation does well for the low pressure data, but overpredicts considerably at higher pressures. Further investigation is recommended to explain this effect.

Figure 7 shows the comparison with Jungnickel et al.'s (1979) R-23/R-13 data. For this set, the present model predicts well for both pressures, with an average error of less than 15 percent. Calus and Leonidopoulos (1974) model is unable to predict the suppression effect in the entire range.

CONCLUSIONS

1. The ideal heat transfer coefficient, $\alpha_{PB,B,id}$, used by earlier investigators does not truly represent the binary pool boiling heat transfer before incorporating the mass diffusion effects. A new pseudo-single component heat transfer coefficient, $\alpha_{PB,B,psc}$, is developed to properly account for the property effects in binary mixtures.

2. A theoretical model is developed to calculate the equilibrium concentrations at the interface of a bubble as it approaches the asymptotic growth condition.

3. The heat transfer coefficient for the binary mixtures is derived, based on theoretical considerations. A transient 1-D model for the heat and mass transfer at the interface of a bubble is applied under diffusion controlled growth conditions. A volatility parameter V_1 given by eq. (23) is utilized to represent the suppression effects.

4. The present model is compared with the other theoretical model by Calus and Leonidopoulos (1974) for five data sets covering a wide range of volatility difference and boiling temperature range. The Calus and Leonidopoulos model is unable to predict the heat transfer coefficients and their trends accurately, generally overpredicting in the entire range. The present model is able to predict the heat transfer coefficient quite well, better than one of the latest empirical methods proposed in the literature by Fujita el al. (1996) in most cases.

5. The final expression for the present model is given by equation (16) in conjunction with eqs. (4), (3), (24), (25), (15), (12) and (13).

6. It is recommended in future to incorporate the effects resulting from the changes in the bubble nucleation characteristics with binary mixtures by conducting experimental as well as analytical work in this area.

ACKNOWLEDGMENTS

The author is thankful to Dr. Vijay Srinivasan of Praxair Inc, Tonawanda, NY for his help in mixture property evaluation using HYSIM property routines.

REFERENCES

Calus, W.F. and Leonidopoulos, D.J., 1974, "Pool Boiling - Binary Liquid Mixtures," International Journal of Heat and Mass Transfer, Vol. 17, pp. 249-256.

Calus, W.F., and Rice, P., 1972, "Pool Boiling - Binary Liquid Mixtures," Chemical Engineering Science, Vol. 27, pp. 1687-1697.

Fujita, Y., Bai, Q., Tsutsui, M., 1996, "Heat Transfer of Binary Mixtures in Nucleate Pool Boiling," 2nd European Thermal Sciences and 14th UIT National Heat Transfer Conference, Celata, G.P., Di Marco, and Mariani, A., editors, pp. 1639-1646.

Fujita, Y., and Tsutsui, M., 1994, "Heat Transfer in Nucleate Pool Boiling of Binary Mixtures,"International Journal of Heat and Mass Transfer, Vol. 37, pp. 291-302.

Gorenflo, D., 1984, Behältersieden, Kap. Ha., VDI-Wärmeatlas, 4. Aufl., VDI-Verlag, Düsseldorf.

Gorenflo, D., Blein, P., Herres, G., Rott, W., Schömann, H., and Solol, P., 1988, "Heat Transfer at Pool Boiling of Mixtures with R-22 and R-114," Rev. Int. Froid, Vol. 11, 257-263.

HYSIM, Version 6, 1996, Hyprotech Ltd., Alberta, Canada.

Jungnickel, H., Wassilew, P., and Kraus, W.E., 1979, "Investigation on the Heat Transfer of Boiling Binary Refrigerant Mixtures," Proceedings of the

XVth International Congress on Refrigeration, Vol. II, pp.525-536.

Kandlikar, S.G., Bijlani, C.A., and Sukhatme, S.P., 1975, "Predicting the Properties of R-22 and R-12 Mixtures - Transport Properties", ASHRAE Transactions, Vol. 81, Part 1, pp. 266-284.

Kandlikar, S.G., "Correlating Heat Transfer Data in Binary Systems," Paper presented at the National Heat Transfer Conference, Minneapolis, July 1991, Phase Change Heat Transfer, E. Hensel et al., eds., ASME HTD-Vol. 159.

Mikic, B.B., and Rohsenow, W.M., 1969, "Bubble Growth Rates in Non-uniform Temperature Field," Progress in Heat and Mass Transfer, Vol. II, pp. 283-293.

NIST, 1995, REFPROP, National Institute for Science and Technology, Washington, D.C.

Rohsenow, W.M., 1952, "A Method of Correlating Heat Transfer Data for Surface Boiling Liquids,"Transactions of ASME, Vol. 74, pp. 969.

Schlünder, E.U., 1982, "Heat Transfer in Nucleate Pool Boiling of Mixtures," Proceedings of the 7th International Heat Transfer Conference, Vol. 4, pp. 2073-2079.

Scriven, L.E., 1959, "On the Dynamics of Phase Growth," Chemical Engineering Science, Vol. 10, pp. 1-13.

Stephan, K., and Körner, M., 1969, "Berechnung des Wärmeübergangs Verdamp Fender Binärer Flüssig-keitsgemische," Chemie Ing. Techn., Vol. 41, pp. 409-417.

Stephan, K., and Abdelsalam, M., 1980, "Heat Transfer Correlations for Natural Convection Boiling,"International Journal of Heat and Mass Transfer, Vol. 23, pp. 73-87.

Taylor, R. and Krishna, P., 1993, Multicomponent Mass Transfer, John Wiley & Sons, Inc., NY.

Thome, J.R., 1983, Prediction of Binary Mixture Boiling Heat Transfer Coefficients using only Phase Equilibrium Data,"International Journal of Heat and Mass Transfer, Vol. 26, pp. 965-974.

Van Stralen, S., "The Mechanism of Nucleate Boiling in Pure Liquids and Binary Mixtures, 1967, "International Journal of Heat and Mass Transfer, 990-1046, 1469-1498.

Thome, J.R., and Shakir, S., 1987, "A New Correlation for Nucleate Pool Boiling of Binary Mixtures," AIChE Symposium Series, 83, pp. 46-51.

Van Stralen, S., 1979, "Growth Rate of Vapor and Gas Bubbles," Chapter 7, Boiling Phenomena, eds. Van Stralen, S., and Cole, R.,Hemisphere Publishing Corp., U.S.A.

Vignes, A., 1966, "Diffusion in Binary Solutions," Industrial and Engineering Chemistry Fundamentals, Vol. 5, 189-199.

Wenzel, U., Balzer, F., Jamialahmadi, M., Müller-Steinhagen, H., 1995, "Pool Boiling Heat Transfer Coefficients for Binary Mixtures of Acetone, Isopropanol, and Water, Heat Transfer Engineering, Vol. 16, 36-43.

Wilke, C.R., and Chang, P., 1955, "Correlation of Diffusion Coefficients in Dilute Solutions," AIChE Journal, Vol. 1, pp. 264-270.

HTD-Vol. 342, National Heat Transfer Conference
Volume 4
ASME 1997

BOILING HEAT TRANSFER WITH BINARY MIXTURES
PART II - FLOW BOILING IN PLAIN TUBES

Satish G. Kandlikar
Mechanical Engineering Department
Rochester Institute of Technology
Rochester, NY 14623
USA
Ph: (716) 475-6728; e-mail: sgkeme@rit.edu

ABSTRACT

Flow boiling heat transfer with pure fluids comprises of convective and nucleate boiling components. In binary mixtures, in addition to the binary suppression effects present in pool boiling, the flow further modifies the nucleate boiling characteristics. In the present work, the flow boiling correlation by Kandlikar (1990) for pure fluids is modified to include the binary effects derived in Part I (Kandlikar, 1996). Three regions are defined on the basis of a volatility parameter, $V_1 = (c_p/\Delta h_{LG})(\alpha/D_{12})^{1/2}|(y_1 - x_1)dT/dx_1|$. They are: region I - near azeotropic, region II - moderate suppression, and region III - severe suppression. The resulting correlation correlates over 2500 data points within 8.3 to 13.3 percent mean deviation for each data set. Furthermore, it represents the α-x_1 trend well for five binary systems.

NOMENCLATURE

A_{12} - thermodynamic factor
Bo - boiling number, $=\dot{q}/(G\Delta h_{LG})$
c_p - specific heat, J/kg K
Co - convection number, $= (\rho_G/\rho_L)^{0.5}((1-x)/x)^{0.8}$
D - diameter of tube, m
D_{12} - Diffusion coefficient of 1 in mixture of 1 and 2
D_{12}^0 and D_{12}^0 - Diffusion coefficient of component 1 present in infinitely low concentration of liquid mixture
f - friction factor
F_D - diffusion factor, $=\alpha/\alpha_{id}$
F_{fl} - fluid-surface parameter, values listed in Table 2
g - $=(x_1 - x_{1,s})/(y_{1,s} - x_{1,s})$
Δh_{LG} - latent heat of vaporization, J/kg
Ja_0 - modified Jakob number, defined by eq. (12)
Pr - Prandtl number, $c_p\eta/\kappa$
$\dot{q}$ - heat flux, W/m^2
Re - Reynolds number, GD/η
T - temperature, K
ΔT_s - $= (T_s - T_{sat})$, K
$v_{m,1}$, $v_{m,2}$ - molar specific volume of 1 and 2, m^3/kg-mol
V_1 - volatility parameter, defined by eq. (6)
x - quality
x_1, x_2 - mass fraction of components 1 and 2 in liquid phase
y_1, y_2 - mass fraction of components 1 and 2 in vapor phase
$\tilde{x}_1$, $\tilde{x}_2$ - mole fraction of components 1 and 2 in liquid phase

Greek Letters

α - heat transfer coefficient, W/m^2 K
η - viscosity, kg/m s
κ - thermal diffusivity, m^2/s
λ - thermal conductivity, W/m K
ρ - density, kg/m^3
σ - surface tension, N/m
ϕ - association parameter

Subscripts

1, 2 - components of a binary system, 1-more volatile component
B - binary
CBD - convective boiling dominant
D - mass diffusion
G - vapor
id - ideal
L - liquid
LO - entire flow as liquid
m - mixture
NBD - nucleate boiling dominant
PB - pool boiling
psc - pseudo-single component
s - liquid-vapor interface of a bubble
sat - saturation
TP- two-phase

INTRODUCTION

Heat transfer in flow boiling of binary mixtures is receiving increasing attention in the refrigeration industry as refrigerant mixtures are being evaluated to replace the conventional refrigerants. The pure refrigerant systems in reality are mixtures of oil and refrigerant. While ternary refrigerants are currently being tested, a fundamental understanding of binary systems is essential before modeling the multi-component heat transfer. Among other applications, chemical, petrochemical and process industry applications of binary mixtures are noteworthy.

Table 1 Summary of Available Methods for Predicting Binary Flow Boiling Heat Transfer

Authors (Year)	Correlation	Comments
Calus et al. (1973)	$(\alpha_{TP} / \alpha_L) = 0.065(1 / X_{tt})(T_{sat} / \Delta T_{sat})(\sigma_{water} / \sigma_L^*)^{0.9} F^{0.6}$; $F = 1 - (y_1 - x_1)(c_{p,L} / h_{LV})(\kappa / D_{12})^{0.5}(dT / dx_1)$	Considers only convective contribution, large errors are seen when nucleate boiling is present.
Bennett and Chen (1980)	$\alpha_{TP} / \alpha_{LO} = \{[(dP / dz)_{2\phi} / (dP / dz)_L](\mathrm{Pr}_L + 1) / 2\}^{0.444}[\Delta T_m / \Delta T_s] +$ $0.00122(\lambda_L^{0.79} c_{p,L}^{0.45} \rho_L^{0.49} g_c^{0.25}) / (\sigma^{0.5} \mu_L^{0.29} h_{LV}^{0.24} \rho_V^{0.24})(\Delta T_{sat})^{0.24}(\Delta P_{sat})^{0.75} S_B \mathrm{Re}_{2\phi}$; $S_B = [1 - (c_{p,L} / h_{LV})(y1 - x_1)(dT / dx_1)(\kappa / D_{12})^{0.5} S$; $[\Delta T_m / \Delta T_s] = 1 - (1 - y_1)(\dot{q} / \Delta T_s)(dT / dx_1)\|_p ; \alpha_m = 0.023(\mathrm{Re}_{2\phi})^{0.8}(Sc)^{0.4} \rho_L D_{12} / d$; $\mathrm{Re}_{2\phi} = \mathrm{Re}_L \{[(dP / dz)_{2\phi} / (dP / dz)_L](\mathrm{Pr}_L + 1) / 2\}^{0.555}$; $S - Chen(1966)\ Suppression\ factor,\ function\ of\ \mathrm{Re}_{2\phi}$	Mass transfer analogy employed to predict the suppression in nucleate boiling. Convective component modified through a temperature difference correction term.
Jung (1988)	$\alpha_{TP} = (N / C_{UN})\alpha_{UN} + C_{me} F_p \alpha_L$ $N = 4048 X_{tt} Bo^{1.13}\ for\ X_{tt} < 1; N = 2 - 0.1\ X_{tt}^{-0.28} Bo^{-0.33}\ for\ 1 \le X_{tt} \le 5$; $\alpha_{SA} = 207(\lambda_L / b_d)[q b_d / (\lambda_L T_{sat})]^{0.674}(\rho_V - \rho_L)^{0.581} \mathrm{Pr}_L^{0.533}$; $b_d = 0.0146\beta \{2\sigma / [g((\rho_V - \rho_L)]\}^{0.5}\ with\ \beta = 35°$; $F_p = 2.37(0.29 + 1 / X_{tt})^{0.85}; C_{UN} = [1 + (b_2 + b_3)(1 + b_4)(1 + b_5)$; $b_2 = (1 - \tilde{x}_1)\ln[(1.01 - \tilde{x}_1) / (1.01 - \tilde{y}_1)] + \tilde{x}_1 \ln(\tilde{x}_1 / \tilde{y}_1) + \|\tilde{x}_1 - \tilde{y}_1\|^{1.5}$; $b_3 = 0\ for\ \tilde{x}_1 \ge 0.01; b_3 = (\tilde{x}_1 / \tilde{y}_1)^{0.1} - 1\ for\ \tilde{x}_1 < 0.01; b_4 = 152(p / p_{cmvc})^{3.9}$; $b_5 = 0.92\|\tilde{x}_1 - \tilde{y}_1\|^{0.001}(p / p_{cmvc})^{0.66}; x_1 / y_1 = 1\ for\ x_1 = y_1 = 0$; $\alpha_{UN} = \alpha_i / C_{UN}; \alpha_i = [\tilde{x}_1 / \alpha_1 + x_2 / \alpha_2]^{-1}; C_{me} = 1 - 0.35\|\tilde{x}_1 - \tilde{y}_1\|^{1.56}$	The ideal mixture α concept for pool boiling is extended for flow boiling. It causes the averaging of the convection contribution as well from the pure component flow boiling α values. A large number of empirically determined constants are introduced.
Kandlikar (1991)	$\alpha_{TP,B} = \alpha_{Conv} + \frac{\alpha_{Nucl}}{\left[1 + \|y_1 - x_1\|(\kappa / D_{12})^{1/2}\right]^{0.7}}$ α_{Conv} and α_{Nucl} obtained from Kandlikar (1990) correlation for flow boiling	The nucleate boiling term was modified to account for the binary effects.

REVIEW OF LITERATURE

There are relatively fewer studies available in the literature on modeling the flow boiling heat transfer compared to those on pool boiling of binary mixtures. Table 1 provides a summary of some of the available correlations. One of the early developments was proposed by Calus et al. (1973) who extended the pool boiling suppression factor derived by Calus and Rice (1972). Calus et al. modified an existing flow boiling correlation, which included only a convective term, and introduced an additional correction factor to account for the rise in the saturation temperature at the interface of a bubble. As their correlation did not include a nucleate boiling term, it considerably underpredicted the results at higher heat flux.

Bennett and Chen (1980) presented a correlation scheme based on the Chen (1966) correlation. The convective component was modified to incorporate the change in the bubble interface temperature. The suppression factor suggested by Calus and Leonidopoulos (1974) for pool boiling was introduced in the nucleate boiling term with some modifications. The interface mass transfer coefficient was calculated from a Dittus-Boelter type correlation. Jung (1988) found that the Chen correlation overpredicted the heat transfer coefficient for pure refrigerants, and the Bennett and Chen correlation was unable to correlate his data on refrigerant mixtures.

Jung (1988) conducted an extensive study on the flow boiling of refrigerant mixtures of R-12/R-152a and R-22/R-114. He developed a correlation using an ideal heat transfer coefficient, similar to that employed for pool boiling by earlier investigators. This approach implicitly incorporates the convective component in the averaging scheme. Jung's correlation utilizes 25 empirical constants determined empirically from their own experimental data.

Kandlikar (1991) extended his earlier pure component flow boiling correlation (Kandlikar, 1990) to binary mixtures. It was postulated that only the nucleate boiling component would be affected due to the diffusion effects, which were modeled after a semi-empirical approach proposed by Calus and Rice (1972) for pool boiling. The results were satisfactory when compared to Jung's (1988) data. However, with the availability of new data sets, it was seen that the Kandlikar (1991) correlation for binary mixtures yielded larger errors, usually underpredicting the suppression effects. One of the main reasons was the inability of the underlying Calus and Rice's suppression model to predict the severe suppression seen for mixtures with large volatility differences as shown in Part I.

OBJECTIVES OF THE PRESENT WORK

From the above discussion it can seen that there is a need for developing a flow boiling correlation applicable to binary systems and azeotropes, particularly with the current interest in refrigerant mixtures. The present work is aimed toward incorporating the binary pool boiling suppression model developed in Part I of this paper in the binary flow boiling model. The new model is tested with experimental data from six different sources covering a broad range of variables. The parametric trends in the binary flow boiling heat transfer are also investigated.

FLOW BOILING MODEL FOR MIXTURES

In developing the present model, the Kandlikar (1990) correlation for pure fluids is used as the starting point. This correlation was able to represent the dependence of α on quality x, mass flux G, and heat flux q. The flow boiling correlation for pure.

$$\alpha_{TP} = larger\ of \begin{cases} \alpha_{TP,NBD} \\ \alpha_{TP,CBD} \end{cases} \tag{1}$$

The subscripts NBD and CBD in eq. (1) refer to the nucleate boiling dominant and the convective boiling dominant regions for which the α_{TP} values are given by:

$$\alpha_{TP,NBD} = 0.6683 Co^{-0.2}(1-x)^{0.8}\alpha_{LO} + 1058.0 Bo^{0.7}(1-x)^{0.8} F_{Fl}\alpha_{LO} \tag{2}$$

$$\alpha_{TP,CBD} = 1.136 Co^{-0.9}(1-x)^{0.8}\alpha_{LO} + 667.2 Bo^{0.7}(1-x)^{0.8} F_{Fl}\alpha_{LO} \tag{3}$$

The first terms in eqs. (2) and (3)represent the convective components, while the second terms, which include Bo, represent the nucleate boiling component. For horizontal tubes with Froude number, Fr_{LO}, less than 0.04, an additional multiplier $(25Fr_{LO})^{0.324}$ is applied to the first terms in eqs. (2) and (3).

F_{fl} is a fluid-surface parameter related to the nucleation characteristics. Table 2 lists its value for several fluids. The single-phase heat transfer coefficient, α_{LO}, is obtained from the following correlations, as suggested later by Kandlikar (1992).

Petukhov and Popov (1963) for $0.5 \le Pr_L \le 2000$ and $10^4 \le Re_{LO} \le 5x10^6$ -

$$Nu_{LO} = \alpha_{LO} D/\lambda_L = Re_{LO} Pr_L (f/2)\left[1.07 + 12.7(Pr_L^{2/3} - 1)(f/2)^{0.5}\right]^{-1} \tag{4a}$$

Gnielinski (1976) for $0.5 \le Pr_L \le 2000$ and $2300 \le Re_{LO} < 10^4$ -

$$Nu_{LO} = \alpha_{LO} D/\lambda_L = (Re_{LO} - 1000) Pr_L (f/2)\left[1.0 + 12.7(Pr_L^{2/3} - 1)(f/2)^{0.5}\right]^{-1} \tag{4b}$$

The friction factor in eqs. (4a) and (4b) is given by

$$f = \left[1.58 \ln(Re_{LO}) - 3.28\right]^{-2} \tag{5}$$

Table 2 Fluid-Surface Parameter F_{fl} for Refrigerants in Copper or Brass Tubes

Fluid	F_{fl}
Water	1.00
R-11	1.30
R-12	1.50
R-13B1	1.31
R-22	2.20
R-113	1.30
R-114	1.24
R-124	1.9
R-134a	1.63
R-152a	1.10

For all fluids in stainless steel tubes, F_{fl}=1.0

The suppression in the nucleate boiling is derived on the basis of the suppression factor F_D presented in Part I of this paper. A volatility parameter V_1 is used to identify different levels of suppression.

$$V_1 = \frac{c_{p,L}}{\Delta h_{LG}}\left(\frac{\kappa}{D_{12}}\right)^{0.5}\frac{dT}{dx_1}(y_1 - x_1) \tag{6}$$

The pool boiling suppression factor F_D is given by the following equation.

$$F_D = C\left[1 + (c_{p,L}/\Delta h_{LG})(\kappa/D_{12})^{1/2}(\Delta T_s/g)\right]^{-1} \tag{7}$$

where C is derived for different ranges of V_1.

$$C = \begin{cases} 2.13/\pi,\ \ or\ 0.678; & for V_1 > 0.005 \\ 1 - \left(\dfrac{1 - 2.13/\pi}{0.05}\right)V_1,\ \ or\ 1 - 64V_1; & for V_1 \le 0.005 \end{cases} \tag{8}$$

The bubble interface concentration is given by:

$$x_{1,s} = x_1 - 0.678 Ja_0 (\kappa/D_{12})^{1/2}(\rho_G/\rho_L)(y_{1,s} - x_1) \tag{9}$$

where

$$g = (x_1 - x_{1,s})/(y_{1,s} - x_{1,s}) \tag{10}$$

$$\Delta T_s = T_{sat,s} - T_{sat} \tag{11}$$

and the Jakob number for the binary system is given by

$$Ja_0 = \frac{(T_W - T_{L,sat})}{(\rho_G/\rho_L)\left[(\Delta h_{LG}/c_{p,L}) + (\kappa/D_{12})^{1/2}(\Delta T_s/g)\right]} \tag{12}$$

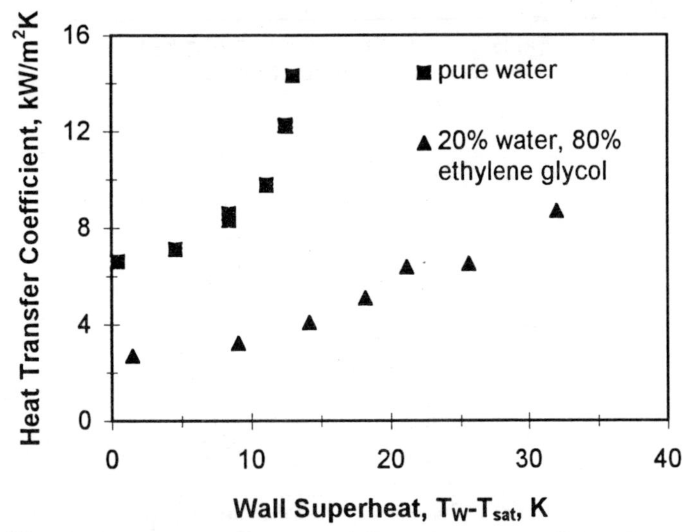

Fig. 1 Heat transfer characteristics of water/ethylene glycol solution under subcooled flow boiling, flow velocity =0.4 m/s in 3 mm x 40 mm rectangular channel

The diffusion coefficient D_{12} is calculated by the Vignes correlation (1971).

$$D_{12} = (D_{12}^0)^{x_2} (D_{21}^0)^{x_1} A_{12} \tag{13}$$

This correlation is found to be very suitable for non-ideal mixtures. D^0_{12} and D^0_{21} are the self diffusion coefficients given by Wilke-Change (1955).

$$D_{12}^0 = 1.1782 \times 10^{-16} \frac{(\phi M_2)^{1/2} T}{\eta_{L,2} V_{m,1}} \tag{14}$$

and A_{12} is the thermodynamic factor to account for the non-ideality of the mixture. Kandlikar et al. (1975) recommended A_{12}= 1.0 since it introduces only a small error. ϕ is the association factor for the solvent (2.26 for water, 1.9 for methanol, 1.5 for ethanol, and 1.0 for unassociated solvents, Taylor and Krishna, 1993).

The suppression factor given by eq. (7) is derived for pool boiling. Before this factor is applied to the nucleate boiling component in flow boiling, a valuable insight is obtained from the experiments conducted by Kandlikar and Raykofff (1997). They employed subcooled water/ethylene glycol solutions over a flat 9.5 mm circular heater placed flush on the lower wall of a 3 mm x 40 mm horizontal flow channel. Figure 1 shows the variation of heat transfer coefficient with wall superheat for pure water, and water/ethylene glycol solution. Nucleation begins at ONB (onset of nucleate boiling), and α increases slowly from its single phase value. This region is similar to the CBD region in which the convective effects are dominant. Beyond a certain value of wall superheat, α begins to rise rapidly due to the increased nucleation activity, and this region corresponds to the NBD region. The binary results for ethylene glycol solution, also shown in Fig. 1, indicate the same trend of slowly increasing α in the CBD region, but the sharp increase in α corresponding to the NBD region is not seen even for a superheat of 33 °C. This shows that the nucleate boiling dominant region is not present, or pushed significantly toward the higher wall superheats for binary mixtures.

From the above discussion, it can be concluded that the CBD region extends into considerably higher values of wall superheats in binary flow boiling. For conditions near azeotropic compositions, the pure component correlation should however hold good. For the conditions where the binary suppression effects are significant, the suppression factor could be applied to the nucleate boiling component in the CBD region correlation. The volatility parameter, V_1, given by eq. (6), derived for binary pool boiling is again utilized here to identify the different levels of suppression. The criteria identifying these regions are obtained by analyzing over 2500 data points for four binary systems reported by five investigators. The following correlation is presented to predict the flow boiling of binary mixtures. The properties of the mixtures at saturation state are employed in the following equations.

Binary Flow Boiling Correlation

Region I: Near-Azeotropic Region; $V_1 < 0.03$

$$\alpha_{TP,B} = larger\ of \begin{cases} \alpha_{TP,B,NBD} \\ \alpha_{TP,B,CBD} \end{cases} \tag{15}$$

In this region, the general correlation for pure fluids is applicable and $\alpha_{TP,NBD}$ and $\alpha_{TP,CBD}$ are obtained from eqs. (2) and (3) respectively using the mixture properties. The fluid-surface parameter for the mixture is given by:

$$F_{fl} = x_1 F_{fl,1} + x_2 F_{fl,2} \tag{16}$$

The near-azeotropic region covers azeotropes and low volatility difference mixtures. Although the nucleate boiling is slightly affected for these mixtures, its effect on the flow boiling heat transfer is insignificant.

Region II: Moderate Suppression Region, $0.03 \le V_1 < 0.2$, and Bo>1E-4

$$\alpha_{TP,B} = \alpha_{CBD,B} = 1.136\, Co^{-0.9} (1-x)^{0.8} \alpha_{LO} + 667.2\, Bo^{0.7} (1-x)^{0.8} F_{fl} \alpha_{LO} \tag{17}$$

F_{fl} for mixtures is obtained from eq. (16). Here, the nucleation is suppressed and the convective heat transfer become dominant. In the CBD region, the bubble growth is primarily limited to the early stages. The correlation for the CBD region without any suppression factor is therefore able to predict this region well.

Region III Severe Suppression Region, (a) For $0.03 \le V_1 < 0.2$ and Bo<1E-4, and (b) $V_1 \ge 0.2$;

$$\alpha_{TP,B} = 1.136 Co^{-0.9} (1-x)^{0.8} \alpha_{LO} + 667.2 Bo^{0.7} (1-x)^{0.8} F_{Fl} \alpha_{LO} F_D \tag{18}$$

F_D is obtained from the following equation:

$$F_D = 0.678 \left[1 + (c_{p,L} / \Delta h_{LG})(\kappa / D_{12})^{1/2} \left| (y_1 - x_1)(dT / dx_1) \right| \right]^{-1} \tag{19}$$

where dT/dx_1 is the slope of the bubble point temperature versus x_1 curve. Equation (19) is simplified from eq. (7) to eliminate the iteration for the interface concentrations. The resulting error in F_D is small, usually less than 5 percent for refrigerant mixtures, and is applied only to the nucleate boiling term. For systems with large volatility parameter, use of eq. (7) is recommended in place of eq. (19).

Table 3 Details of the Experimental Data Available for Binary Flow Boiling

Source (Year)	Binary System	Press (bar)	Tube/ Orient.	Mass Fraction	Quality	Heat Flux kW/m^2	Mass Flux, kg/m^2s
Jung et al (1988)	R-12/R-152a R-22/R-114, R-500	3.1-4.8	9 mm, SS, Hor.	0-1	0-0.9	10-45	250-720
Hihara et al. (1989)	R-12/R-22 R-22/R-114	0.23-0.83	8 mm, SS, Hor.	0-1	0-1	5.8-28.5	100-350
Takamatsu et al. (1993)	R-22/R-114	4 - 8.1	7.9 mm, Copper, Hor.	0-1, Only 0.51 used	0-0.9	1.8-72.8	214-393
Celata et al. (1993)	R-12/R-114	10 - 30	7.57 mm, SS, Hor.	0 - 1	0 - 1	10 - 45	300 - 1800
Murata and Hashizume (1993)	R-134a/R-123	2.2 and 2.4	10.3 mm, Copper, Hor.	90/30 mole-fraction	0.1-1	10 - 30	100 - 300

Table 4 Parameter Ranges of Data Sources and Comparison with the Present Correlation

Data source	Binary System	Bo x 10e-5	Co	V_1	Present Correlation % Mean Absolute Deviation
Jung et al. (1988)	R-12/R-152a	6.1-6.3	0.52-1.8	0.025-0.044	8.3 %
	R-12/R-152a	8.8 - 71	0.01 - 1.45	0.013 - 0.022	10.4 %
	R-22/R-114	7.1- 77	0.04 - 1.39	0.1 - 0.72	13.0 %
	R-500	7.5 - 77	0.01 - 1.83	0.0006 - 0.002	11.4 %
Hihara et al. (1989)	R-12/R-22	47 - 61	0.01 - 1.91	0.015 - 0.064	13.3 %
	R-22/R-114	37 - 60	0.023- 1.64	0.07 - 0.67	9.0 %
Takamatsu et al. (1993)	R-22/R-114	19 - 76	0.018 - 3.2	0.28 - 0.54	9.2 %
Celata et al. (1993)	R-12/R-114	9.4 - 88	0.52 - 1.83	0.06 - 0.15	8.9 %
Murata and Hashizume (1993)	R-134a/R-123	20 - 185	0.004 - 0.05	0.28 - 0.34	12.1 %

The severely suppressed region is dominated by the convective effects. The nucleate boiling contribution in this region is reduced due to the large difference in composition between the two phases, and the resulting mass diffusion resistance at the liquid-vapor interface of a growing bubble.

RESULTS AND DISCUSSION

Table 3 shows the details of the experimental data sets used to compare the results from the correlation. The data cover a broad range of the volatility parameter, V_1. Table 4 shows the mean deviation from each data set for different binary systems. The ranges of Bo, Co and V_1 covered in each data set are also listed in Table 4. The correlation is applicable for qualities from near zero to about 0.8. The properties of the mixtures are evaluated using REFPROP by NIST (1995).

It can be seen from Table 4 that the mean absolute error for each data set is between 8.3 and 13.3 percent. Additional discussion on the comparison for specific ranges of V_1 covering different suppression regions is presented in the following paragraphs.

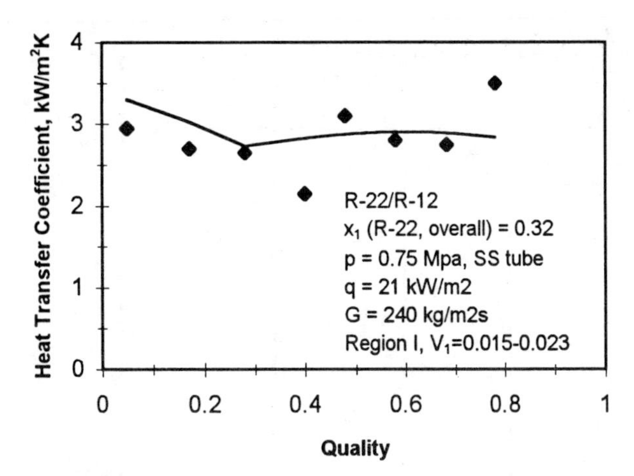

Fig. 2 Comparison of model predictions with the experimental data of Hihara et al. (1989) for R-22/R-12 in region I, near azeotropic region.

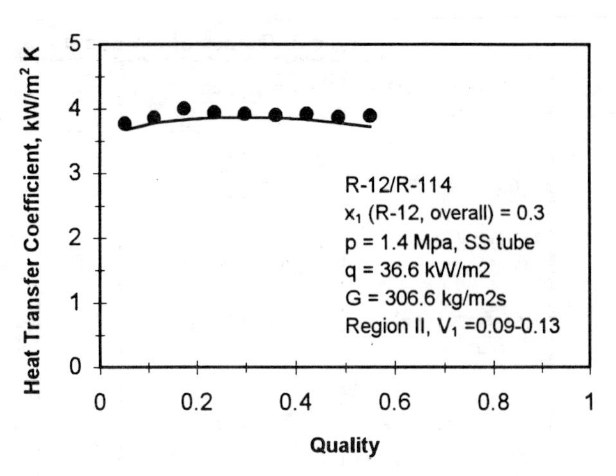

Fig. 4 Comparison of model predictions with the experimental data of Celata et al. (1993) for R-12/R-114 in region II, moderate suppression region.

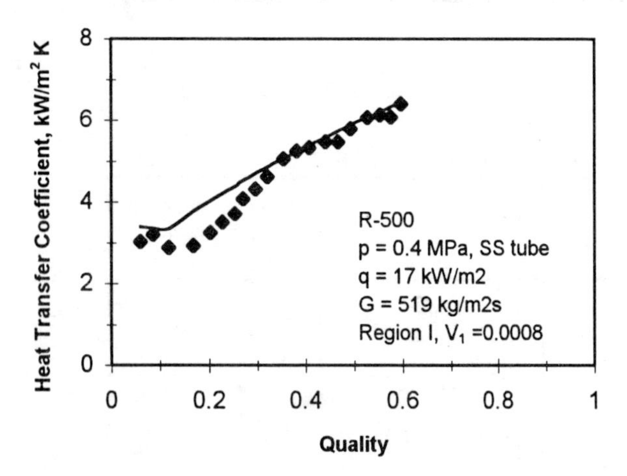

Fig. 3 Comparison of model predictions with the experimental data of Jung (1989) for R-500 azeotrope in region I, near azeotropic region.

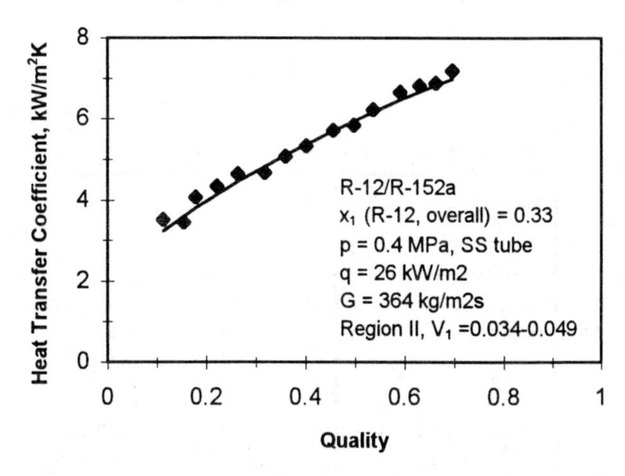

Fig. 5 Comparison of model predictions with the experimental data of Jung (1989) for R-12/R-152a in region II, moderate suppression region.

Region I - Near Azeotropic Region

In this region, the compositions of the two phases are nearly equal. The data by Jung et al. (1988) on azeotrope R-500 is representative of this region. R-500, an azeotrope of R-12 and R-152a, shows a small volatility difference with V_1 in the range of 0.0006-0.002. Figure 2 shows a comparison between the predicted and the experimental variation of α with quality x. It can be seen that the nucleate boiling dominant region at lower x and convective boiling dominant region at higher x are well represented by the correlation. The mean absolute error for this data set is 11.4 percent.

Part of Hihara et al.'s (1989) data also falls in this region for R-22/R-12 system. Figure 3 shows a comparison between the predicted and the experimental values. It can be seen that the trend in α vs. x is well represented by the correlation showing the transition between the NBD and the CBD regions. No suppression is observed in the entire region and the pure fluid correlation correctly represents α and its trends with x.

Region II - Moderate Suppression

This region covers the range $0.03<V_1<0.2$ with Bo>1E-4. In this region, the nucleate boiling dominant region is not present, and the heat transfer is mainly in the CBD region. However, the suppression is not severe, and does not affect the nucleate boiling term in the CBD region.

Figure 4 shows Celata et al.'s (1993) data for R-12/R-114 obtained in an electrically heated stainless steel test section. F_{fl} for this case is 1.0 throughout. As seen from Fig. 4, the correlation is able to predict the data well. The mean absolute error with this data set is 8.9 percent. Similar observations are made with Jung's (1988) data sets for R-22/R-114 and R-22/R-152a. Figure 5 shows the comparison with Jung's R-12/R-152a data falling in the moderate suppression region, with a good agreement between the observed and the predicted trends.

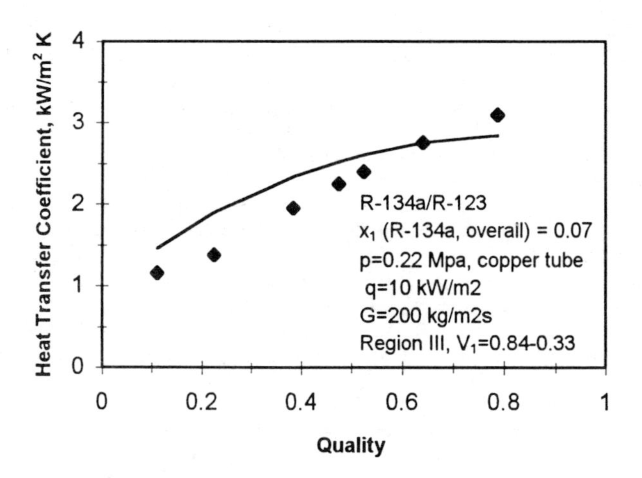

Fig. 6 Comparison of model predictions with the experimental data of Murata et al. (1989) for R-134a/R-123 in region III, severe suppression region.

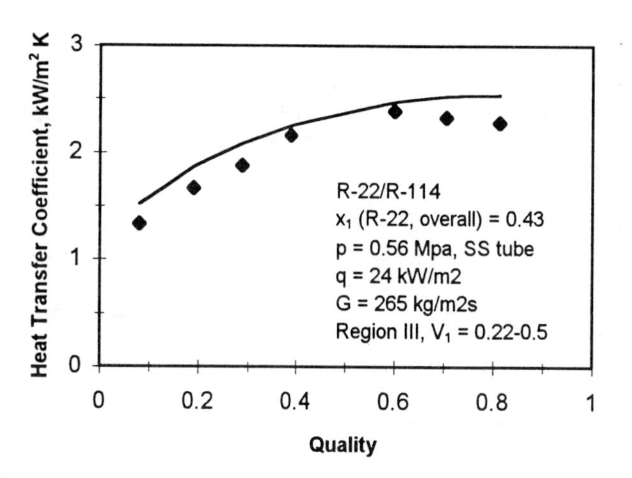

Fig. 8 Comparison of model predictions with the experimental data of Hihara et al. (1989) for R-22/R-114 in region III, severe suppression region

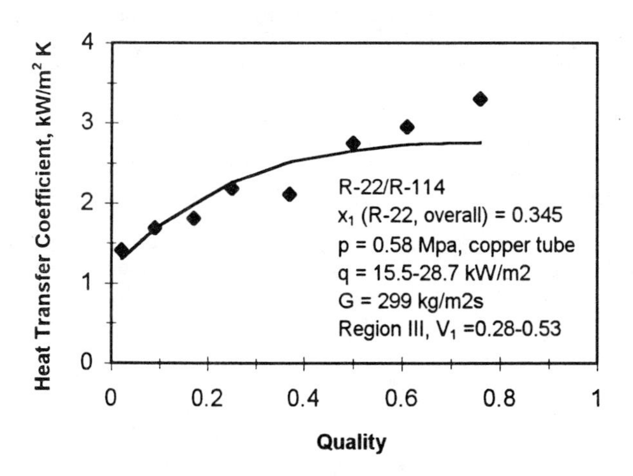

Fig. 7 Comparison of model predictions with the experimental data of Takamatsu et al. (1993) for R-22/R-114 in region III, severe suppression region

Region III - Severe Suppression

The nucleate boiling mechanism is strongly affected by the mass diffusion effects, and the nucleate boiling component in the CBD region is suppressed considerably. A large number of data investigated in the present work falls in this region.

Figure 6 shows Murata and Hashizume's (1989) data for R-134a/R-123 system. This data set is obtained using a copper tube wrapped with an electrical heater element around it. The fluid-surface parameter F_{fl} given by eq. (16) is applied for this case. Again the correlation does a good job in predicting the heat transfer coefficient in this region as well, with a mean error of 12.1 percent for the entire data set. Figure 7 shows Takamatsu et al.'s (1993) data obtained using a copper test section exchanging heat with hot water. This corresponds to a fluid heated test section. The fluid dependent parameter F_{fl} is applicable to copper and brass tubes. The correlation is able to represent this data set also quite well and the trend in α vs. x is also accurately represented. The mean absolute error is seen from Table 4 to be 9.2 percent for the entire data set.

Figure 8 shows Hihara et al.'s (1989) data set for R-22/R-114 under the severe suppression region. The data was obtained in an electrically heated stainless steel test section for which F_{fl}=1.0 applies. The agreement between the correlation and the data is very good, with the mean absolute error seen from Table 4 as 9.0 percent.

Additional Comments:

The experimental error in a binary system is higher than that with pure components due to additional uncertainties associated with equilibrium phase compositions. The saturation temperature along the length of the evaporator tube varies due to changing liquid and vapor compositions as well as changing pressure. Additional factors are introduced by the fluid heated or electrically heated test sections, and stainless steel or copper tubes. A study of the experimental set up and the procedure suggests that the errors in the experimental data are around 10 to 15 percent for the sets considered in this study. The present correlation is able to predict well within the experimental uncertainty for all these cases.

Another comment may be made regarding a somewhat larger error associated with the first one or two thermocouple locations in many experimental data sets. The heat transfer coefficients measured at these locations are much lower than at the subsequent locations. The main reason for this is believed to be the delay in initiating nucleation with binary mixtures as the flow enters the heated test section. The nucleation characteristics are also believed to be affected for mixtures and additional work in this area is warranted to quantify these effects.

CONCLUSIONS

1. A flow boiling correlation is developed for binary systems. It incorporates the nucleate boiling suppression factor developed in Part I of this paper. The basic correlation by Kandlikar (1990) for pure liquids is extended to cover the binary systems under three regions depending on the volatility parameter: region I - near-azeotropic, region II - moderate suppression, and region III - severe suppression.

2. The correlation is compared with the experimental data for six binary refrigerant systems reported in literature. The overall absolute mean deviation for over 2500 data points is around 10 percent.

3. The trend int α versus x is well represented by the correlation in all three regions.

4. The correlation performs equally well for fluid-heated as well as electrically heated test sections. Also, the tube material effect, stainless steel or copper/brass, is well accounted by the fluid dependent parameter, which reduces to 1.0 for stainless steel test sections.

ACKNOWLEDGMENT

The author gratefully acknowledges the support provided by Taavo Raykoff in the programming for data analysis.

REFERENCES

Bennett D.L., and Chen, J.C., 1980, "Forced Convective Boiling in Vertical Tubes for Saturated Pure Components and Binary Mixtures," AIChE Journal, Vol. 26, No. 3, pp. 454-461.

Calus, W.F. and Leonidopoulos, D.J., 1974, "Pool Boiling - Binary Liquid Mixtures," International Journal of Heat and Mass Transfer, Vol. 17, pp. 249-256.

Calus, W.F., di Montegnacco, A., and Kenning, D.B.R., 1973, "Heat Transfer in a Natural Circulation Single Tube Reboiler, Part II: Binary Liquid Mixtures," The Chemical Engineering Journal, Vol. 6, pp. 251-264.

Calus, W.F., and Rice, P., 1972, "Pool Boiling - Binary Liquid Mixtures," Chemical Engineering Science, Vol. 27, pp. 1687-1697.

Celata, G.P., Cumo, M., and Setaro, T., 1993, "Forced Convective Boiling in Binary Mixtures," International Journal of Heat and Mass Transfer, Vol. 36, No. 13, pp. 3299-3309.

Chen, J.C. 1966, "A correlation for boiling heat transfer to saturated fluids in convective flow," Industrial and Engineering Chemistry, Process Design and Development, Vol. 5, No. 3, pp. 322-329.

Gnielinski, V., 1976, "New Equations for Heat and Mass Transfer in Turbulent Pipe and Channel Flow," International Chemical Engineer, Vol. 16, pp. 359-368.

Hihara, E., Tanida, K., and Saito, T., 1989, "Forced Convective Boiling Experiments of Binary Mixtures," JSME International Journal, Ser. II, Vol. 32, No. 1, pp. 98-106.

Jung, D.S., McLinden, M., Radermacher, R., and Didion, D., 1988, "Horizontal Flow Boiling Experiments with a Mixture of R-22/R-114," International Journal of Heat and Mass Transfer, Vol. 32, No. 1, pp. 131-145.

Jung, D.S., 1988, "Horizontal Flow Boiling Heat Transfer Using Refrigerant Mixtures," Ph.D. dissertation, University of Maryland.

Kandlikar, S. G., 1990a, "A General Correlation for Saturated Two-Phase Flow Boiling Heat Transfer Inside Horizontal and Vertical Tubes," ASME Journal of Heat Transfer, Vol. 112, pp. 219-228.

Kandlikar, S.G., Bijlani, C.A., and Sukhatme, S.P., 1975, "Predicting the Properties of R-22 and R-12 Mixtures - Transport Properties", *ASHRAE Transactions*, Vol. 81, Part 1, pp. 266-284.

Kandlikar, S.G., and Raykofff, T., 1997, "Investigating Bubble Characteristics and Convective Effects on Flow Boiling of Binary Mixtures," Paper presented at the Engineering Foundation Conference on Convective Flow and Pool Boiling, Irsee, Germany, May 1997.

Kandlikar, S.G., 1991, "A Model for Predicting the Two-Phase Flow Boiling Heat Transfer Coefficient in Augmented Tube and Compact Heat Exchanger Geometries," ASME Journal of Heat Transfer, Vol. 113, pp. 966-972.

Kandlikar, S.G., 1997, "Boiling Heat Transfer with Binary Mixtures, Part I: A Theoretical Model for Pool Boiling," Paper presented at the ASME National Heat Transfer Conference, Baltimore, appears in the same bound volume.

Murata, K., and Hashizume, K., 1989, "Forced Convection Boiling of Nonazeotropic Refrigerant Mixtures Inside Tubes," ASME Journal of Heat Transfer, Vol. 115, pp. 680-689.

NIST, 1995, REFPROP, National Institute for Science and Technology, Washington, D.C.

Petukhov, B.S., and Popov, V.N., 1963, "Theoretical Calculation of Heat Exchange and Frictional Resistance in Turbulent Flow in Tubes of an Incompressible Fluid with Variable Physical Properties," Teplofiz.Vysok.Temperatur (High Temperature Heat Physics), Vol. 1, No. 1.

Takamatsu, H., Momoki, S., and Fujii, T., 1993, "A Correlation for Forced Convection Boiling Heat Transfer of Nonazeotropic Refrigerant Mixture of HCFC22/CFC114 in a Horizontal Smooth Tube," Int. Journal of Heat and Mass Transfer, Vol. 36, pp. 3555- 3563.

HTD-Vol. 342, National Heat Transfer Conference
Volume 4
ASME 1997

Modeling of Conduction Effects on Pool Boiling Critical Heat Flux of Dielectric Liquids

A. A. Watwe[1] and A. Bar-Cohen
Thermodynamics and Heat Transfer Division
University of Minnesota, Minneapolis

[1]Currently at Fluent Inc., Evanston, Illinois

ABSTRACT

Experimental studies have shown that the geometric and thermophysical properties of the heater significantly influence the critical heat flux (CHF). While the thermal activity parameter $\delta\sqrt{\rho_h C_h k_h}$ has been found to correlate this variation very effectively for pool boiling, a physical explanation for its success is only now beginning to emerge. A conceptual CHF model, emphasizing both the role of transient radial conduction within the heater and the relationship between the heater surface temperature and CHF, is proposed to account for this dependence. A finite element analysis is used to investigate the effect of heater and substrate thermal properties on the temperature distribution at CHF. The numerical results show that the thermal activity parameter is associated with radial transient conduction in the heater and confirm the empirical observation that the presence of a substrate will augment CHF on thin heaters with low thermal activity values. The effects of wettability on CHF are also highlighted.

Nomenclature

A	Surface area [m^2]
C	Specific Heat [J/kg K]
CHF	Critical heat flux [W/m^2]
g	Gravitational acceleration [m/s^2]
h	Heat transfer coefficient [W/m^2 K]
k	Thermal conductivity [W/m K]
q	Heat Flux [W/m^2]
R	Heater radius [m]
r	Radial coordinate [m]
T	Temperature [°Cor K]
ΔT	Temperature difference $(T - T_{sat})$ [°Cor K]
t	time [s]
z	Vertical coordinate [m]

Greek Symbols

α	Thermal diffusivity [m^2/s]
β	Contact angle [°]
δ	Heater thickness [m]
λ_D	Most dangerous taylor wavelength [m]
ρ	Density [kg/m^3]
σ	Surface tension [N/m]
τ_d	Hovering period [s]
ξ	Fraction of liquid dragged by the vapor mushroom

Subscripts

$boil$	Boiling
cr	Critical
d	Dry spot
f	Liquid
gen	Volumetric heat generation
h	Heater
hom	Homogeneous nucleation
max	Maximum dry spot
rad	Radial
sat	Saturation
sub	Subcooled
v	Vapor
zub	As per the Zuber model

INTRODUCTION

High performance microprocessor chips are expected to dissipate 30 to 50 W/cm^2 by the end of the next decade (SIA Report, 1994). Reliability and performance considerations often dictate that these chips be maintained at temperatures below 85 ° C. Direct immersion cooling, with nucleate pool boiling of a dielectric coolant on the bare chip, has been explored as a cooling technique for such chips. (Bar-Cohen, 1991). However, the departure from nucleate boiling, or "Critical Heat Flux" (CHF), places an upper limit on this highly efficient heat transfer mechanism. At CHF, an insulating film of vapor forms on the heated surface, leading to a large increase ($\approx$ 100 ° C for the dielectric coolants) in surface superheat.

The central theme in most existing CHF models is the "choking off", or interruption, of the liquid supply to the heater, due to a hydrodynamic instability. Predictions from these models for saturated liquids at atmospheric pressure usually lie within $\pm 20\%$ of experimental measurements. However, the models fail to account for the parametric effects, on CHF, of liquid subcooling and more importantly of heater thermal properties. These effects are usually incorporated through empirical corrections to the basic equations.

In a revision of long held beliefs on the role played by hydrodynamic instabilities, the current understanding of CHF is based on the formation of large coalesced vapor bubbles (also called vapor mushrooms) and dry patches on the heater surface. These dry patches are re-wetted by the liquid following the periodic departure of the vapor mushrooms. If the dry patch temperature exceeds some critical value, the dry patch cannot be re-wetted and it grows leading to CHF. The thermal properties of the heater are intrinsic to such a model and it is believed that only a model which links vapor mushroom behavior to conduction within the heater could provide a satisfactory description of CHF.

The primary goal of this study is to extend the hydrodynamic vapor-mushroom based CHF model to include the effects of heater properties. In doing so the study evaluates the importance of thermal conduction within the heater via a dry-spot based analysis.

THEORETICAL BACKGROUND

Noting the similarity between "flooding" in distillation columns and CHF, Kutateladze (1951) obtained an expression for CHF based on a similitude analysis of the momentum and energy equations governing two-phase flow near the heated surface. Subsequently, Zuber (1959) derived an analytical equation for CHF assuming that it was controlled by a hydrodynamic instability. He postulated that vertical coalescence of vapor bubbles resulted in counter-flowing liquid and vapor streams. Assuming that CHF occurred when the interface between the liquid and vapor streams became unstable and completely choked off the flow of liquid to the surface, he obtained the following relation for CHF:

$$CHF_{zub} = \frac{\pi}{24} h_{fv} \sqrt{\rho_v} \left[\sigma_f(\rho_f - \rho_v)g\right]^{1/4} \tag{1}$$

Equation (1) was derived specifically for saturated pool boiling of a liquid on an infinite, upward-facing horizontal plate. Despite these limitations, it has been applied to a wide variety of situations, by introducing correction factors to account for the effects of liquid subcooling and heater geometry. Photographic studies by Gaertner and Westwater (1960) and Gaertner (1965) have demonstrated the absence of the vapor columns, on large horizontal flat plates, postulated by Zuber (1959). The recent experiments of Williamson and El-Genk (1991) indicate that in addition to possible vertical coalescence, bubbles often merge laterally, forming large mushrooms of vapor along the heated surface.

The periodic formation and departure of such vapor slugs is at the heart of the macrolayer model, proposed by Haramura and Katto (1983). The vapor slug grows, due to the vaporization of the liquid trapped between the vapor stems, and periodically departs from the heater surface. The heater surface is then re-wetted by the surrounding liquid, and the entire process of vapor mushroom formation and departure is repeated. The time period, between the start of growth of a vapor mushroom and its departure from the surface, is called the hovering period, τ_d. Assuming that all the available heat is used to evaporate the macrolayer, Haramura and Katto (1983) proposed that at CHF the macrolayer completely evaporated in one hovering period. They evaluated the vapor area fraction on the heater by using the Zuber (1959) model. Consequently, the CHF values predicted using this model are identical to the predictions from Equation (1). The vapor area fraction at CHF evaluated using this method for FC-72 is approximately 2% and appears to be rather small. Moreover, visual observations of boiling by Kirby and Westwater (1965), and van Ouwerkerk (1972) indicate that dry patches on the heater can be re-wetted by the liquid after the departure of the vapor slug. Thus, drying out of the macrolayer is not sufficient to cause burnout. Like the Kutateladze-Zuber model, the macrolayer model ignores heater related effects on CHF. The primary contribution of this model lies in the modified vapor-liquid configuration, and in the concept of periodic formation and removal of vapor mushrooms from the heated surface.

Experimental studies by several researchers, spanning nearly four decades, have shown that the heater thickness and thermal properties do have a significant influence on CHF and that for very thin, thermally-poor surfaces CHF drops precipitously. Various parameters such as $k_h\delta$, $\rho_h C_h \delta$, $\sqrt{\rho_h C_h k_h}$, $\alpha_h \delta$ and $\delta\sqrt{\rho_h C_h k_h}$ have been proposed to correlate this effect. A detailed review of these experimental studies and the various correlating parameters has been previously provided in Watwe and Bar-Cohen (1994).

Carvalho and Bergles (1992) recently reviewed experimental data reported in the literature and concluded that this effect was best correlated by the thermal activity parameter $\delta\sqrt{\rho_h C_h k_h}$, originally proposed by Saylor (1989). This paper explores the physics underpinning the correlation of this effect by performing extensive simulations of the conduction effects in the heater.

CONDUCTION MODELING

Conceptual Model

In the present model it is postulated that lateral coalescence of individual vapor bubbles or vapor-columns produces a large vapor mushroom. The vaporization of the liquid macrolayer in the regions where it is relatively thin, or the lateral coalescence of vapor bubbles in regions of high nucleation site density, is expected to create local dry patches on the heater surface. The heat in the dry spot region must be absorbed locally and/or conducted to parts of the heater still experiencing nucleate boiling. If the resulting temperature of the dry patch exceeds the critical re-wetting temperature, during the residence time (hovering period) of the vapor mushroom, the liquid is unable to quench the overheated surface, local dryout proceeds to global dryout and CHF is said to occur.

Ramilison and Lienhard (1987) have shown that the limiting value of the re-wet temperature is approximately equal to the homogeneous nucleation temperature T_{hom} for the highly wetting liquids ($\beta \to 0$). Lienhard (1976) has correlated T_{hom} with the following equation,

$$\frac{T_{hom}}{T_{cr}} = 0.923 + 0.077 \left(\frac{T_{sat}}{T_{cr}}\right)^9, \tag{2}$$

This equation gives a T_{hom} value of 145 °C for FC-72 at 101.3 kPa. The corresponding wall superheat is 89 °C. In applying the dry-patch model to FC-72, CHF is assumed to occur if the maximum temperature in the dry spot exceeds 145 °C.

If the heater is very thin and has poor thermal properties, then the dry patch temperature may exceed the critical re-wetting temperature soon after the formation of the first vapor mushroom and dry patch. The critical heat flux on such heaters will then be given by the transition heat flux at which discrete vapor bubbles merge to form large vapor slugs. This can be thought of as the lower limit on CHF. At the other end of the spectrum, a very thick heater, with good thermal properties, would be able to conduct the heat away from the dry patch, thus keeping the dry patch temperature from exceeding the critical re-wet temperature. In this case, the heater may be able to sustain the sequential formation and departure of many vapor mushrooms and the heater temperature, even under periodic steady state conditions, may not exceed the critical re-wetting temperature. The critical heat flux would then occur primarily due to an instability phenomenon when the liquid flow to the heater would be completely interrupted. The hydrodynamic instability models would, therefore, provide the upper or asymptotic limit for CHF.

It may be postulated that for each heater thickness between these values, there exists a heat flux that would result in the temperature of the dry spot exceeding the re-wet temperature after the departure of one or more vapor mushrooms. The critical heat flux on these heaters would then vary between the lower limit, given by the transition heat flux for lateral coalescence, and the upper limit predicted by the Kutateladze-Zuber relation given by Equation (1).

Numerical Simulation

A model consisting of a heater with a dry spot at its center was used to evaluate the influence of heater properties on CHF for heaters which lie between the two limits discussed above. Like the Haramura and Katto (1983) model, the characteristic length of a vapor mushroom was assumed to be approximately one Taylor wavelength λ_D. For simplicity, an axisymmetric geometry was used to model the one Taylor wavelength region of the heater, as shown in Figure 1.

The transient conduction equation for an axisymmetric domain within the heater can be written as:

$$\frac{\partial T}{\partial t} = \alpha_h \left(\frac{\partial^2 T}{\partial r^2} + \frac{1}{r}\frac{\partial T}{\partial r} + \frac{\partial^2 T}{\partial z^2} \right) + \frac{q_{gen}}{\rho_h C_h} \tag{3}$$

This equation must be solved, with the following boundary conditions, during the hovering period of the vapor mushroom:

$$\frac{\partial T}{\partial r} = 0, \; r = 0, \; r = R, \; t > 0 \tag{4}$$

$$\frac{\partial T}{\partial z} = 0, \; 0 \leq r \leq R, \; z = 0, \; t \geq 0 \tag{5}$$

$$\frac{\partial T}{\partial z} = 0, \; 0 < r < R_d, \; z = \delta, \; t > 0 \tag{6}$$

$$-k_h \frac{\partial T}{\partial z} = h_{boil}[T(r,z,t) - T_{sat}], \; R_d < r < R, \; z = \delta, \; t > 0 \tag{7}$$

The presence of nearly-insulated and very well cooled regions on the heater surface, and the need for very short time transient temperature distributions, necessitated a numerical solution for the temperature field in the heater. The numerical solution of Equations (3) through (7) was obtained using the TOPAZ2D finite element code (Shapiro, 1986), on an Onyx SGI workstation. The numerical analysis was carried out for 16 different heater materials, corresponding to those in the experimental study of Golobic and Bergles (1992), with thicknesses ranging from 2 to 1000 microns. About 1333 elements in the radial direction and 2 elements in the thickness direction were used to resolve the temperature field for thin heaters ($< 13\ \mu$m). For heater thicknesses ranging between 13 and 75 μm, 750 radial elements and 5 thickness elements were used. For very thick heaters ($> 75\ \mu$m) 350 elements in the radial direction and 21 elements in the thickness direction were used. In order to capture the very small time transient temperature distribution, a time step of 0.05 ms was used for the computations. Increasing the mesh density or decreasing the time step was found to result in negligible change in the computed temperature field.

It was assumed that each heater dissipated the same heat flux q, which was set to a value of 17 W/cm^2 based on the CHF measurements in fluorocarbons reported by Danielson et al. (1987). The volumetric heat generation q_{gen} was then estimated by dividing q with the heater thickness δ. The initial temperature in the heater was obtained by performing a steady state calculation with a nucleate boiling heat transfer coefficient, h_{boil}, of 6800 W/m^2 K, applied to the entire top surface of the heater. This value for was estimated from the experimental wall superheat data presented in Watwe et al. (1996). Following the initial temperature computation, a dry spot region was established on the heater surface by applying a zero heat flux boundary condition in the numerical model. The heat transfer coefficient, h_{boil}, was applied over the remainder of the heater surface. The transient calculation of the dry patch temperature was conducted for a hovering period τ_d of 40 ms, estimated using the following equation proposed by Haramura and Katto (1983):

$$\tau_d = \left(\frac{3\,qA}{4\pi \rho_v h_{fv}} \right)^{1/5} \left[\frac{4(\xi \rho_f + \rho_v)}{g(\rho_f - \rho_v)} \right]^{3/5} \tag{8}$$

where, $\xi = 11/16$, is the volumetric ratio of the liquid accompanying the moving vapor bubble. Typically 15 min of CPU time, on an SGI Onyx workstation, were needed to obtain the transient temperature distribution in each heater at the end of the hovering period of 40 ms.

If the dry spot temperature at the end of the hovering period is too high, as in the case of thin, poorly conducting heaters, the CHF would be very low. On the other hand if the dry spot temperature is very small, the heater can sustain a much larger heat flux before the dry spot exceeds the critical re-wetting temperature. This relationship between the dry spot temperature and CHF is employed in this paper by using the magnitude of the maximum dry spot temperature as a measure of the critical heat flux.

NUMERICAL RESULTS

Two sets of computations were conducted. In the first set, the diameter of the dry spot was set to $\lambda_D/2$, which is the diameter of the vapor columns in the Zuber (1959) model. The dry spot diameter for FC-72, based on this assumption, is approximately 3 mm. In experiments with a moderately non-wetting liquid like water boiling on a glass plate, the maximum dry spot size, observed by van Ouwerkerk (1972), was approximately 6 mm in diameter as compared to the $\lambda_D/2$ value of 13.6 mm. The dry spot size for a highly-wetting liquid like FC-72 can be expected to be smaller. To assess the influence of the dry spot size on heater conduction, the second set of computations was conducted with a larger dry spot, 6 mm in diameter.

Results:Smaller Dry Spot

The numerically determined maximum dry spot superheat was plotted against two parameters namely $\delta\sqrt{\rho_h C_h k_h}$ and $\rho_h C_h \delta$. These plots are shown in Figures 2 and 3. Comparison of the results with the other parameters can be found in Watwe and Bar-Cohen (1994). The wall superheat of 89 °C corresponding to the critical re-wetting temperature is also shown in each of the above figures. In order to estimate the correlating capability of the two parameters, an equation having the form

$$T_{max} - T_{sat} = \frac{1}{a + bC_{par}^{c}} \tag{9}$$

was fitted to the numerically calculated dry spot superheats. In Equation (9), C_{par} represents the relevant correlating parameter. The correlation coefficient resulting from the above fitting procedure provides a good estimate of the ability of various parameters to correlate the maximum dry spot temperature and hence the critical heat flux.

The figures clearly show that the best correlation (r^2 = 0.96) is obtained by using the thermal activity parameter $\delta\sqrt{\rho_h C_h k_h}$ (see Figure 2). Figures 2 and 3 also show that the correlation of the dry patch superheat with the parameter $\delta\sqrt{\rho_h C_h k_h}$ is definitely better when compared to the correlation with the parameter $\rho_h C_h \delta$ (r^2 = 0.68) proposed by Houchin and Lienhard (1966) and Tachibana et al. (1967). These results are in agreement with those of Carvalho and Bergles (1992) who found that plotting the experimental CHF values, measured using a variety of heater geometries and fluids, against $\delta\sqrt{\rho_h C_h k_h}$ resulted in the least scatter, as compared to the correlation of CHF with the other parameters. Bar-Cohen and McNeil (1992) found the asymptotic upper limit of CHF to occur at $\delta\sqrt{\rho_h C_h k_h}$ greater than 20. Carvalho and Bergles (1992) reported that CHF attained 90% of the asymptotic value for $\delta\sqrt{\rho_h C_h k_h} =$ 5. The present analysis shows that, for the conditions examined, the maximum superheat and hence the CHF is nearly asymptotic for $\delta\sqrt{\rho_h C_h k_h} > 5$.

Based on the presence of the property group $\sqrt{\rho_h C_h k_h}$, Bar-Cohen and McNeil (1992) attempted to explain the successful correlation of CHF data with $\delta\sqrt{\rho_h C_h k_h}$ via transient one-dimensional conduction into the heater. However, it appears that most of the heaters used in previous experimental studies were relatively thin compared to the depth of penetration of the thermal wave during the hovering period. This is apparent from Figure 4 which compares the heater thicknesses δ and the thermal penetration depth δ_{pen}, calculated as $\sqrt{12\alpha_h \tau_d}$ (Eckert and Drake, 1972) for a hovering period τ_d of 40 ms. Consequently, an explanation different from that proposed by Bar-Cohen and McNeil (1992) must be sought for the success of the correlating parameter $\delta\sqrt{\rho_h C_h k_h}$. For very thin heaters, thermal conduction occurs primarily in the radial (or lateral) direction away from the hot spot. When the heaters are very thick, there is also significant heat flow in the thickness direction. The one-dimensional transient temperature of a semi-infinite surface to which heat flux is applied suddenly is given as (Eckert and Drake, 1972):

$$T_{surface} - T_{initial} = \frac{2q}{\sqrt{\pi}}\sqrt{\frac{t}{\rho_h C_h k_h}} \tag{10}$$

By analogy, it might be anticipated that for conduction in the radial direction the temperature would be expressible as:

$$T_{surface} - T_{initial} = f\left(\frac{q_{rad}}{\sqrt{\rho_h C_h k_h}}, \frac{r}{R}, \sqrt{t}\right) \tag{11}$$

The parameter group $\sqrt{\rho_h C_h k_h}$ in the thermal activity parameter, thus, represents the semi-infinite radial conduction in the heater. The radial flow of uniformly generated heat away from the dry spot, applies a heat flux on the walls of a hollow cylindrical region of radius R_d according to:

$$q_{rad} = \frac{qR_d}{\delta} \tag{12}$$

Thus, the thickness δ scales the applied heat flux q to the radial heat flux q_{rad}, and the maximum wall superheat can be expected to vary inversely with $\delta\sqrt{\rho_h C_h k_h}$.

Results: Larger Dry Spot

The computation of the maximum dry patch superheat was repeated for a dry spot having twice the diameter. A comparison of Figures 5 and 6 shows that in this case, the thermal capacity $\rho_h C_h \delta$ correlates the maximum dry spot superheat better (r^2 = 0.93) than the thermal activity parameter $\delta\sqrt{\rho_h C_h k_h}$ (r^2 = 0.82). The larger dry spot makes it more difficult for the heat to conduct from the center of the dry spot to the wetted regions at its periphery in the short hovering time of 40 ms. Thus, although a strong correlation exists between the surface temperature and the thermal activity parameter, the temperature rise in the dry patch is more directly governed by the thermal capacity of the heater.

It is clear from the above results that the size of the dry patch has a significant influence on how the heater properties affect the critical heat flux. The size of the dry spot is a function of the heater surface characteristics as well as its wettability. Wettablility has been shown to affect CHF significantly (Liaw and Dhir, 1986). Thus the conduction model also highlights the importance of heater surface effects.

The value of λ_D for FC-72 at 101.3 kPa is approximately 0.8 cm. Consequently, only one vapor mushroom would be expected to form on a 1 cm wide silicon chip. However, the photographic evidence in Watwe (1996) suggests that several small vapor mushrooms with their own dry patches may co-exist on the heater surface. Therefore, it appears appropriate to assume that relatively small dry spots are formed on the heaters. In the remainder of this paper, the numerical methodology used to obtain the maximum dry spot temperature for the case of a small dry spot will be used to investigate the influence on CHF of other effects such as substrate properties and contact resistance.

Effect of Substrate Properties

The previous discussion examined a "stand-alone" heater which is perfectly insulated on the bottom surface. However, most heaters in practical electronics cooling applications are attached to a substrate. Since most heaters are thinner than the thermal penetration depth, as seen in Figure 4, the thermal wave can be expected to traverse the heater thickness and penetrate the underlying substrate during the relevant hovering period. Thus, the substrate provides an additional parallel path for radial heat conduction.

Two different substrates, Plexiglas and Pyrex, were used to investigate the influence of a substrate on CHF. The thermal conductivity of Pyrex is an order of magnitude higher than that of Plexiglas. Thus, the choice of these two substrate materials helps to illustrate the effect of substrate properties on the dry spot temperature and hence on the critical heat flux. In order to minimize the number of finite elements required in the numerical computation, the thickness of the substrate was limited to 1 mm. All numerical parameters for these computations were identical to that used in previous computations for the smaller dry spot size.

Figure 7 compares the maximum dry patch superheats for heaters with and without a Plexiglas substrate. The filled symbols in the figure represent the superheats for the no-substrate cases and the open symbols represent the corresponding superheats for cases with the Plexiglas substrate. As expected, the addition of the substrate substantially lowers the peak temperature for heaters with low thermal activity values. The presence of a substrate does not appear to have any effect on the maximum dry spot temperature for heaters with $\delta\sqrt{\rho_h C_h k_h}$ greater than 5. This is because these heaters have a relatively low resistance to the radial penetration of heat and do not require any additional heat transfer path.

Figure 7 also shows that $\delta\sqrt{\rho_h C_h k_h}$ does not correlate the two different data sets. The data are re-plotted in Figure 8 using a modified thermal activity parameter calculated as the sum of the thermal activities of the heater and the substrate. Since the thermal wave may not reach the full substrate thickness during the hovering period, the thermal penetration depth in the substrate, δ_s, can be used to define the thermal activity of the substrate. The thermal wave penetrates the heater depth in the first few miliseconds of the hovering period. The substrate penetration depth δ_s can be estimated using the time available to the thermal wave during the remainder of the hovering period. Figure 8 also shows that even when the substrate is included, the modified thermal activity must exceed 5 for CHF to attain its asymptotic value. Table 1 shows the substrate thickness required to produce a modified thermal activity parameter of 5 for a variety of heaters. Similar results were obtained using Pyrex for the substrate material.

Table 1. Lexan thickness needed for asymptotic CHF for FC-72

Heater specifications		Lexan
Material	Thickness	Thickness
Silicon	0.3 mm	0.6 mm
Silicon	0.2 mm	3.8 mm
Platinum	0.005 mm	10.1 mm

Effect of Contact Resistance

In the previous section it was assumed that there was perfect contact between the heater and the substrate. However, in most practical applications the interface between the heater and substrate gives rise to a thermal contact resistance. The finite element code TOPAZ2D allows for the specification of a contact resistance at an interface through the introduction of an element with zero volume. In the numerical model this was done through the use of a contact conductance value which was applied to the zero-volume elements located at the heater-substrate interface. The substrate material for these computations was Plexiglas. The dry patch superheat was computed for selected heater materials and thicknesses chosen such that the parameter $\delta\sqrt{\rho_h C_h k_h}$ ranged approximately from 0.01 to 10.

For each material and thickness pair, the dry spot temperature was calculated for a range of contact resistances varying from 10^{-6} °C m^2/W to 10^{+6} °C m^2/W. The no-substrate results reflect the situation where the contact resistance is infinitely large. The dry spot temperatures obtained for the heater in perfect contact with a Plexiglas substrate reflect the situation where the contact resistance is zero.

The maximum dry spot temperature from the no-substrate case is used to non-dimensionalize the temperature values computed in the presence of contact resistance. These results are shown in Figure 9. The figure shows that for heaters with a very small value of $\delta\sqrt{\rho_h C_h k_h}$, the presence of thermal contact resistance can have a significant influence on the maximum dry spot temperature and thus the CHF. On the other hand, the presence of interfacial thermal contact resistance has little or no effect on the CHF on heaters with $\delta\sqrt{\rho_h C_h k_h} > 5$.

Effect of hovering period τ_d

The results presented above were obtained by assigning a value of 40 ms to the bubble residence time (hovering period). This section presents a discussion of the influence of the hovering period on the maximum dry spot temperature and how this affects the variation of CHF with heater properties. Figures 10 and 11 show the variation of the maximum dry patch superheat with the correlating parameters $\delta\sqrt{\rho_h C_h k_h}$ and $\rho_h C_h \delta$ respectively for a τ_D of just 20 ms. A similar comparison is presented in Figures 12 and 13 which show the results under steady state conditions (which represent a very long hovering period). These results show that for a very large range of hovering periods (20 ms to ∞) the parameter $\delta\sqrt{\rho_h C_h k_h}$ appears to correlate the variation of dry patch superheat, and hence the CHF, better than the thermal capacity parameter. This indicates that the parameter $\delta\sqrt{\rho_h C_h k_h}$ would probably correlate the heater property effects well for most pool boiling situations.

The time scales in flow boiling are expected to be much smaller than the hovering periods in pool boiling. Moreover, the time scales are expected to decrease with increasing flow velocity. Using the data and model for CHF during flow boiling in narrow channels, proposed by Galloway et al. (1993a,b), the flow boiling time scales were estimated as 5 ms and 1.85 ms at flow velocities of 1 m/s and 2 m/s respectively. The maximum dry patch superheats from these computations are plotted against the correlating parameters $\delta\sqrt{\rho_h C_h k_h}$ and $\rho_h C_h \delta$ in Figures 14 and 17. It is obvious from these figures that, for the short hovering periods, the thermal capacity $\rho_h C_h \delta$ correlates the maximum dry patch superheat better than the parameter $\delta\sqrt{\rho_h C_h k_h}$. Hovering periods less than 5 ms do not appear to allow enough time for the conduction to be established within the heater. Consequently, the temperature rise in the dry spot region is dictated primarily by the thermal capacity of the heater. Table 2 summarizes the effects of the hovering period.

Table 2. Effectiveness of correlating parameters vs. τ_d

τ_d	Correlation Coefficient r^2	
(ms)	with $\delta\sqrt{\rho_h C_h k_h}$	with $\rho_h C_h \delta$
1.85	0.68	0.98
4.5	0.72	0.98
20.0	0.88	0.84
40.0	0.96	0.68
∞	0.87	0.36

Figures 15 and 17 show that the dry spot temperature and hence the critical heat flux is nearly constant for $\rho_h C_h \delta > 200$. For a Copper heater this corresponds to a heater thickness of approximately 58 μm. From Figure 2 for the pool boiling case the asymptotic CHF values is expected to be attained for $\delta\sqrt{\rho_h C_h k_h} > 5$. The thickness of a Copper heater corresponding to this $\delta\sqrt{\rho_h C_h k_h}$ value is approximately 135 μm. Thus, the dry spot based conduction model of CHF reflects the practical observations that a thinner heater is required to attain the asymptotic critical heat flux in flow boiling.

CONCLUSION

This paper has presented a detailed analysis of the conduction effects on CHF under a variety of scenarios. Based on photographic evidence and the excellent correlation of experimental data with the parameter $\delta\sqrt{\rho_h C_h k_h}$, it appears that radial transient conduction primarily governs the effect of heater properties on CHF. The value of $\delta\sqrt{\rho_h C_h k_h}$ sets the lower limit on the CHF on a given heater. The asymptotic critical heat flux predicted by the hydrodynamic models of Zuber (1959) and Haramura and Katto (1983) cannot be exceeded simply by increasing $\delta\sqrt{\rho_h C_h k_h}$. The results clearly demonstrate that substrate properties can have a significant influence on CHF for heaters with $\delta\sqrt{\rho_h C_h k_h} < 5$. Contact resistance is also expected to significantly influence CHF for very thin or poorly conducting heaters. Thermal capacity, $\rho_h C_h \delta$, plays a significant role in cases where the dry spot is very large (as in heaters with poor surface wettability) or when the time scales are too small (as in flow boiling). In both these situations lateral conduction is insignificant. The results clearly indicate that the heater-side and fluid-side effects on CHF cannot be isolated and that together they will determine the time scales and the dry patch geometry.

References

Bar-Cohen, A., 1991, "Thermal Management of Electronic Components with Dielectric Liquids," *Proc. 3rd ASME/JSME Thermal Engineering Joint Conference*, Reno, Nevada, Vol. 2, pp. xv–xxxviii.

Bar-Cohen, A., McNeil, A., 1992, "Parametric Effects on Pool Boiling Critical Heat Flux in Highly-Wetting Liquids," *Proc. Engineering Foundation Conference on Pool and External Flow Boiling*, Santa Barbara, California, pp. 171–175.

Carvalho, R. D. M., Bergles, A. E., 1992, "The Effects of the Heater Thermal Conductance/Capacitance on the Pool Boiling Critical Heat Flux," *Proc. Engineering Foundation Conference on Pool and External Flow Boiling*, Santa Barbara, California, pp. 203–212.

Danielson, R. D., Tousignant, L., Bar-Cohen, A., 1987, "Saturate Pool Boiling Characteristics of Commercially Available Perfluorinated Inert Liquids," *Proc. 1987 ASME/JSME Thermal Engineering Joint Conference*, Honolulu, Hawaii, Vol. 3, pp. 419–430.

Eckert, E. R. G., Jr., R. M. D., 1972, *Analysis of Heat and Mass Transfer,* McGraw Hill Kogakusha, Tokyo.

Gaertner, R. F., 1965, "Photographic Study of Nucleate Boiling on a Horizontal Surface," *J. Heat Transfer*, Vol. 87, pp. 17–29.

Gaertner, R. F., Westwater, J. W., 1960, "Population of Active Sites in Nucleate Boiling Heat Transfer," *Chem. Engg. Prog. Symp. Ser.*, Vol. 56, pp. 39–48.

Galloway, J. E., Mudawar, I., 1993a, "CHF Mechanism in Flow Boiling From a Short Heated Wall - Part I Examinatin of Near-Wall Conditions with the Aid of Photomicrography and High Speed Video Imaging," *Int. J. Heat Mass Transfer*, Vol. 36, pp. 2511–2526.

Galloway, J. E., Mudawar, I., 1993b, "CHF Mechanism in Flow Boiling From a Short Heated Wall - Part II Theoretical CHF Model," *Int. J. Heat Mass Transfer*, Vol. 36, pp. 2527–2540.

Golobic, I., Bergles, A. E., 1992, "Effects of Thermal Properties and Thickness of Horizontal Vertically Oriented Ribbon Heaters on the Pool Boiling Critical Heat Flux," *Proc. Engineering Foundation Conference on Pool and External Flow Boiling*, Santa Barbara, California, pp. 213–218.

Haramura, Y., Katto, Y., 1983, "A New Hydrodynamic Model of Critical Heat Flux Applicable Widely to both Pool and Forced Convective Boiling on Submerged Bodies in Saturated Liquids," *Int. J. Heat Mass Transfer*, Vol. 26, pp. 389–399.

Houchin, W. R., Lienhard, J. H., 1966, "Boiling Burnout in Low Thermal Capacity Heaters," ASME Paper 66-WA/HT-40.

Kirby, D. B., Westwater, J. W., 1965, "Bubble and Vapor Behavior on a Heated Horizontal Plate During Pool Boiling Near Burnout," *Chem. Engg. Prog. Symp. Ser.*, Vol. 57, pp. 238–248.

Kutateladze, S. S., 1951, "A Hydrodynamic Theory of Changes in the Boiling Process Under Free Convection Conditions," *Izv. Akad. Nauk SSSR, Otd. Tekhn. Nauk No. 4 (Translation in AEC-TR-1441)*, pp. 529.

Lienhard, J. H., 1976, "Correlation of Limiting Liquid Superheat," *Chem. Engg. Sci.*, Vol. 31, pp. 847–849.

Ramilison, J. M., Lienhard, J. H., 1987, "Transition Boiling Heat Transfer and the Film Transition Regime," *J. Heat Transfer*, Vol. 109, pp. 746–752.

SIA Report, 1994, "The National Technology Roadmap for Semiconductors," Semiconductor Industry Association.

Saylor, J. R., 1989, "An Experimental Study of the Size Effect in Pool Boiling Critical Heat Flux on Square Surfaces," Master's thesis, Mechanical Engineering, University of Minnesota, Minneapolis.

Shapiro, A. B., 1986, "TOPAZ2D - A Two-Dimensional Finite Element Code for Heat Transfer Analysis, Electrostatic, and Magnetostatic Problems," *Lawrence Livermore National Laboratory.*

Tachibana, F., Akiyama, M., Kawamura, H., 1967, "Non-hydrodynamic Aspects of Pool Boiling Burnout," *J. Nuclear Science and Technology*, Vol. 4, pp. 121–130.

Van Ouwerkerk, J. H., 1972, "Burnout in Pool Boiling: The Stability of Boiling Mechanisms," *Int. J. Heat Mass Transfer*, Vol. 15, pp. 25–34.

Watwe, A. A., 1996, "Measurement and Prediction of the Pool Boiling Critical Heat Flux in Highly Wetting Liquids," Ph.D. thesis, Mechanical Engineering, University of Minnesota, Minneapolis.

Watwe, A. A., Bar-Cohen, A., 1994, "The Role of Thickness and Thermal Effusivity in Pool Boiling in Highly-Wetting Liquids," *Proc. 10th International Heat Transfer Conference*, Brighton, UK, Vol. 5, pp. 183–188.

Watwe, A. A., Bar-Cohen, A., McNeil, A., 1996, "Combined Pressure and Subcooling Effects on Pool Boiling from a PPGA Chip Package," *Proc. Will be published in the June 1997 issue of ASME J. Electronics Packaging*.

Williamson, C. R., El-Genk, M. S., 1991, "High-Speed Photographic Analysis of Saturated Nucleate Pool Boiling at Low Heat Flux," ASME Paper 91-WA-HT-8.

Zuber, N., 1959, "Atomic Energy Commission Technical Information Service," Atomic Energy Commission Report AECU-4439.

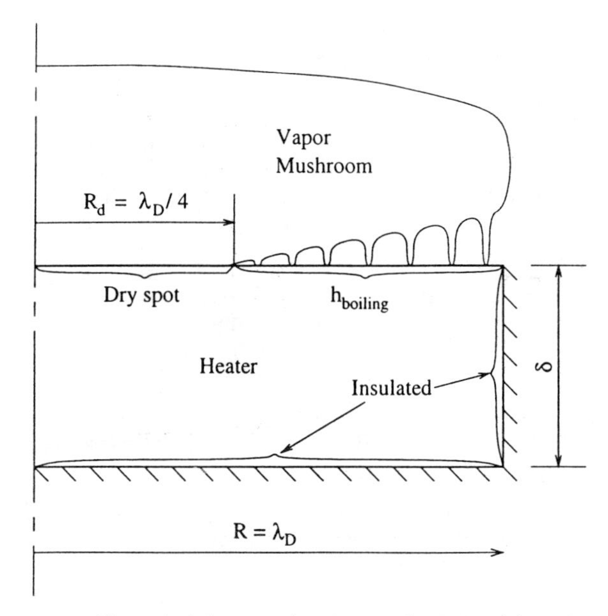

Figure 1. Axisymmetric geometry for the model

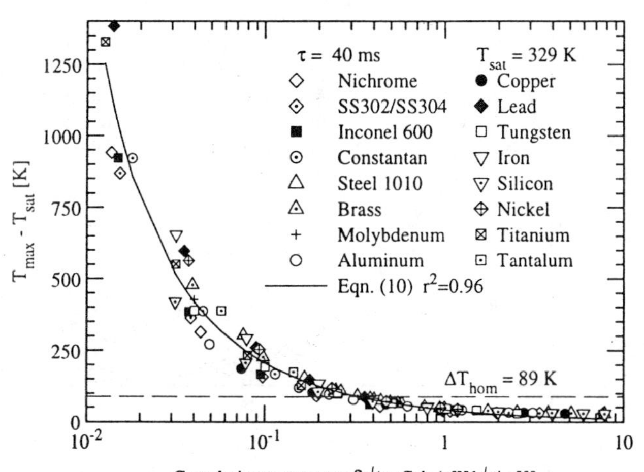

Figure 2. Plot of dry patch superheat against $\delta\sqrt{\rho_h C_h k_h}$

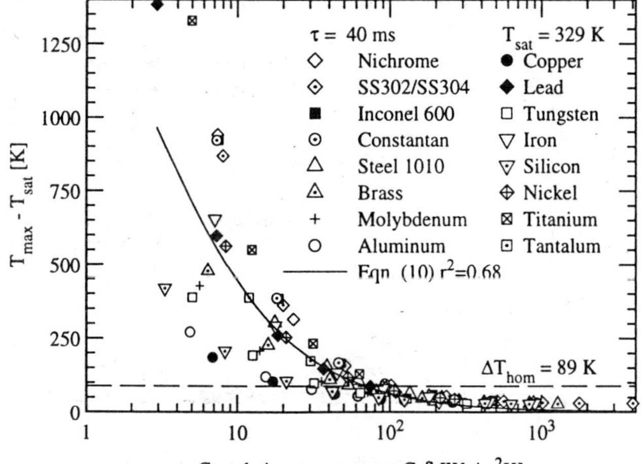

Figure 3. Plot of dry patch superheat against $\rho_h C_h \delta$

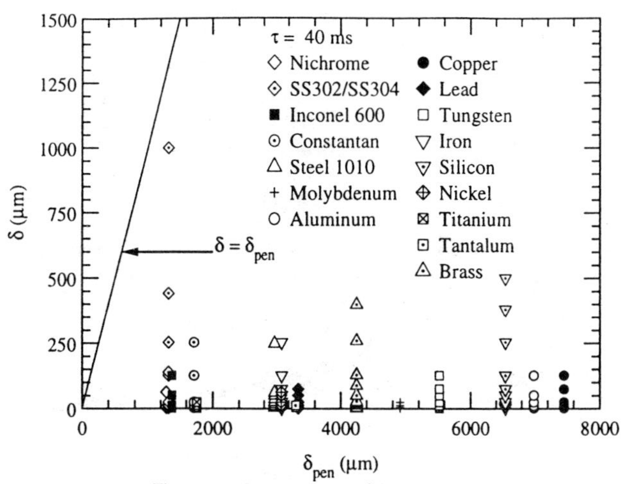

Figure 4. Comparison of δ with δ_{pen}

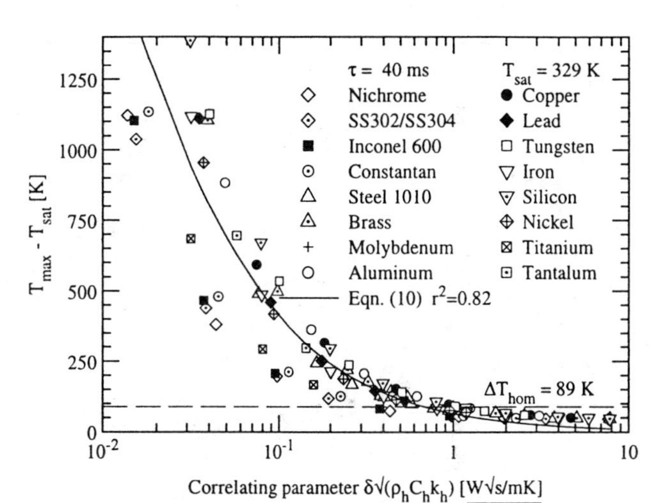

Figure 5. Plot of dry patch superheat against $\delta\sqrt{\rho_h C_h k_h}$: Larger dry spot

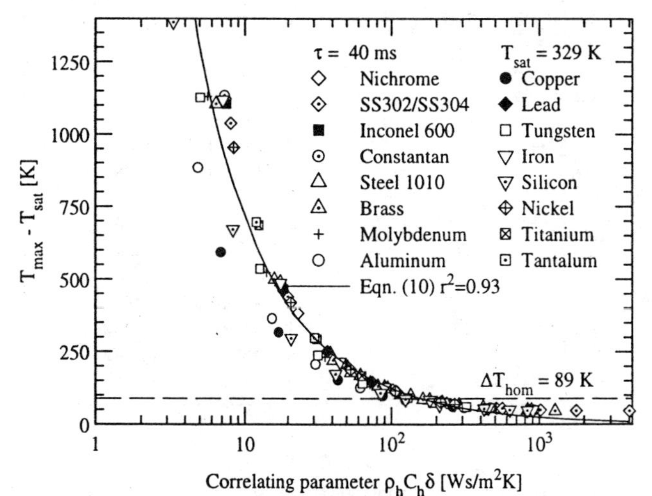

Figure 6. Plot of dry patch superheat against $\rho_h C_h \delta$: Larger dry spot

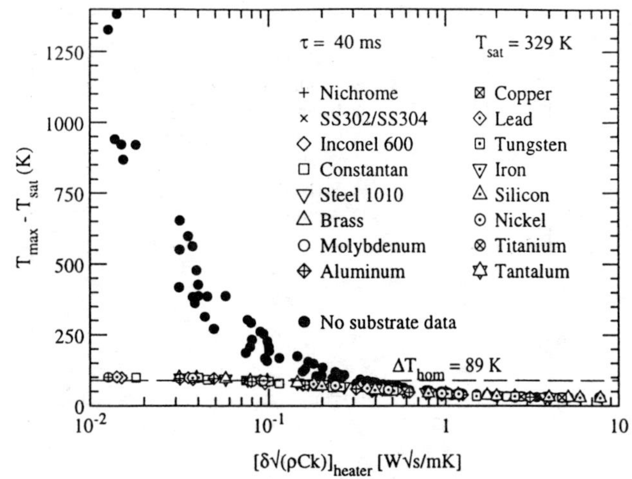

Figure 7. The effect of a Plexiglas substrate

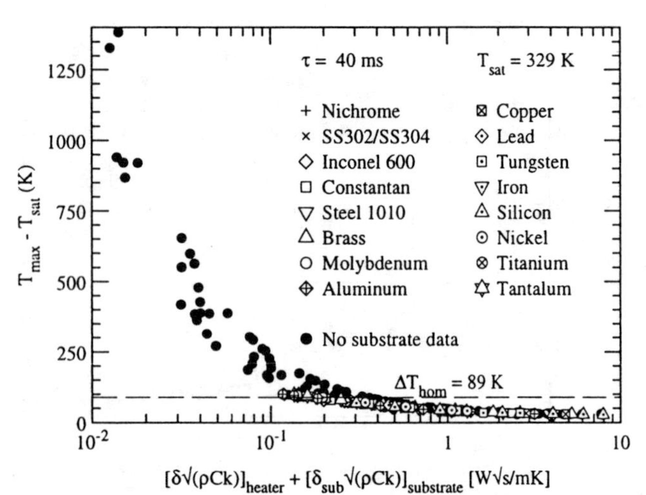

Figure 8. Modified thermal activity parameter for a Plexiglas substrate

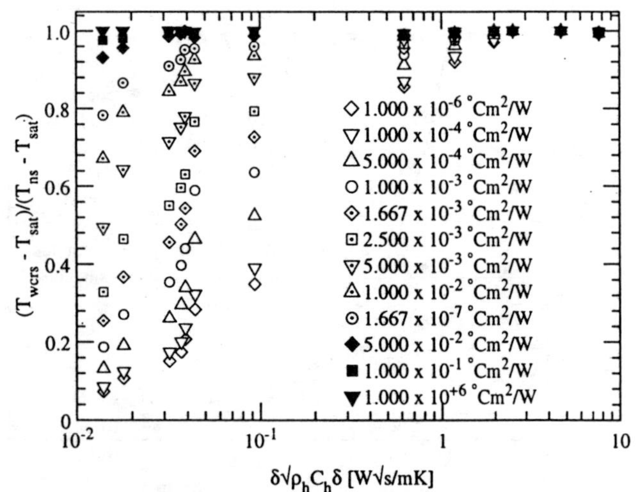

Figure 9. Maximum dry spot superheat variation with $\delta\sqrt{\rho_h C_h k_h}$ for various contact resistances

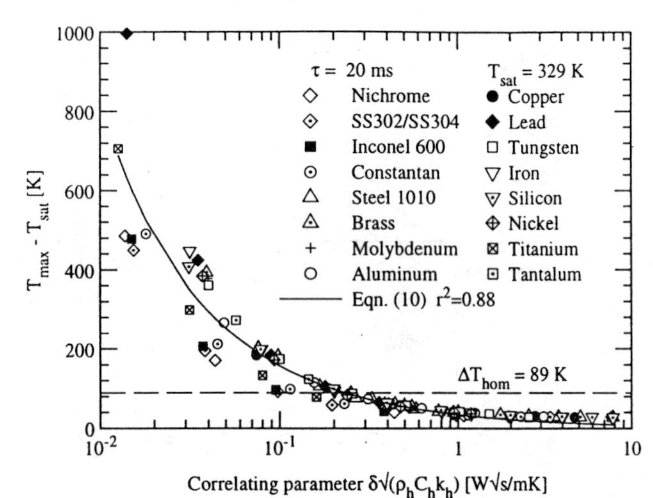

Figure 10. Plot of dry patch superheat against $\delta\sqrt{\rho_h C_h k_h}$: τ = 20 ms

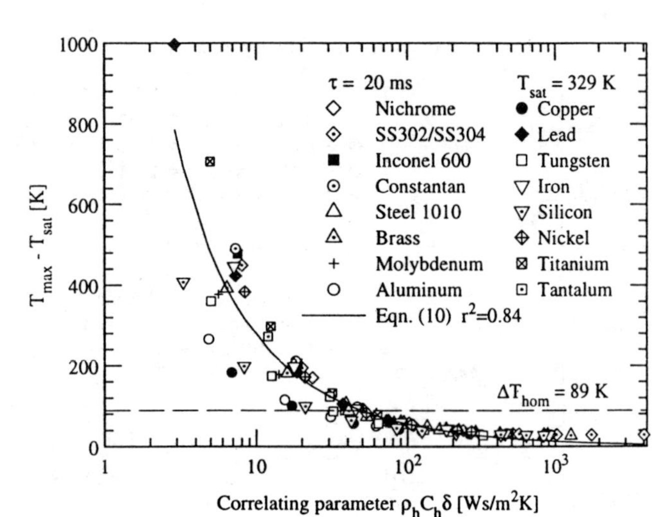

Figure 11. Plot of dry patch superheat against $\rho_h C_h \delta$: τ = 20ms

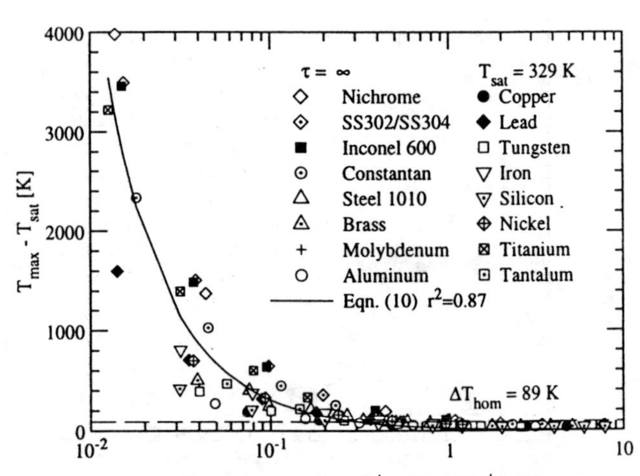

Figure 12. Plot of dry patch superheat against $\delta\sqrt{\rho_h C_h k_h}$: $\tau = \infty$

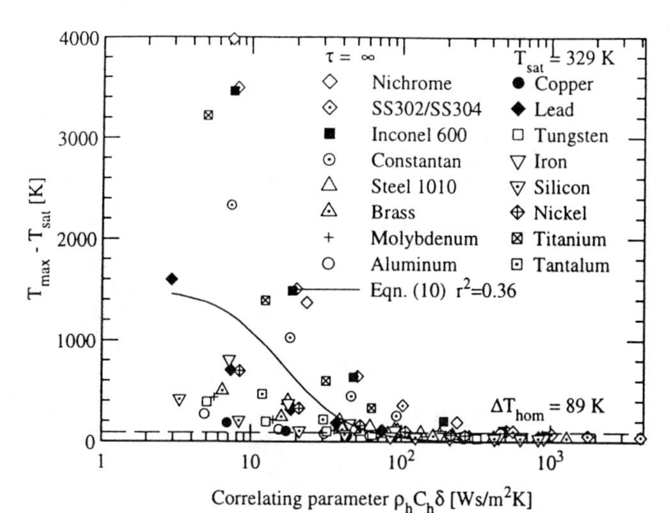

Figure 13. Plot of dry patch superheat against $\rho_h C_h \delta$: $\tau = \infty$

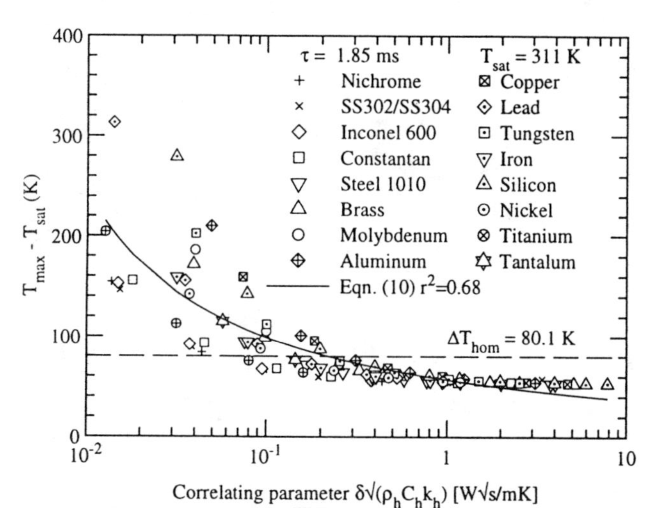

Figure 14. $T_{max} - T_{sat}$ vs. $\delta\sqrt{\rho_h C_h k_h}$ in flow boiling of FC-87 ($\bar{u}$ = 2 m/s)

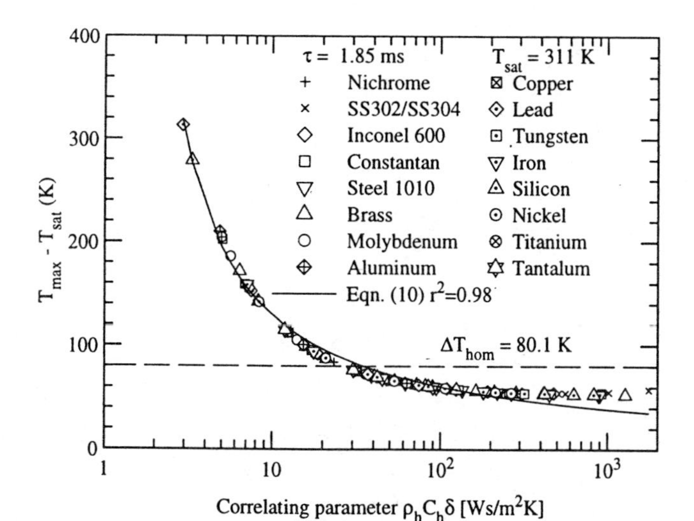

Figure 15. $T_{max} - T_{sat}$ vs. $\rho_h C_h \delta$ in flow boiling of FC-87 ($\bar{u}$ = 2 m/s)

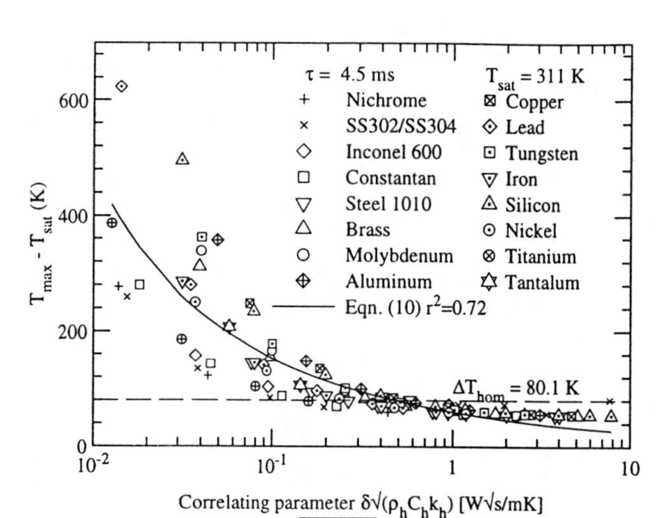

Figure 16. $T_{max} - T_{sat}$ vs. $\delta\sqrt{\rho_h C_h k_h}$ in flow boiling of FC-87 ($\bar{u}$ = 1 m/s)

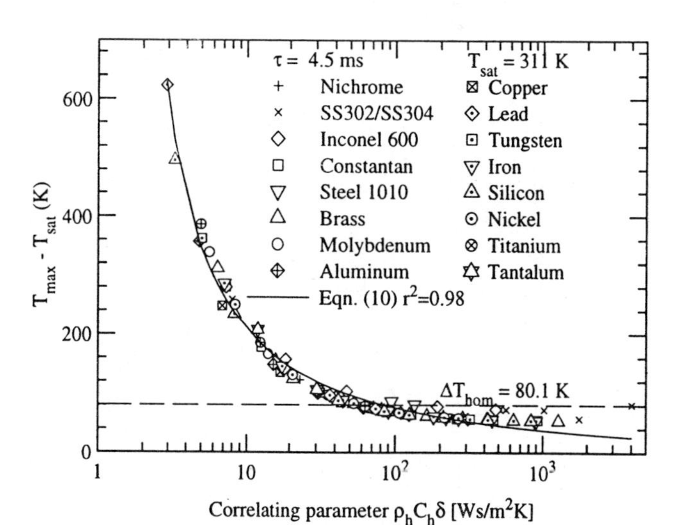

Figure 17. $T_{max} - T_{sat}$ vs. $\rho_h C_h \delta$ in flow boiling of FC-87 ($\bar{u}$ = 1 m/s)

HTD-Vol. 342, National Heat Transfer Conference
Volume 4
ASME 1997

EVAPORATION OF A SMALL WATER DROPLET CONTAINING AN ADDITIVE

Michelle D. King, Jiann C. Yang[1], Wendy S. Chien, and William L. Grosshandler

Building and Fire Research Laboratory
National Institute of Standards and Technology
Gaithersburg, Maryland 20899, U.S.A.

ABSTRACT

An experimental study on the evaporation of a small water droplet containing an additive on a heated, polished stainless-steel surface was performed. Solutions of water containing 30 % (w/w) and 60 % (w/w) of potassium acetate and sodium iodide were used in the experiments. Surface temperatures used in the experiments ranged from 50 °C to 100 °C. The average evaporation rates for the potassium acetate and sodium iodide solutions were found to be lower than that of pure water at a given surface temperature. A simple evaporation model was developed to interpret the experimental results.

INTRODUCTION

Fine water systems have several advantages over conventional fire protection sprinklers in certain applications when water supply is limited and collateral damage by water is a concern. These systems have recently been considered as a potential replacement for halon fire-protection systems in shipboard machinery spaces and crew compartments of armored vehicles. However, below 0 °C water will freeze, thus posing a limitation in low temperature operations. Certain additives, if selected properly, not only can suppress the freezing point of water but also can improve its fire suppression effectiveness. Some water-based agents have recently been proven to be more effective than pure water when applied in the form of mist to suppress a small jet fuel pool fire (Finnerty *et al.*, 1996). Among the thirteen agents tested, potassium lactate (60% w/w), potassium acetate (60% w/w), and sodium bromide (10% w/w) were found to be superior as fire-extinguishing sprays than pure water and other candidate solutions. The suppression benefit of adding a solute to a water spray was only noted when the spray was applied directly toward the fire; however, the effect of these additives on the overall fire suppression effectiveness when the spray is not directed toward the base of the fire remains unclear.

When a fine mist is formed in a nozzle, the majority of the droplets are unlikely to penetrate to the base of the fire because the droplet momentum is small enough that they are deflected away by the rising plume (Downie *et al.*, 1995). The deflected mist droplets subsequently experience a cooler environment outside the hot gas plume, thus resulting in slow droplet evaporation. Some of the slowly vaporizing droplets will impinge upon the enclosure surfaces wherein the fire is located, or upon obstacles within the enclosure. These droplets will eventually be vaporized on the heated surfaces. Droplets that impinge on these surfaces in the vicinity of the fire zones can still play many indirect roles in facilitating fire suppression through (1) surface cooling, thus mitigating flame spread and (2) entrainment of water vapor from the evaporating droplets into the flame. Therefore, rapid evaporation of water droplets from the surrounding heated surfaces may be desirable. The evaporation of water/additive droplets may not be an important issue if the droplets can penetrate the hot plume and reach the base of the fire. However, the role of the deflected droplets and their subsequent evaporation becomes significant if only a small amount of droplets with significant momentum can penetrate the flame.

While the addition of a solute to water may improve the suppression effectiveness within the fire through chemical or physical means, it may also affect the droplet vaporization and generation processes. The addition of a solute lowers vapor pressure and the mass transfer driving force for evaporation, elevates the boiling point of the solvent, and modifies other physical properties of water. Furthermore, the addition of solute decreases the relative amount of water in a droplet (*e.g.*, 60% w/w potassium lactate solution). For droplets that fall outside the flame zone, the solid residuals of the solute are left deposited on the heated surface after the water has been evaporated and may not contribute to the chemical suppression process.

The evaporation of suspended droplets containing dissolved solids in a hot ambience was first studied by Charlesworth and Marshall (1960). The formation of a solid crust and various appearance changes during the course of evaporation under a wide range of experimental conditions were observed. Three major evaporation stages were identified: (1) evaporation before the formation of the solid phase, (2) progressive formation of the solid phase about the droplet, and (3) evaporation during the solid phase

[1]Author to whom correspondence should be addressed.

formation. Several studies on droplets with dissolved solids have since been conducted (see *e.g.*, Nešić and Vodnik, 1991; Kudra *et al.*, 1991; Taniguchi and Asano, 1994). The formation of dried solids in a droplet impedes the evaporation process of the droplet. Liquid in the interior of the droplet must reach the surface in order to evaporate. Increasing the amount of solute increases the resistance to mass transfer inside the droplet, slowing the movement of moisture out of the droplet (Masters, 1985). However, to the best of our knowledge, no research has been performed on the evaporation of a water droplet with a dissolved salt on a heated surface.

The objective of the present work is to examine the evaporation characteristics of some water-based fire suppressing agents on a heated surface at temperatures below nucleate boiling. Because of its potentially superior fire suppression ability and its use in suppressing cooking grease fires, water with dissolved potassium acetate was used in the experiments. Sodium iodide was selected as another additive in lieu of sodium bromide (recommended by Finnerty *et al.*, 1996) because it is believed that the iodine compound may be more effective in fire suppression than its bromine counterpart (Pitts *et al.*, 1990). Previous studies (*e.g.*, diMarzo and Evans, 1989; Chandra and Avedisian, 1991; Qiao and Chandra, 1996) were focused on relatively large drops (above 1 mm in diameter) of pure solvents (water or hydrocarbons) or water with a small amount of surfactant added. For the present work, smaller droplets of highly concentrated electrolyte solutions with diameters between 0.3 mm and 0.6 mm (to simulate mist droplets) were used.

EXPERIMENTAL METHOD

Figure 1 shows a schematic diagram of the experimental apparatus. It consists of a droplet generator, a solution reservoir, a nickel-plated copper block equipped with two small cartridge heaters, a stainless steel surface, a temperature controller, and a CCD camera. The droplet generator has a chamber, a piezoelectric ceramic disc, and a glass nozzle (Yang *et al.*, 1997) and is based on the drop-on-demand ink-jet technique. A small droplet is ejected from the nozzle as a result of the deflection of the piezoelectric ceramic disc upon application of a squared pulse with controlled amplitude and duration to the disc. The use of this droplet generator enables the production of smaller droplets and repeatable operation. The surface on which the droplet is vaporized is a 5 cm x 3 cm x 0.5 cm polished stainless steel (SS 304) block fastened to the nickel-plated copper block. The 5 cm x 3 cm x 1.25 cm nickel-plated copper block is used to heat the surface to the desired temperature (between 50 °C and 100 °C). Surface temperature is maintained within ± 1 °C by using a temperature controller. The CCD camera is used to record the evaporation histories of the droplets. The evaporation times (see discussion below) of the droplets can be determined by using frame-by-frame analysis of the video records. The basic experimental procedure involved triggering the generator to deliver a single droplet from the nozzle to the heated surface located 6.5 cm below.

Single-shot stroboscopic photography (see *e.g.*, Chandra and Avedisian, 1991) was used to record the droplet formation at the nozzle exit of the droplet generator, droplet diameters before impact, and average droplet impact velocities. A 35 mm SLR camera equipped with a 105 mm lens and extended bellows, an electronic strobe for backlighting, and an electronic delay timer were used. The average impact velocity, which was taken to be the time between the instant when an electronic pulse was sent to the droplet generator and the instant when the droplet impacted the surface, was found to be 72 cm/s ± 2 cm/s. No shattering of droplets due to impact on the surface was observed within the range of surface temperatures tested. The Weber number ($We = \rho_l V^2 D_o/\sigma$) of the droplets was less than 80.

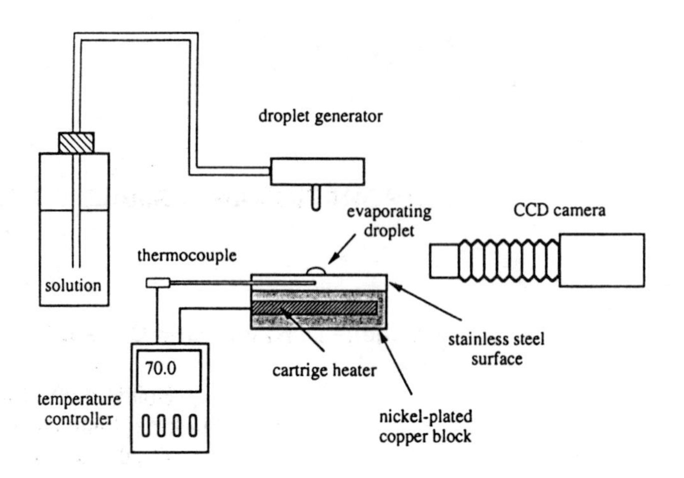

Figure 1. Schematic diagram of experimental apparatus

Since the mass loss of the evaporating droplet in the present experiment was not monitored continuously, it was not possible to determine the time for *complete* evaporation of water from the water/dissolved-solid droplet on the heated surface from the video records because of subsequent formation of solid residual. Therefore, an *apparent* evaporation time of a water droplet with dissolved solid is defined as the time when the solid residual first appears during evaporation with the assumption that the amount of water vaporized before solid formation constitutes the bulk of the initial water content. Such an assumption appears to be reasonable for small droplets with $D_o < 1.3$ mm (Charlesworth and Marshall, 1960). The average water evaporation rate is defined as the initial amount of water in the droplet divided by the apparent evaporation time. Knowing the initial solution density, droplet diameter, and solute concentration, the initial mass of water in the droplet can be determined.

Table 1 summarizes the physical properties of the aqueous solutions studied. The surface tensions of the aqueous solutions were determined by using a tensiometer which measured the force required to withdraw a platinum-iridium ring from the surface of the liquid. The surface tensions of sodium iodide solutions are greater than that of distilled water and increase as the concentration increases. For the potassium acetate solutions, it is interesting to note that the 30 % w/w and the 60 % w/w have surface tensions greater than and less than that of pure water, respectively. The densities of the solutions were taken from Söhnel and Novotný (1985), and the solution viscosities and normal freezing points were obtained from Washburn (1929).

RESULTS AND DISCUSSION

Since droplet formation depends on the initial salt concentration, the resulting initial droplet diameters will differ somewhat. In addition, the initial mass of water in the droplet decreases as the concentration of the dissolved salt increases. For the purpose of comparison, it is more meaningful to plot apparent or average mass evaporation rate than evaporation time as a function of surface temperature. The experimental average water mass evaporation rates at different surface temperatures for 0 %, 30 %, and 60 % potassium acetate and sodium iodide solutions are shown in Figure 2. The error bars in the figures are the 2-σ (standard deviation) of at least six runs. The large scatter in some of the experimental data reflects the difficulty in determining the first appearance of solid

Table 1. Physical properties of the solutions

Fluid	Density @ 22 °C (g cm^{-3})	Viscosity (g s^{-1} cm^{-1}) x 10^3	Normal Freezing point (°C)	Surface tension (@ 22 °C) (dyne cm^{-1}) ± 2 dyne cm^{-1}
Distilled water	1.00	9.6 (@ 22 °C)	0	72
30% potassium acetate	1.16	22.8 (@ 18 °C)	-23	73
60% potassium acetate	1.34	50.9 (@ 18 °C)	---- §	68
30% sodium iodide	1.29	10.9 (@ 20 °C)	-18	76
60% sodium iodide	1.80	23.4 (@ 20 °C)	---- §	78

§Not available from literature

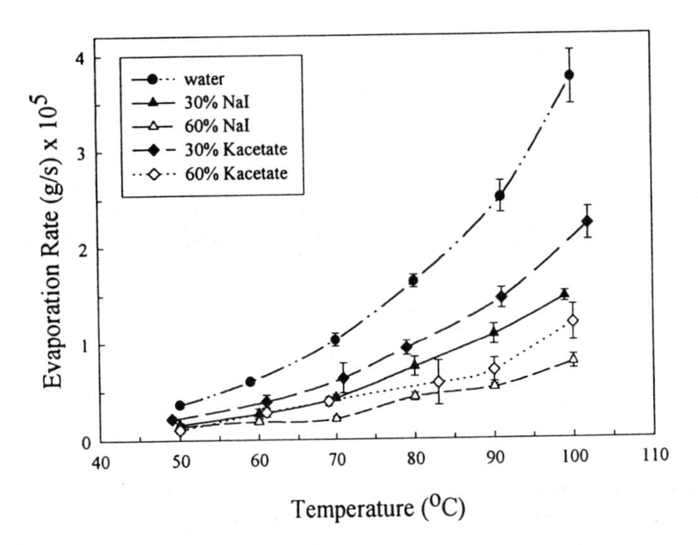

Figure 2. Experimental average mass evaporation rates at different surface temperatures

formation. The addition of potassium acetate or sodium iodide to water decreases the evaporation rate below that of pure water as shown in the figure. As the concentration of dissolved salt increases, the average evaporation rate becomes slower.

Heat and Mass Transfer Analysis

A simple model for predicting the average evaporation rate of a water droplet containing a dissolved solid on a heated surface is formulated in an attempt to compare with the experimental observations. The problem description is as follows. A droplet impinges on the heated surface and spreads. The spread droplet then evaporates due to heat transfer from the heated surface. The droplet evaporation model is based on the following simplifying description of the process:

1. After impact on the heated surface, the droplet immediately assumes the shape of a truncated sphere, whose diameter of the contact circle with the surface is taken to be the maximum spread diameter of the droplet, as shown in Figure 3.
2. The time for the droplet to attain its maximum spread diameter is negligible compared to the droplet evaporation time on the heated surface.
3. The droplet maintains its initial maximum spread diameter during evaporation; therefore, only H is a function of time (see Figure 3).
4. The heated surface is treated as an infinite thermal reservoir.
5. Molecular diffusion dominates at the vapor-liquid interface.
6. The instantaneous concentration of the dissolved solid in the droplet is spatially uniform during evaporation, and liquid-vapor phase equilibrium is maintained at the droplet surface.

The calculated evaporation time is defined as the time between impact and when the dissolved solute mole fraction, $X_d(t)$, becomes equal to the solubility of the solute, $X_{d,sat}(t)$; *i.e.*, a phase transition from dissolved solute to a solid phase will occur when $X_d(t) = X_{d,sat}(t)$. Note that the solubility of the solute is a function of time because the droplet temperature is changing.

After the droplet has impinged on the surface, its maximum spread diameter can be estimated by using the following empirical correlation (Scheller and Bousfield, 1995):

$$\frac{D_{max}}{D_o} = 0.61\,(Re\sqrt{We})^{0.166} \tag{1}$$

where the Reynolds number Re is based on the initial droplet diameter before impact. The assumption of maximum spread is reasonable because it was observed in this study that during most of the evaporation period the contact diameter between a water droplet containing a dissolved salt and the heated surface remained relatively constant. The spread diameter is required for the calculation of heat transfer from the heated surface to the droplet. The instantaneous droplet height H of the spherical segment can be calculated by equating the mass of the initial droplet (before impact) to that of the truncated sphere.

$$\frac{1}{6}\pi H\left(\frac{3}{4}D_{max}^2 + H^2\right) = \frac{\pi}{6}D_o^3 \tag{2}$$

Once H is determined, the radius of curvature R_c of the truncated sphere can be obtained by the following equation (Beyer, 1981):

$$R_c = \frac{1}{6H}\left(\frac{3}{4}D_{max}^2 + H^2\right) + \frac{H}{3} \tag{3}$$

The instantaneous heat transfer rate from the heated surface to the droplet is approximated by

$$\dot{Q} \approx A_b\frac{k_l\,(T_w - T_s)}{\delta} \tag{4}$$

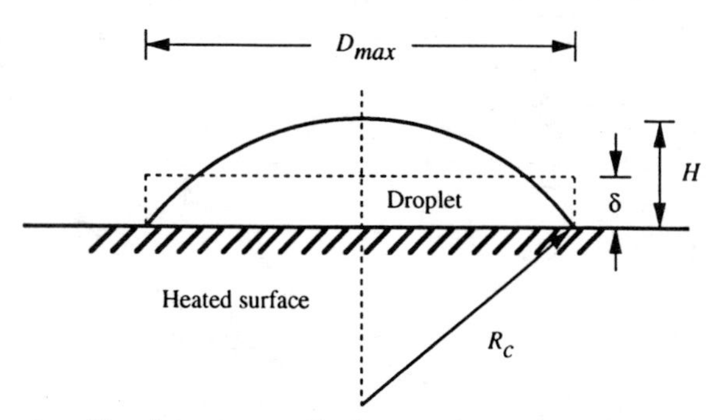

Figure 3. Droplet geometry for heat and mass transfer analysis

In writing Equation (4), the heat transfer is modeled as a thin right circular cylinder with a thickness δ (see Figure 3) and an equivalent volume equal to the spherical segment, and the temperature profile within the droplet is assumed to be linear (Bonacina *et al.*, 1979). If the heat transferred to the droplet from the surface is used solely to evaporate the liquid from the liquid-vapor interface, then an energy balance on the evaporating droplet can be expressed as

$$\dot{m} \Delta H_v = \dot{Q} \tag{5}$$

Following the treatment given in Bird *et al.*(1960), the instantaneous evaporation rate can be written as

$$\dot{m} = k_m A_f \ln \left[1 + \frac{Y_{ws} - Y_{w\infty}}{1 - Y_{ws}} \right] \tag{6}$$

In order to calculate the evaporation rate $\dot{m}$, the value of T_s or Y_{ws} is needed. Using Equations (4), (5), and (6), one gets

$$k_m A_f \ln \left[1 + \frac{Y_{ws} - Y_{w\infty}}{1 - Y_{ws}} \right] = A_b \frac{k_l (T_w - T_s)}{\delta \Delta H_v} \tag{7}$$

If equilibrium is assumed at the liquid-vapor interface and the vapor phase is assumed to be an ideal gas mixture, then equating the component fugacities in both phases (Prausnitz *et al.*, 1986) results in

$$Y_{ws} \approx \frac{P_{solution}}{P_t} \tag{8}$$

where $P_{solution}$ is the vapor pressure of the aqueous solution. The reduction of vapor pressure due to the presence of the dissolved solid can be seen in Equation (8) since without the dissolved solid, $P_{solution}$ equals to $P_{w,sat}$, assuming the solubility of air in water is negligible under atmospheric conditions. Knowing the vapor pressure of the solution to be a function of droplet surface temperature, Equation (7) can be solved for T_s. Once T_s is known, the evaporation rate $\dot{m}$ can be calculated by using Equations (4) and (5).

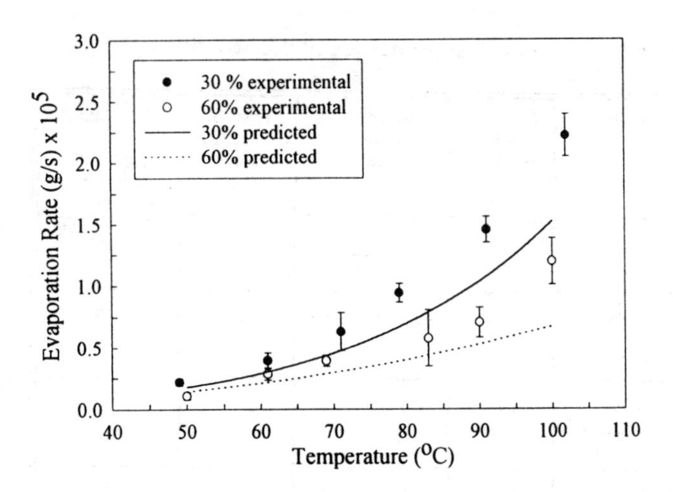

Figure 4. Experimental and predicted average evaporation rates of potassium acetate at various surface temperatures

Since mass transfer coefficients of spherical droplets are readily available in the literature, the mass transfer process is treated as if the truncated droplet were evaporating as a spherical droplet with a radius equivalent to the radius of curvature R_c of the truncated sphere. In this case, the mass transfer coefficient k_m can then be estimated by using the Chilton-Colburn analogy (Bird *et al.*, 1960).

$$\frac{k_m 2 R_c}{c_f D_{wa\,f}} = 2 \tag{9}$$

Calculations were performed for two different concentrations of potassium acetate and sodium iodide solutions (see the Appendix for calculation procedure). Figures 4 and 5 compare the predicted and measured average evaporation rates of the two aqueous solutions at different surface temperatures. The predicted values were obtained based on the average initial droplet diameters used in all the experiments with the *same* initial dissolved salt concentration. Since most of the estimation methods for the physical properties of aqueous electrolyte solutions are primarily applicable to dilute or moderately concentrated solutions at 25°C, a certain degree of extrapolation had to be used in the calculations for concentrated solutions at higher temperatures when no literature values were available. The discrepancies between the predictions and measurements in Figures 4 and 5 may be largely attributed to the estimation methods. However, the disagreements are equally likely due to the simplified heat and mass transfer models used in the analysis and the time-averaging of the evaporation rate.

Direct comparison of the results between the two aqueous solutions with the same initial salt concentration proves to be not straightforward. The degree of vapor pressure lowering due to the presence of a salt depends on the ionic strength of the solution. The two aqueous solutions used in the experiments are not at the same ionic strength despite the same initial mass fraction of salt in the solutions. Even if solutions with same ionic strength and same initial droplet diameters were used, the spread of the liquid droplet after impact would still differ because of different Reynolds and Weber numbers. This would change the heat and mass transfer processes between the droplet and the surface. Therefore, it is difficult to interpret the results

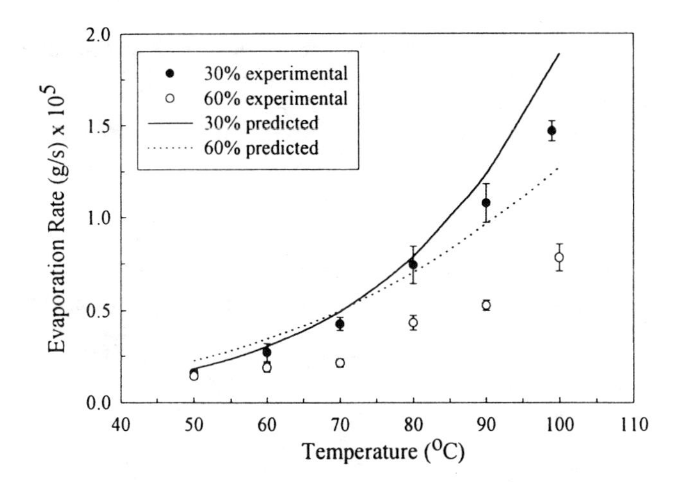

Figure 5. Experimental and predicted average mass evaporation rates of sodium iodide at various surface temperatures

in Figures 4 and 5 solely based on the effect of vapor pressure lowering on the evaporation rate, although for solutions with the same ionic strength, the degree of vapor pressure lowering at a given temperature is less for potassium acetate than for sodium iodide (Washburn, 1929).

The evaporation model described above is rudimentary compared to those previously developed for pure water evaporation (*e.g.*, DiMarzo and Evans, 1989; Qiao and Chandra, 1997); however, the model does not require *a priori* information (*i.e.*, convective heat transfer correlations or spread diameter and contact angle measurements) other than initial droplet diameters, droplet impact velocities, and initial physical properties of the solutions. Since the heat transfer rates to the droplets were directly or indirectly measured and used as input in their analysis, it is not surprising that the models developed by DiMarzo and Evans (1989) and Qiao and Chandra (1997) correlate the experimental data very well. However, the present model overpredicts the evaporation times of pure water. This is probably partially due to the simplified heat transfer model and partially due to the breakdown of Equation (9) in the mass transfer analysis. Since the mass transfer coefficient depends on the radius of the curvature in Equation (9), it will approach zero as $R_c \rightarrow \infty$. As the droplet is evaporating, R_c is increasing as a result of decreasing H, and the instantaneous mass evaporation rate is asymptotically approaching zero at later times, thus resulting in a relatively long droplet evaporation lifetime. If the average mass evaporation rate of pure water were calculated based on a final mass of 20 % (rather than 0 %) water remaining in order to circumvent this asymptotical behavior, the calculated evaporation rates would correlate well with the measurements. This situation is not encountered in the calculation for a droplet with a dissolved salt because the solute concentration reaches its saturation value before H is significantly reduced.

CONCLUSIONS

Experiments on the evaporation of a small droplet with a dissolved salt on a heated surface have been performed as part of an assessment of the performance of water with an additive as a fire suppressant. The surface temperatures varied from 50 °C to 100 °C. Two salts, potassium acetate and sodium iodide, were used. The addition of potassium acetate and sodium iodide decreases the average evaporation rates. At a given surface temperature, the average mass evaporation rate decreases as the concentration of the dissolved salt increases. A simple evaporation model was developed. Despite the assumptions, idealization, and the uncertainties associated with the thermophysical property estimations of the concentrated electrolyte solutions, this simple analysis captures most of the essential characteristics of the evaporation process of a droplet with a dissolved solid on a heated surface and agrees qualitatively with the trend of the experimental data.

Several issues regarding the application of dissolved salts in water in fire suppression should be carefully considered. If the aqueous solution droplets penetrate the fire plume and reach the fire, the chemical suppression action of the salt could be fully utilized, and the reduction in water evaporation rate due to the added salt would not be a governing factor. On the other hand, if most of the droplets were deflected away from the fire, the chemical suppression effectiveness of the salt would be minimized or lost, and the reduction in water evaporation rate would impose an additional penalty because for a given time less water vapor would be generated and entrained into the adjacent fire for suppression.

ACKNOWLEDGMENTS

The authors would like to thank Professors. S. Chandra of the University of Toronto and S. C. Yao of Carnegie-Mellon University for many helpful suggestions and discussions.

NOMENCLATURE

A_b = contact surface area of the truncated sphere = $\pi D_{max}^2/4$, cm^2
A_f = droplet free surface area = $2\pi R_c H$, cm^2
c = vapor concentration, mole/cm^3
D_{max} = maximum spread diameter of the droplet after impact, cm
D_o = initial droplet diameter (before impact), cm
D_{wa} = diffusion coefficient of water vapor in air, cm^2/s
H = height of the truncated sphere, cm
ΔH_v = latent heat of vaporization of water, J/mole
k_l = thermal conductivity of the liquid, J/cm s K
k_m = mass transfer coefficient, mole/ s cm^2
$\dot{m}$ = mass transfer rate, mole/s
P = pressure, dyne/cm^2
$\dot{Q}$ = heat transfer rate, J/s
R_c = radius of curvature of the truncated sphere, cm
Re = Reynolds number = $\rho_l D_o V/\mu_l$
T = temperature, K
T_w = temperature of the heated surface, K
V = impact velocity of the droplet, cm/s
We = Weber number = $(\rho_l V^2 D_o/\sigma)$
X = mole fraction in the liquid phase
Y = mole fraction in the vapor phase
δ = equivalent heat conduction distance, cm
μ = viscosity, g/cm s
ρ = density, g/cm^3
σ = surface tension of the liquid, dyne/cm
$\Delta\tau$ = time step, s

Subscripts
d = dissolved solid
f = at film temperature [= $(T_w + T_\infty)/2$] or at film mole fraction [= $(Y_{ws} + Y_{w\infty})/2$]

l = liquid phase
o = initial
s = droplet surface
t = total
sat = saturation
w = water or heated surface
∞ = ambience

REFERENCES

Beyer, W. H., *CRC Standard Mathematical Tables*, 27th ed., CRC Press, Boca Raton, 1981.

Bird, R. B., Stewart, W. E., and Lightfoot, E. N., *Transport Phenomena*, John Wiley & Sons, New York, 1960.

Bonacina, C., Del Giudice, S., and Comini, G., "Dropwise Evaporation," *ASME Journal of Heat Transfer*, Vol. 101, August 1979, pp. 441-446.

Chandra, S., and Avedisian, C. T., "On the Collision of a Droplet with a Solid Surface," *Proceedings of the Royal Society of London A*, Vol. 432, 1991, pp. 13-41.

Charlesworth, D. H., and Marshall, W. R., Jr., "Evaporation from Drops Containing Dissolved Solids," *A.I.Ch.E. Journal*, Vol. 6, No. 1, March 1960, pp. 9-23.

Cisternas, L. A., and Lam, E. J., "An Analytic Correlation of Vapour Pressure of Aqueous and Non-Aqueous Solutions of Single and Mixed Electrolytes," *Fluid Phase Equilibria*, Vol. 53, 1989, pp. 243-249.

Cisternas, L. A., and Lam, E. J., "An Analytic Correlation of Vapour Pressure of Aqueous and Non-Aqueous Solutions of Single and Mixed Electrolytes. Part II. Application and Extension," *Fluid Phase Equilibria*, Vol. 62, 1991, pp. 11-27.

Daubert, T. E., and Danner, R. P., *Physical and Thermodynamic Properties of Pure Chemicals, Data Compilation*, Hemisphere, New York, 1992.

Dean, J. A. (editor), *Lange's Handbook of Chemistry*, 12th ed., McGraw-Hill, New York, 1979.

Downie, B., Polymeropoulos, C., and Gogos, G., "Interaction of a Water Mist with a Buoyant Methane Diffusion Flame," *Fire Safety Journal*, Vol. 24, 1995, pp. 359-381.

Finnerty, A. E., McGill, R. L., and Slack, W. A., "Water-Based Halon Replacement Sprays," ARL-TR-1138, U.S. Army Research Laboratory, Aberdeen Proving Ground, July 1996.

Horvath, A. L., *Handbook of Aqueous Electrolyte Solutions, Physical Properties, Estimation and Correlation Methods*, Ellis Horwood Ltd., Chichester, 1985.

Kudra, T., Pan, Y. K., and Mujumdar, A. S., "Evaporation from Single Droplets Impinging on Heated Surfaces," *Drying Technology*, Vol. 9, No. 3, 1991, pp. 693-707.

diMarzo, M., and Evans, D. D., "Evaporation of a Water Droplet Deposited on a Hot High Thermal Conductivity Surface," *ASME Journal of Heat Transfer*, Vol. 111, February 1989, pp. 210-213.

Masters, K., *Spray Drying Handbook*, 4th ed., CRC Press, Boca Raton, 1985.

Nešić, S., and Vodnik, J., "Kinetics of Droplet Evaporation," *Chemical Engineering. Science*, Vol. 46, No.2, 1991, pp. 527-537.

Pitts, W. M., Nyden, M. R., Gann, R. G., Mallard, W. G., and Tsang, W., *Construction of an Exploratory List of Chemicals to Initiate the Search for Halon Alternatives*, NIST Technical Note 1279, U. S. Department of Commerce, Washington D.C., August 1990.

Prausnitz, J. M., Lichtenthaler, R. N., and Gomes de Azevedo, E., *Molecular Thermodynamics of Fluid-Phase Equilibria*, 2nd ed., Prentice-Hall, Englewood Cliffs, 1986.

Qiao, Y. M., and Chandra, S., "Experiments on Adding a Surfactant to Water Droplets Boiling on a Hot Surface," to be published in *Proceedings of the Royal Society of London A*, 1997.

Reid, R. C., Prausnitz, J. M., and Poling, B. E., *The Properties of Gases and Liquids*, 4th ed., McGraw-Hill, New York, 1987.

Scheller, B. L., and Bousfield, D. W., "Newtonian Drop Impact with a Solid Surface," *A.I.Ch.E. Journal*, Vol. 41, No. 6, June 1995, pp. 1357-1367.

Söhnel, O., and Novotný, P., *Densities of Aqueous Solutions of Inorganic Substances*, Elsevier, Amsterdam, 1985.

Taniguchi, I., and Asano, K., "Evaporation of an Aqueous Salt-Solution Drop in Dry Air During Crystallization of Salt," *Proceedings of the Sixth International Conference on Liquid Atomization and Spray Systems*, July 18-22, 1994, Palais des Congrès, Rouen, France, pp. 859-866.

Washburn, E. W. (editor-in-chief), *International Critical Tables of Numerical Data, Physics, Chemistry and Technology*, Vol. 5, 1st ed., McGraw-Hill, New York, 1929.

Yang, J. C., Chien W., King, M. D., and Grosshandler, W. L., "A Simple Piezoelectric Droplet Generator," submitted to *Experiments in Fluids*, 1997.

APPENDIX

This appendix describes the procedure to calculate the evaporation time and the average evaporation rate of a droplet with dissolved solid on a heated surface. For given D_o, V, μ_l, σ, and $\Delta\tau$, the following computational scheme may be used:

1. Calculate D_{max} using Equation (1).
2. Solve for H and R_c using Equations (2) and (3) respectively.
3. Calculate k_l using X_w and $(T_s + T_w)/2$.
4. Calculate c_f and $D_{wa,f}$ using $(T_s + T_\infty)/2$.
5. Calculate k_m using Equation (9).
6. Calculate T_s using Equations (7) and (8) by iteration.
7. Calculate $\dot{m}$ using Equation (5).
8. Calculate the amount of water evaporated (Δm) during $\Delta\tau$ which is equal to $\dot{m}\,\Delta\tau$.
9. Calculate new X_w and X_d.
10. If $X_d \approx X_{d,sat}$, then stop; else calculate ρ_l, go to Step 2 to recalculate H and R_c knowing that Δm has been evaporated over a period of $\Delta\tau$, and repeat Steps 3 to 10.

The calculated evaporation time is then the sum of $\Delta\tau$'s required to reach $X_{d,sat}$, and the average evaporation rate is the ratio of the difference between the initial water content and the final water content at $X_{d,sat}$ to the total $\Delta\tau$s. A time-step $\Delta\tau$ of 0.1 s was found to be adequate and was used in all the calculations.

The method of Riedel (Horvath, 1985) is used to estimate the thermal conductivity of the solution. The solution density is estimated by the method of Koptev (Horvath, 1985), and the data on $X_{d,sat}$ are obtained from Dean (1979). The latent heat of vaporization of water as a function of temperature is obtained from Daubert and Danner (1992). Strictly speaking, the latent heat of vaporization of the solution should be used. The vapor pressures of the aqueous solutions are calculated by the method of Cisternas and Lam (1989, 1991). The diffusion coefficient of water vapor in air is estimated by using the method of Fuller *et al.* (Reid *et al.*, 1987), and the vapor concentration c_f is calculated by assuming the vapor mixture (at film temperature T_f and P) to be ideal.

HTD-Vol. 342, National Heat Transfer Conference
Volume 4
ASME 1997

Influence of Bubble Formation on Pressure Drop in a Uniformly Heated Vertical Channel

Amy S. Fleischer, E.V. McAssey, Jr. and G.F. Jones
Villanova University, Dept. of Mechanical Engineering
Villanova, PA 19085

ABSTRACT

An analytical model to predict the influence of bubble formation on pressure drop in uniformly heated vertical channels, and on the resulting onset of flow instability has been developed and evaluated. The model is general in nature and covers the complete range of boiling regimes from single phase flow to fully saturated boiling, regardless of flow orientation. The pressure drop in the channel is determined by dividing the channel into regions of similar bubble behavior. The extent of each region is determined by the local fluid temperature. The theoretical results are compared to an extensive set of experimental data covering a range of channel diameters and operating conditions. Two different theoretical models for subcooled boiling fluid mass fraction are used, to gauge the influence of model selection on the accuracy. The model also predicts a threshold heat flux below which flow instability may not occur during flowrate reductions.

NOMENCLATURE

C_p	Fluid specific heat at constant pressure
D	Channel diameter
f	Friction factor
g	Acceleration due to gravity
h	Enthalpy
k	Conductivity
Nu	Nusselt Number
P	Pressure
Re	Reynolds number
R2	Martinelli-Nelson (1948) saturated boiling parameter
St	Stanton number
T	Temperature of fluid
V	Fluid velocity
v	Specific volume
X	Thermodynamic quality
X_b	Thermodynamic quality at onset of significant voiding
X_t	True mass fraction
Y_b	Bubble diameter in the nucleate boiling region
Z	Length of region
α	Void fraction
ϕ	Heat flux
ϕ^2_{fo}	Martinelli-Nelson (1948) saturated boiling two phase multiplication factor
θ	Angle of channel (90° for downflow, -90° upflow)
ρ	Density
σ	Surface tension

subscripts

b	bubble
f	fluid
f	Fanning friction factor
fdb	fully developed subcooled boiling region
g	gas
pdb	partially developed subcooled boiling region
sat	saturated boiling region
sp	single phase
tp	two-phase

Introduction

The formation of bubbles as a consequence of uniformly heating flow through a vertical channel has a significant effect on the overall pressure drop along the channel. When the fluid is completely single phase, the pressure drop decreases as the fluid velocity decreases. This is because in single phase flow, the pressure drop is comprised primarily of effects due to wall friction associated with tube surface roughness, which decrease with velocity. However, as the velocity continues to decrease, the applied heat flux begins to cause vapor formation in the fluid. Boiling causes an increase in frictional drag due to the presence of bubbles, and also acts to increase pressure drop through acceleration and buoyancy effects. At some velocity, the increase in pressure drop due to boiling completely offsets the decrease in pressure drop due to channel frictional components. Further velocity reductions now cause the pressure drop to rise. This minimum point in the pressure drop versus velocity curve (demand curve) is referred to as the onset of flow instability (OFI) point. This type of excursive instability is referred to as Ledinegg instability.

If parallel flow paths exist, this increase in pressure drop due to bubble presence in one channel may cause flow to be diverted to alternate channels. The reduction in flow in the channel of interest will result in even more bubble formation. The situation continues to escalate until the channel experiences burnout.

Prediction of Ledinegg flow instability is an important design consideration for any high flux heat exchanger application in which parallel flow paths exist, and particularly in nuclear reactor design due to the possibility of a flow excursion during a postulated accident.

For a given channel geometry and operating conditions (exit pressure, inlet temperature) the velocity at which Ledinegg instability occurs is proportional to the applied heat flux. However, in certain upflow situations, the problem becomes even more complicated as the bubble formation does not lead to Ledinegg instability, but can lead instead directly to critical heat flux conditions as the flowrate drops. This is due to the influence of buoyancy in upflow. In certain upflow cases, the increase in pressure drop due to boiling effects is more than offset by the decrease in gravity head from buoyancy, and no minimum occurs in the demand curve. These are low heat flux applications and the flux below which Ledinegg instability will no longer occur is referred to as the threshold flux.

Several experimental and analytical studies have been performed to determine the influence of bubble formation on channel pressure drop. These studies have been limited by the determination of the bubble behavior over the entire length of the flow regime and the method of determining bubble influence on fluid mass fraction in the subcooled boiling region. The present study seeks to develop a general model which covers the complete range of boiling regimes from single phase flow to fully saturated boiling, regardless of flow orientation. In addition, the model for subcooled boiling fluid mass fraction is varied in order to compare the accuracy of two different models. Finally, the model predicts a threshold heat flux below which Ledinegg instability may not occur for upflow situations. The theoretical results are compared to an extensive set of experimental data covering a range of channel diameters and operating conditions.

Background

A substantial body of research exists on the calculation of pressure drop in two-phase flow and on the prediction of flow instability for vertical flow in heated channels. Some of the earliest work on two-phase flow pressure drop calculations was that of Martinelli and Nelson (1948). Martinelli and Nelson (1948) calculated the two-phase frictional pressure drop for saturated boiling conditions simply by applying empirical multipliers to the widely accepted methods for calculating single phase pressure drop. Despite its simplicity, this method was the basis for most of the future work in this field and its ease of use leads to its continued application as an estimating method.

Soon, researchers began to more thoroughly investigate the direct effect that the bubble formation had on the friction factor. Sabersky and Mulligan (1955) and Thom (1964) were among the first to account for effects due to vapor presence in their pressure drop correlations.

In the late 1950s and early 1960s, using these methods of calculating two-phase flow pressure drop, researchers began to identify the onset of flow instabilities caused by two-phase flow pressure drop. This early research, by Marshek (1958), Dormer and Bergles (1964) and Maulbetsch and Griffith (1966), identified that flow excursions occurred when the pressure gradient versus mass flow reached a minimum point, and that the location of the minimum point was a function of the applied heat flux and the geometry of the channel.

Whittle and Forgan (1967) collected experimental data for various channel designs under subcooled boiling conditions. They determined that the minimum in the demand curve occurred at fixed values of the ratio of the channel temperature rise to the inlet subcooling. Levy (1967) examined the formation of vapor during subcooled boiling and developed a model to predict the vapor volumetric fraction. He determined the point of bubble separation from the channel wall by a force balance on the bubble.

With the work of Owens and Schrock (1960) and Saha and Zuber (1974), researchers began to differentiate the different regions of bubble behavior along the length of a heated tube, and identify bubble dynamic influence on pressure drop. Owens and Schrock (1960) theorized that the flow would transition from single phase to subcooled boiling when the channel wall temperature exceeded the boiling point and Saha and Zuber (1974), proposed that the subcooled boiling region could be further divided into two regions: the nucleate boiling region, characterized by the formation of bubbles that remain on the channel wall, and the fully developed subcooled region, in which the bubbles become entrained in the general flow before collapsing in the subcooled bulk fluid. The bubble behavior in each region would affect the pressure drop calculation differently. Both these hypothesis were confirmed with experimental results.

Block et al. (1990) collected experimental data identifying OFI for annulus geometries and compared that data to that from an analytical model. The analytical model was based on three distinct boiling regimes, (single phase, nucleate boiling and fully developed subcooled boiling) as identified by the fluid temperature. The method uses the work of Levy (1967) and Saha and Zuber (1974). Stelling et. al (1996) presented a modeling approach similar to Block et. al (1990) but with an improved subcooled boiling quality model. The work of Stelling et. al (1996) was limited to the downflow of a subcooled boiling fluid.

Dougherty and Yang (1995) developed a non-equilibrium, homogenous flow pressure drop model based on a novel rate equation for net steam generation. The method makes use of a quantity defined as true mass fraction, which is zero at the onset of significant voiding and increases from that point. Unlike the conventionally defined quality for boiling heat transfer, where quality is a function of saturation enthalpy values, true mass fraction is always positive in the regions in which vapor content is substantial.

The methods of Block et al. (1990), Stelling et. al (1996), Martinelli-Nelson (1948), and Dougherty-Yang (1995) are used to develop the theoretical models presented in this study. This study seeks to develop a general model which covers the complete range of boiling regimes and flow orientations. In addition, the model for subcooled boiling fluid mass fraction is varied in order to compare the accuracy of two different models. The results are compared to a database of experimental results collected under controlled flow conditions for upflow and downflow.

Pressure Drop Models

As a fluid flows through a uniformly heated tube, the fluid may pass through four regions of flow. The fluid enters the channel as single-phase flow. As the flow progresses down the channel, the constant heat addition causes the fluid to begin nucleate boiling, continue through fully developed subcooled boiling, and finally reach saturated boiling. The transition points between these regions of flow can be identified by the bulk temperature of the fluid. The pressure drop in each region is found by solving a set of equations appropriate for the fluid behavior in that region. The regional pressure drops are then summed to determine the overall channel pressure drop.

In the single phase fluid region, the pressure drop is determined primarily by the flow speed and a friction factor based on Reynolds number and channel wall roughness. The partially developed subcooled boiling region (nucleate boiling) is characterized by the presence of small vapor bubbles on the walls of the channel. These bubbles, which do not have enough energy to depart from the channel wall, are formed as the channel wall temperature rises past the saturation temperature. The transition point between single phase flow and partially developed subcooled boiling is referred to as the onset of nucleate boiling (ONB). The pressure drop model in this region is similar to that in the single phase region with the exception that the characteristics of the bubbles lining the walls affect the friction factor.

Fully developed subcooled boiling begins when the bubbles obtain enough energy to detach from the channel wall. This point is known as the onset of significant voiding (OSV). Although the vapor bubbles are detaching from the channel wall, there is still no net vapor generation due to the fact that the bubbles collapse soon after becoming entrained in the still subcooled fluid. However, the bubble presence does introduce a significant acceleration term into the pressure drop model.

The final region of fluid behavior is saturated boiling. This region is reached when the bulk fluid temperature reaches the saturation temperature. The region is characterized by significant net vapor generation. The pressure drop model used in the saturated region must take into account the vapor presence and the pressure drop becomes a function of the fluid quality.

The two theoretical models presented in this paper treat the fully developed subcooled and saturated boiling regions differently. The first model (Model A) is based on the method proposed by Stelling et. al (1996) for the fully developed subcooled boiling region and is supplemented by the Martinelli and Nelson (1948) method for saturated boiling. The second model (Model B), based on ideas proposed by Dougherty and Yang (1995), does not distinguish between the fully developed subcooled boiling region and the saturated flow region, instead treating the entire region after OSV as a single two phase region.

For the single phase and partially developed subcooled boiling regions, the developed model uses the same pressure drop equations as Stelling et. al (1996), but modified to include the effects of flow direction. Additional refinements have been made in the fluid properties, in that the fluid properties are based on a regional average temperature rather than the fluid inlet temperature, and the fluid saturation temperature is adjusted for the effect of pressure drop. A detailed description is presented in Fleischer (1996).

The extent of the single phase and partially developed subcooled boiling regions can be found by the fluid temperature at the transitions points. These are based on the Davis and Anderson (1966) and Saha and Zuber (1974) correlations respectively.

$$T_{ONB} = T_{SAT} + \sqrt{\frac{8\sigma\phi T_{SAT}}{h_{fg} k \rho_g}} - \frac{\phi D}{Nuk} \tag{1}$$

$$T_{OSV} = T_{SAT} - \frac{\phi}{St\rho_f C_p V} \tag{2}$$

With these temperatures, energy balances can be performed to find the length of each region, and the total pressure drop calculated. The friction factor in the single phase region is a Fanning type friction factor and the friction factor in the partially developed subcooled region, f_{PDB}, is a function of the friction caused by the bubbles lining the walls.

$$\Delta P_{SP} = \Delta Z_{SP}\left(\frac{2 f_{SP}}{D} V^2 \rho_f + \rho_f g \sin(\theta)\right) \tag{3}$$

$$\Delta P_{PDB} = \Delta Z_{PDB}\left(\frac{2 f_{PDB}}{D} V^2 \rho_f + \rho_f g \sin(\theta)\right) \tag{4}$$

$$f_{pdb} = \frac{1}{4}\left(-2\log_{10}\left(\frac{Y_b}{3.7D} - \frac{5.02}{Re}\log_{10}\left(\frac{Y_b}{3.7D} + \frac{13}{Re}\right)\right)\right)^{-2} \tag{5}$$

For Model A, the pressure gradient in the fully developed subcooled boiling region must account for the vapor present in the flow. A subcooled boiling model developed by Bowring (1962) is used to calculate the fluid mass fraction vapor, X. The pressure drop in the saturated flow region is calculated according to a model developed by Martinelli and Nelson (1948). This method predicts pressure drop in saturated forced convective boiling based on a separated flow model.

$$\Delta P_{FDB} = \Delta Z_{FDB}\left(\frac{2 f_{FDB} V^2 \rho_f}{D} + V^2 \rho_f^2 (v_g - v_f)\frac{dX}{dZ} + \rho_f g \sin(\theta)\right) \tag{6}$$

$$\Delta P_{sat} = \frac{2\Delta Z_{SAT} f_{SAT} v_{fsat} \rho_f^2 V^2 \phi_{fo}^2}{D} + \rho_f^2 V^2 v_{fsat} R2 + \frac{\Delta Z_{SAT} g \sin(\theta)}{X}\left(\frac{1}{v_g}\alpha + \rho_f(1-\alpha)X\right) \tag{7}$$

For Model B, the pressure gradient in the two phase region is calculated according to a non-equilibrium homogenous model proposed by Dougherty and Yang (1995). The key to the method is the calculation of a quantity referred to as true mass fraction, X_t. X_t is zero at the onset of significant voiding and increases from that point. Unlike conventionally defined quality, which is a function of

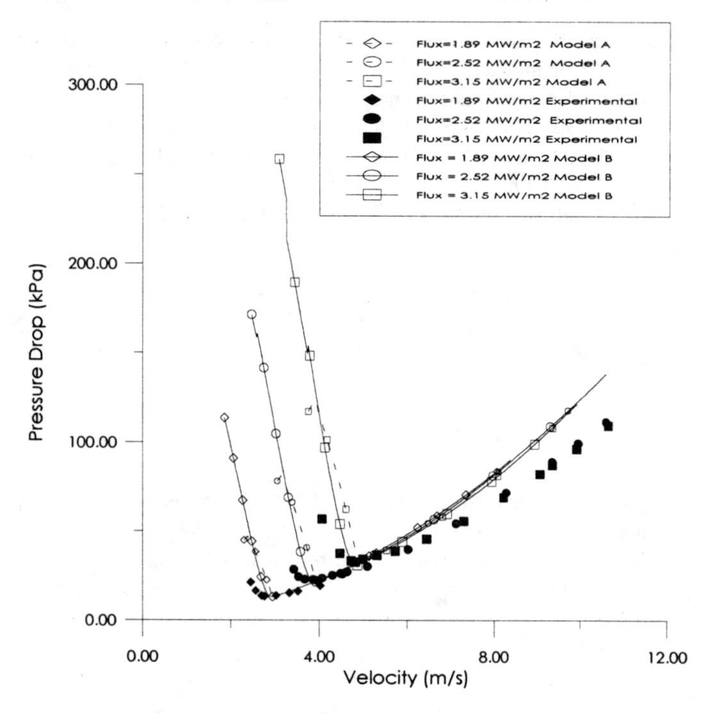

FIGURE 1 Downflow in a 15.5 mm Diameter Tube
TI = 25 C PE = 446 kPa

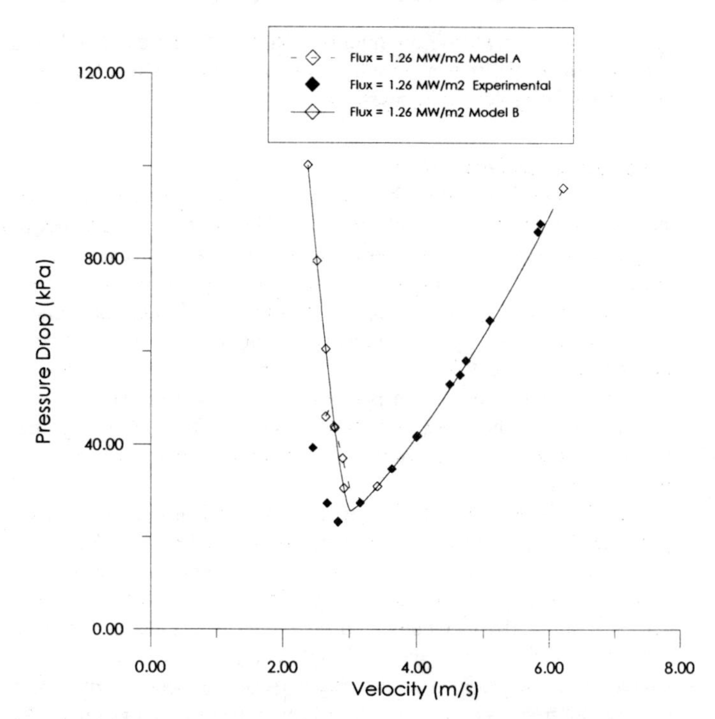

FIGURE 2 Downflow in a 9.1 mm Diameter Tube
TI= 25 C PE= 446 kPa

saturation enthalpy values, true mass fraction is always positive in the regions in which vapor is entrained in the flow. The concept of true mass fraction arose from the need to correlate low pressure critical heat flux data.

True mass fraction is calculated from eq.(8), and used to calculate specific volume, void fraction and the axial derivative of specific volume from eqs. (9)-(11).

$$X_b \ln\left(\frac{X - X_t}{X_b}\right) + \ln\left(1 - X + X_b - X_b X_t\right) = 0 \quad (8)$$

$$v = v_f + X_t\left(v_g - v_f\right) \quad (9)$$

$$\alpha = \frac{\beta}{1+\beta} \quad \text{where} \quad \beta = \frac{X_t \rho_f}{\rho_g\left(1 - X_t\right)} \quad (10)$$

$$\frac{dv}{dZ} = \left(v_g - v_f\right)\left(\frac{1 - R}{1 + R X_b}\right)\left(\frac{4\phi}{\rho_f h_{fg} V D}\right)$$

$$\text{where } R = \frac{X - X_t}{\left(1 - X + X_b\left(1 - X_t\right)\right) X_b} \quad (11)$$

These variables can be then used in eq. (12) to find pressure drop in the two phase region.

$$\Delta P_{TP} = \Delta Z_{TP}\left(V^2 \rho_f^2 \frac{dV_S}{dZ} - \frac{g\sin(\theta)}{V_S} + \frac{2 f_f}{D} V^2 \rho_f^2 v\right) \quad (12)$$

Results and Discussion

The proposed pressure drop models are compared to two large experimental data bases for vertical upflow and downflow in uniformly heated tubes. Dougherty et. al (1989), (1990), presented a database for vertical downflow for a wide range of channel diameters and operating conditions. Yang et. al (1993) and Dougherty et al. (1991) presented a data set for vertical upflow, limited to one tube diameter, but over a large range of applied flux values.

Figure 1 presents a comparison of pressure drop versus inlet velocity results for both Model A and Model B, for downflow conditions in a 15.5 mm tube over a range of surface fluxes. For both models, the agreement between the experimental results and the theoretical prediction is very good. It can be seen that the minimum point in the curve is more accurately predicted by Model B. However, both methods tend to overpredict the pressure drop prior to the OFI point. A possible explanation for this is the lack of correction in the models for the horizontal temperature variations at any given cross-sectional point. The horizontal variation in temperature will affect the fluid properties and thus the pressure drop. Also, the method of calculating the friction factor in the nucleate boiling region, as based on the bubble diameter, may need to be investigated further. Looking at Fig. 2, downflow in a 9.1 mm tube, is seen again that Model B more accurately predicts the minimum point, but also that both models make a more accurate prediction of pressure drop prior to OFI. It is expected that the horizontal property variations will have less effect in a smaller diameter tube.

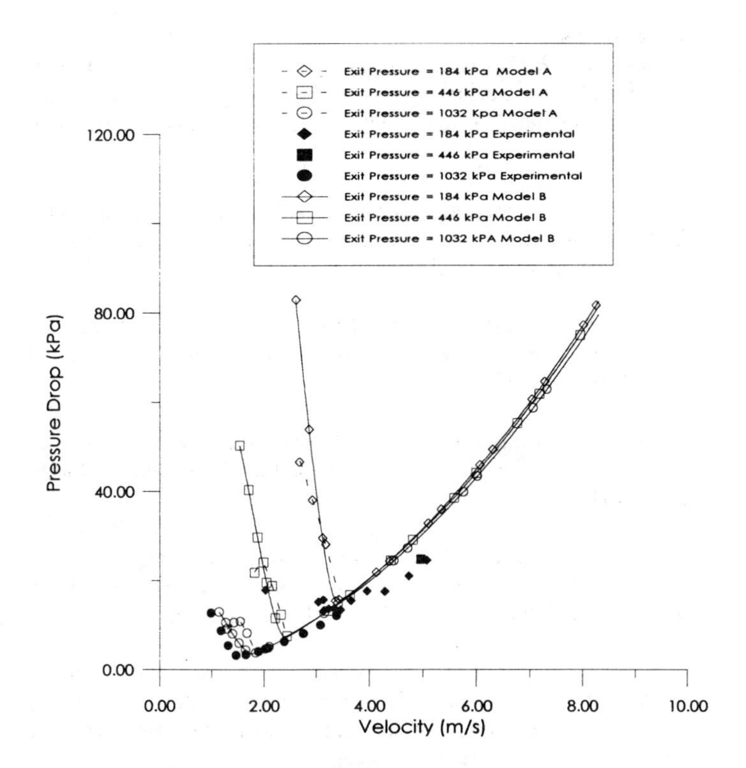

FIGURE 3 Upflow in a 15.5 mm Diameter Tube
1.26 MW/m2 Applied Flux TI = 50 C

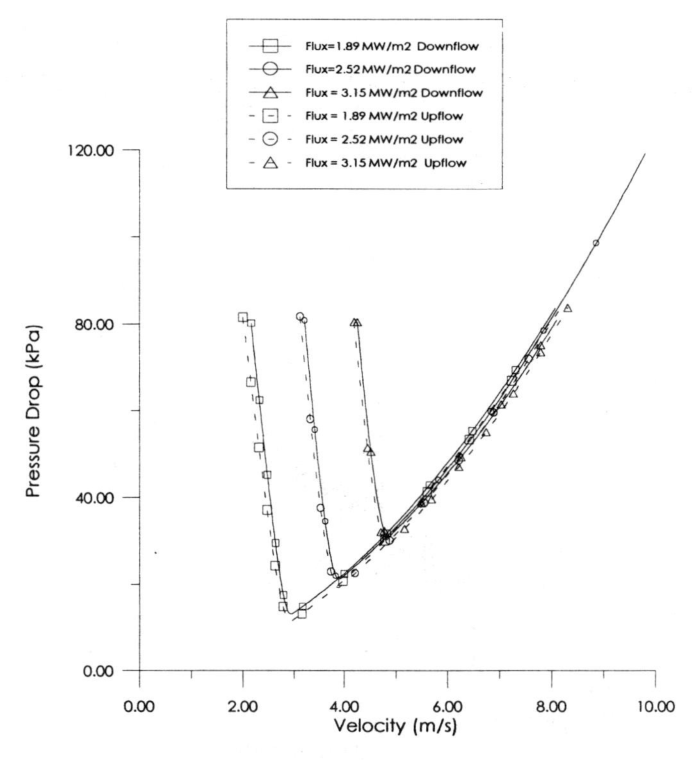

FIGURE 4 Upflow vs. Downflow in a 15.5mm Diameter Tube
Model B TI = 25 C PE = 446 kPa

A comparison of pressure drop versus inlet velocity for upflow conditions is presented in Fig. 3. Once again, Model B more accurately predicts the minimum point. This plot also shows the effect of channel exit pressure on the minimum point. An increase in exit pressure suppresses subcooled boiling and reduces the OFI velocity. For both models, the agreement between the experimental results and the theoretical prediction is very good.

Direct comparisons between upflow and downflow theoretical pressure drop for the same operating conditions are shown in Fig. 4. For identical operating conditions, upflow pressure drops were consistently lower than the corresponding downflow pressure drops. This is a result of the fundamental differences between upflow and downflow in forced convective boiling. For upflow, the increase in pressure drop due to boiling effects is offset by the decrease in gravity head since buoyancy works with, instead of against fluid flow, resulting in lower pressure drop values for upflow. It is this competition between frictional/accelerational effects and decreasing gravity head in upflow that leads to cases where no minimum occurs in the demand curve.

Both Model A and Model B will identify operating cases where no minimum point will occur. In Figs. 5 and 6, the applied flux was dropped in stages and the theoretical demand curve calculated. The theoretical demand curves predict the OFI points for the high flux situations, but as the flux is lowered, the threshold flux is passed and no minimum point occurs. In these instances, there will be no warning, such as a sudden increase in channel pressure drop before the onset of critical heat flux.

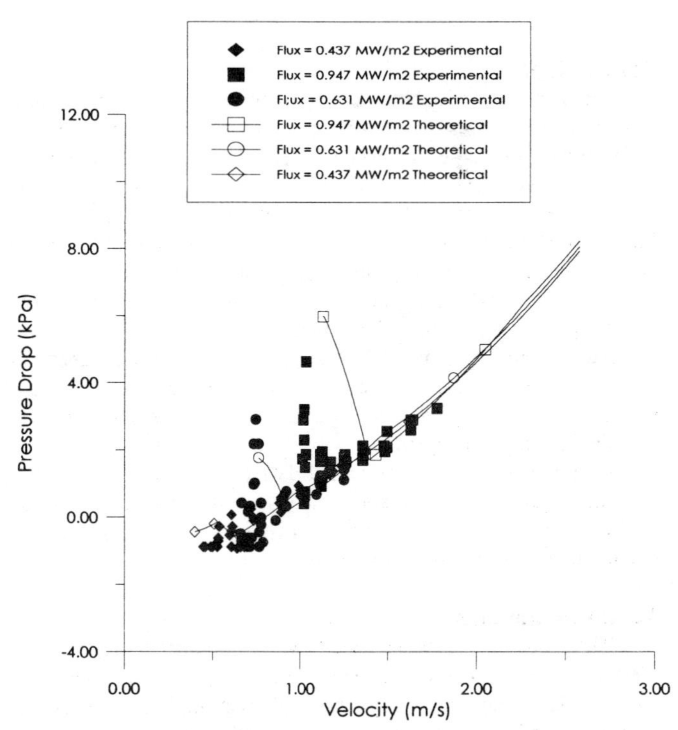

FIGURE 5 Threshold Flux, Model A
Exit Pressure =1032 kPa 15.8 mm Tube Diameter TI = 50 C

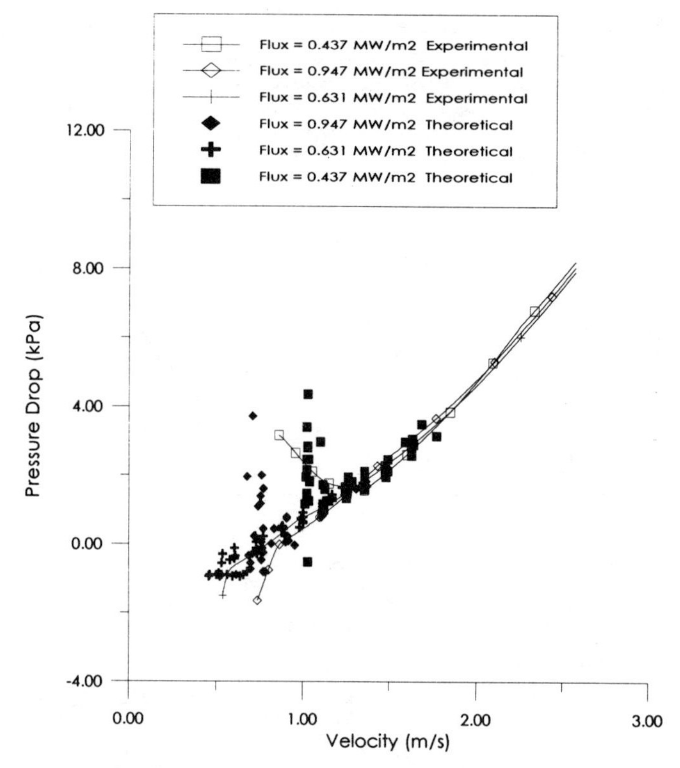

FIGURE 6 Threshold Flux, Model B
Exit Pressure = 1032 kPa 15.8 mm Tube Diameter Tl=50 C

Conclusions

The formation of bubbles as a consequence of uniformly heating flow through a vertical channel has a significant effect on the overall pressure drop along the channel. Two simple models to predict this effect have been developed. The models can handle flow conditions from single phase through saturated boiling and are valid for any flow orientation. In addition, the models predict a threshold heat flux below which flow instability may not occur for upflow situations.

The model which calculates the pressure gradient in the two phase region as a function of true mass fraction is consistently more accurate at predicting the OFI velocity. This model does not distinguish between the fully developed subcooled boiling region and the saturated flow region, instead treating the entire region after OSV as a single two phase region.

Comparisons with experimental data for a variety of operating conditions and flow orientations show good agreement with the predicted results. The model provides both qualitative and quantitative pressure drop information.

Acknowledgments

This material is based upon work completed while A. S. Fleischer was supported under a National Science Foundation Graduate Research Fellowship. Any opinions, findings, conclusions or recommendations expressed in this publication are those of the authors and do not necessarily reflect the views of the National Science Foundation.

References

Bowring, R.W., 1962, "Physical Model Based on Bubble Detachment and Calculation of Steam Voidage in the Subcooled Region of a Heated Channel," OECD Halden Reactor Project Report HPR-10.

Block, J.A., Crowley, C.J., Dolan, F.X., Sam, R.G., Stoedefalke, B.H., 1990, "Nucleate Boiling Pressure Drop in an Annulus," Creare, Inc., Hanover, N.H.

Davis, E.J and Anderson, G.H., 1966, "The Incipience of Nucleate Boiling in Forced Convection Flow," AIChE Journal, Volume 12 pp.774-780.

Dormer, J. Jr. and Bergles, A.E., 1964, "Pressure Drop with Surface Boiling in Small Diameter Tubes," Report 8767-31, Dept. of Mech. Eng. MIT.

Dougherty,T., Fighetti, C., McAssey, E., Reddy, G., Yang, B., Chen, K. and Qureshi, Z.C., 1989, "Flow Instability in Vertical Down-Flow at High Heat Fluxes," ASME HTD-Vol. 199, pp.17-23.

Dougherty, T., Fighetti, C., Reddy, D., Yang, B., Jafri,T., McAssey,E. and Qureshi, Z., 1990, "Flow Boiling in Vertical Downflow," Proceedings of the Ninth International Heat Transfer Conference, Vol. 2, Jerusalem, Israel.

Dougherty, T., McAssey, E., Reddy, D., Yang, B., Maciuca, C., Fighetti, C., Cheh, H., 1991, "HWNPR Two-Phase Cooling Experiments with Multiple Heated Test Sections," Report Number CU-HTRF-91-09, Columbia University.

Dougherty, T., McAssey, E., Reddy, D., Yang, B., Maciuca, C., Fighetti, C., Cheh, H., 1990, "Flow Excursion Experimental Program- Single Tube Uniformly Heated Tests," Report Number CU-HTRF-T4 CU-90-01, Columbia University.

Dougherty, T., McAssey, E., Reddy, D., Yang, B., Maciuca, C., Fighetti, C., Cheh, H., Carrano, V., Kokolis, S., Ouyang, W., 1992, "HWRF-NPR Onset of Flow Instability and Critical Heat Flux Experiment,." Report Number CU-HTRF-92-06, Columbia University.

Dougherty, T.J. and Yang, B.W. 1995, " Flow Instability and Critical Heat Flux in a Natural Circulation Loop at Low Pressure," Proceedings of the 1995 ASME National Heat Transfer Conference, HTD-Volume 316.

Fleischer, A.S., 1996, "Prediction of the Onset of Ledinegg Flow Instability in Uniformly Heated Constant Cross Section Channels," Master Thesis, Villanova University, Villanova, Pa.

Levy, S., 1967, "Forced Convective Subcooled Boiling- Prediction of Vapor Volumetric Fraction," Journal of Heat and Mass Transfer, Volume 10, pp. 951-965, Great Britain.

Marshek, S., 1958, "Transient Flow of Boiling Water in Heated Tubes," Savannah River Laboratory Report PP-301TL.

Martinelli, R.C. and Nelson, D.B., 1948, "Prediction of Pressure Drop During Forced Circulation Boiling of Water," Transactions of the ASME. Volume 70, pp.695.

Maulbetsch, J.S and Griffith, P., 1966, "A Study of System Induced Instabilities in Forced Convective Flow with Subcooled Boiling," Proceedings of the 3rd International Heat Transfer Conference. pp. 247-257.

Owens, W.L and Schrock, V.E., 1960 "Local Pressure Gradients for Subcooled Boiling of Water in Vertical Tubes," ASME Paper 60-WA-249.

Sabersky, R.H. and Mulligan, H.E., 1955, "On the Relationship Between Fluid Friction and Heat Transfer in Nucleate Boiling," Jet Propulsion, Volume 25, pp. 9.

Saha, P. and Zuber, N. 1974, "Point of Net Vapor Generation and Vapor Void Fraction in Subcooled Boiling." Proceedings of the 5th International Heat Transfer Conference, B4.7, Tokyo, Japan.

Stelling, R., McAssey, E.V., Dougherty, T., Yang, B., 1996, "The Onset of Flow Instability for Downward Flow in Vertical Channels," Journal of Heat Transfer, Volume 118, pp.709-714.

Thom, J.R.S., 1964, "Prediction of Pressure Drop During Forced Circulation Boiling of Water," International Journal of Heat and Mass Transfer, Volume 7, pp. 709-724, Great Britain.

Whittle, R.H., and Forgan, R., 1967, "A Correlation of the Minima in the Pressure Drop versus Flowrate Curves for Subcooled Water Flowing in Narrow Heated Channels," Nuclear Engineering and Design, Volume 6, pp.89-99.

Yang, B.W., Dougherty, T.J., Ouyang, W., McAssey, E.V., Smalec, L., 1993, "Flow Instability in Vertical Upflow Under Low Heat Flux Conditions," Ninth Proceedings of Nuclear Thermal Hydraulics, American Nuclear Society Winter Annual Meeting, San Francisco, CA

HTD-Vol. 342, National Heat Transfer Conference
Volume 4
ASME 1997

NUMERICAL INVESTIGATION OF HEAT TRANSFER TO A NON-SPHERICAL DROP SUSPENDED IN AN ELECTRIC FIELD: INTERNAL PROBLEM

M. A. Hader, A. Verma, and M. A. Jog
Department of Mechanical, Industrial, and Nuclear Engineering
University of Cincinnati, Cincinnati, OH 45221-0072.

ABSTRACT

Heat transfer to a drop of a dielectric fluid suspended in another dielectric fluid in the presence of an electric field is numerically investigated. The internal heat transfer problem is considered where the bulk of the resistance to the heat transfer is assumed to be in the dispersed phase. We have analyzed the effect of drop deformation on the heat transport to the drop. The deformed drop shape is assumed to be a spheroid and is prescribed in terms of the ratio of drop major and minor diameter. Both prolate and oblate shapes are considered with a range of diameter ratio b/a from 2.0 to 0.5. The electrical field and the induced stresses are obtained analytically. The resulting flow field is determined by numerically solving the Navier-Stokes equations in the continuous and the dispersed phase. An alternating-direction-implicit (ADI) method is used to obtain the transient temperature field for drop Peclet number from 5 to 1500. Heat transfer results for a nearly spherical drop ($b/a = 0.99$ and $b/a = 1.01$) show excellent agreement with the results available in published literature. Results indicate that the drop shape significantly affects the flow field and the heat transport to the drop. At very low and very high Peclet numbers, the steady state Nusselt number is higher for a deformed drop than that for a sphere. However, at intermediate Peclet number, the Nusselt number for a sphere may be higher than that for a deformed drop. For both prolate and oblate drops, the steady state Nusselt number increases with increasing Peclet number, and at high Peclet number, becomes increasingly independent of the Peclet number. The maximum steady state Nusselt numbers for an oblate drop are higher than that for a prolate drop.

NOMENCLATURE

a	half of drop diameter perpendicular to the axis of symmetry
A_s	drop surface area
b	half of drop diameter along the axis of symmetry
E	electric field
i	unit vector
k	dielectric constant
M	viscosity ratio (μ_1/μ_2)
Nu	Nusselt number based on drop major diameter (Equation (16))
$\hat{\text{Nu}}$	equivalent Nusselt number (Equation (23))
p	pressure
Pe	Peclet number based on drop major diameter
$\hat{\text{Pe}}$	equivalent Peclet number (Equation (22))
P_n	Legendre Polynomial of order n.
Q	heat flux
R	electric conductivity ratio (σ_1/σ_2).
Re	Reynolds number
$(r,\ z,\ \theta)$	cylindrical polar co-ordinates.
S	ratio of dielectric constants (k_1/k_2).
t	time
T	temperature
u	velocity
U	maximum surface velocity
V	electric potential
$\hat{V}$	volume
w	vorticity

Greek Symbols

α thermal diffusivity
γ surface tension
μ viscosity
ρ density
σ electrical conductivity
τ stress
ψ stream function
Φ Taylor's discriminating function (Equation (1))
(ξ, η, ϕ) oblate spheroidal co-ordinates.

Subscripts

0 initial
1 continuous phase
2 dispersed phase
b bulk
E electrically induced
s surface
∞ far field

Superscripts

$*$ dimensional quantities.

INTRODUCTION

When a uniform electric field is applied to a drop of dielectric liquid suspended in another immiscible fluid, the electric field induces stresses at the fluid interface. The normal stresses may deform the drop and the tangential stresses produce a circulatory motion inside the drop. This electrically induced motion can be used to enhance the heat/mass transfer from the drop and finds applications in development of compact direct-contact heat exchangers. If the ratio of thermal conductivity of the dispersed to the continuous phase is much smaller than unity, then the resistance to the heat transport can be considered to be mainly in the dispersed phase. This is the internal heat transfer problem. Oliver et al.(1985) analyzed the internal heat transfer to a spherical liquid drop suspended in an electric field. They showed that for large Peclet numbers the Nusselt number for purely electrically driven flow becomes increasingly independent of the Peclet number. The value of the maximum Nusselt number for the internal heat transfer to a spherical drop suspended in a uniform electric field is shown to be 29.8 by Oliver and DeWitt (1993). This value is significantly higher than 17.9, the Nusselt number for a drop translating in gravitational field in the absence of an electric field (Clift et al., 1978; Ayyaswamy, 1995).

Heretofore, the studies of heat/mass transport from a liquid drop in a uniform electric field consider the drop shape to be spherical. For a moving drop, in the absence of electric field, the assumption of sphericity is appropriate if the Weber number and the Eötvös numbers are small (Sadhal and Johnson, 1986; Jog et al., 1996). However, for a drop suspended in a uniform electric field, drop deformations are possible even for small Eötvös numbers. This is due to the existence of non-uniform normal stresses induced by the electric field. Only for a certain specific combination of the electro-thermo-physical properties of the dispersed and the continuous phase, the drop may remain spherical (Taylor, 1966). Under these conditions the viscous stresses normal to the drop surface due to the induced circulatory flow exactly cancel the effect of the normal stress variation due to the electric field. In general, the non-uniformity in the normal stress at the drop surface leads to deformation of the drop. As the drop deforms, the electric field at the drop surface changes and the electrically induced surface stresses are altered (Feng and Scott, 1996). This results in a change in the flow field in the dispersed and the continuous phase. Consequently, the deformation of a drop can significantly impact the heat transport characteristics. A computational model to calculate drop deformations under the influence of electric field has been recently reported by Feng and Scott(1996). They found that the field strength usually exhibits a turning point when it reaches a critical value. Physically those turning points indicate the stability limits of drops stressed by an electric field. No steady solution can be obtained if the externally applied electric field is increased beyond the critical field strength. However, in some special cases with fluid conductivities closely matched, they found that the drop deformations can grow indefinitely as the electric field strength increases. Therefore, with closely matched fluid electrical conductivities very large drop deformations are possible without becoming unstable.

Recently, we have studied the electrohydrodynamics and heat transfer for an oblate spheroidal drop (Jog and Hader, 1997). The electric field and the electrically induced stresses were calculated numerically and only the oblate drop shapes were considered. The maximum steady state Nusselt number for oblate drops were found to be significantly higher than that for a sphere. This work is extended in the present paper to include both the prolate and the oblate drop shapes. We have analyzed the effect of drop deformation on the heat transport to a drop suspended in uniform electric field. Analytical solutions are presented for the electric field and the electrically induced stresses. We have assumed the drop shape to be a spheroid. Taylor (1966) analytically showed that for small deformations the deformed drop shape is a spheroid. Numerous experimental studies (Allan and Mason, 1962; Taylor, 1966; Torza et al., 1971; Vizika and Saville, 1992) and computational study (Feng and Scott, 1996) have shown that the drop shape is very nearly spheroidal even for moderate deformations. The simplification of a spheroidal drop shape allows us use an orthogonal co-ordinate system which conforms to the drop surface thereby reducing the computational efforts without much loss of accuracy. The electrically induced flow field is numerically calculated in the continuous and the dispersed phase. The transient temperature variation and the Nusselt number variations are numerically obtained in the limit of bulk of the resistance to the heat transport being in the drop. Results indicate that the drop shape significantly affects the flow field and the heat transport to the drop. At very low and very high Peclet numbers, the steady state Nusselt number is higher for a deformed drop than that for a sphere. However, at intermediate Peclet number, the Nusselt number for a sphere may be higher than that for a deformed drop. For both prolate and oblate drops, the steady state Nusselt number increases with increasing Peclet number, and at high Peclet number, becomes increasingly independent of the Peclet number. The maximum steady state Nusselt numbers for an oblate drop are higher than that for a prolate drop.

PROBLEM FORMULATION

Consider a drop of a dielectric liquid suspended in another dielectric liquid. Application of a uniform electric field to such a system results in development of stresses on the surface of the drop. The non-uniform stress at the droplet surface results in a circulatory motion inside the drop and may cause the drop to deform. We note that both prolate and oblate drop deformations are possible and can be determined by Taylor's discriminating function

$$\Phi = S(R^2+1) - 2 + 3(SR-1)\frac{2M+3}{5M+5}. \tag{1}$$

$\Phi < 0$ indicates oblate deformations whereas $\Phi > 0$ results in prolate deformations. $\Phi = 0$ corresponds to a spherical drop. For most practical situations, it has been shown that prolate deformations will result when the drop is more conductive than the surrounding fluid and oblate deformations are likely when the drop is less conductive than the surrounding fluid (Feng and Scott, 1996). For sake of brevity, the formulation is presented in terms of the oblate spheroidal coordinates (see Figure 1). The corresponding expressions in terms of the prolate spheroidal coordinates can be readily obtained by substituting $\cosh\xi$ for $i\sinh\xi$ and $\sinh\xi$ for $i\cosh\xi$. The formulation is axisymmetric.

Governing equations and boundary conditions

The electric field in both the phases is governed by the Laplace's equation and the flow is described by the Navier-Stokes equations. The temperature in the drop interior can be calculated by the solution of energy conservation equation. The timescales for momentum and heat diffusion in the drop are $a^2/(\mu/\rho)$ and a^2/α, respectively. The ratio of the two timescales is O(Pr). Therefore, for liquids with moderate to high Prandtl number, in the timescale of drop

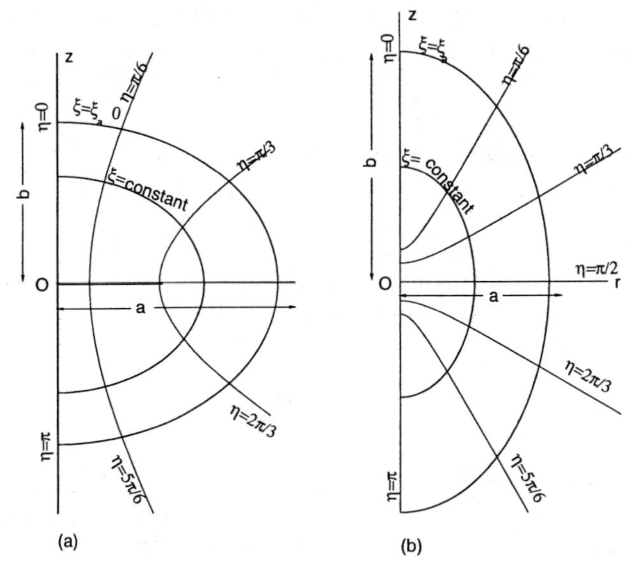

Figure 1: (a)Oblate spheroidal co-ordinates, (b) Prolate spheroidal co-ordinates

heat transport, the momentum transport can be regarded as quasi-steady. We consider that the bulk of the resistance to the heat transfer is in the droplet. Therefore the drop surface temperature corresponds to the free stream temperature and the energy equation in the continuous phase need not be solved.

$$\nabla \cdot \mathbf{u}_i^* = 0 \tag{2}$$

$$\mathbf{u}_i^* \cdot \nabla \mathbf{u}_i^* = -\frac{1}{\rho_i}\nabla p_i^* + \nu_i \nabla^2 \mathbf{u}_i^* \tag{3}$$

$$\nabla \cdot \mathbf{E}_i^* = 0 \tag{4}$$

$$\frac{\partial T_2^*}{\partial t^*} + \mathbf{u}_2^* \cdot \nabla T_2^* = \alpha_2 \nabla^2 T_2^* \tag{5}$$

where $\mathbf{E}^* = -\nabla V^*$ and $i = 1,\ 2$.

Initially the drop is uniformly at temperature $T_{2,0}^*$. At the interface the following conditions are satisfied (Melcher and Taylor, 1969)

$$\begin{aligned} \mathbf{n}\cdot\mathbf{u}_i^* &= 0 \\ \mathbf{u}_1^* &= \mathbf{u}_2^* \\ \mathbf{n}\cdot(\mathcal{T}_E^* + \mathcal{T}_1^* - \mathcal{T}_2^*) &= \mathcal{T}_\gamma^* \\ V_1^* &= V_2^* \\ \sigma_1 \mathbf{n}\cdot\nabla V_1^* &= \sigma_2 \mathbf{n}\cdot\nabla V_2^* \\ T_2^* &= T_s^* \end{aligned} \tag{6}$$

Here $\mathcal{T}_E$ is the electrically induced stress at the drop surface, $\mathcal{T}_\gamma$ is the stress due to surface tension, and $\mathbf{n}$ is unit vector normal to the drop surface.

The symmetry condition is used along the z axis and far away from the drop we have

$$\mathbf{u}^* \rightarrow 0, \quad p^* \rightarrow p^*_\infty, \quad \text{and } V^* \rightarrow E^* z^*. \tag{7}$$

We use the oblate spheroidal co-ordinates (ξ, η, ϕ) for computational convenience. The origin of the co-ordinate system is at the drop center as shown in Figure 1. The drop surface corresponds to $\xi = \xi_a$. The oblate spheroidal coordinates are related to the cylindrical polar co-ordinates (r, z, θ) as(Happel and Brenner, 1965)

$$\begin{aligned} z &= c \sinh\xi \cos\eta \\ r &= c\cosh\xi \sin\eta \\ \theta &= \phi \end{aligned} \tag{8}$$

The governing equations are made dimensionless as follows. Velocity is made dimensionless by the maximum surface velocity U and distances are nondimensionalized by a. The stresses and pressure are normalized by $\mu U/a$ and the Reynolds number is $Re = \rho U\ 2a/\mu$. We define the dimensionless temperature as $T = (T_2^* - T_s^*)/(T_{2,0}^* - T_s^*)$ and dimensionless time as $t = \alpha_2 t^*/a^2$. The Peclet number is given by $\mathrm{Pe} = 2Ua/\alpha_2$. The electric field is made dimensionless by $[\mu_2 U/(\epsilon_0 a)]^{1/2}$. We introduce stream function ψ (nondimensionalized by Ua^2) such that

$$u_\xi = -\frac{\cosh^2\xi_a}{(\cosh^2\xi - \sin^2\eta)^{1/2}\cosh\xi\sin\eta}\frac{\partial\psi}{\partial\eta} \tag{9}$$

$$u_\eta = \frac{\cosh^2\xi_a}{(\cosh^2\xi - \sin^2\eta)^{1/2}\cosh\xi\sin\eta}\frac{\partial\psi}{\partial\xi} \tag{10}$$

and vorticity $\mathbf{w}$ (nondimensionalized by U/a) as

$$\mathbf{w} = \boldsymbol{\nabla}\times\mathbf{u} \tag{11}$$

As the formulation is axisymmetric, only the ϕ component of vorticity is nonzero. $\mathbf{w} = w\mathbf{i}_\phi$.
The governing equations can be written in the oblate spheroidal coordinates as

$$\begin{aligned} &\frac{\partial}{\partial\xi}\left(-\cosh\xi_a\frac{\partial\psi}{\partial\eta}w\right) + \frac{\partial}{\partial\eta}\left(\cosh\xi_a\frac{\partial\psi}{\partial\xi}w\right) - \frac{2}{Re}\times \\ &\left\{\frac{\partial}{\partial\xi}\left(\cosh\xi\sin\eta\frac{\partial w}{\partial\xi}\right) + \frac{\partial}{\partial\eta}\left(\cosh\xi\sin\eta\frac{\partial w}{\partial\eta}\right)\right\} \\ &= w\left[-\tanh\xi\frac{\partial\psi}{\partial\eta}\cosh\xi_a + \cot\eta\frac{\partial\psi}{\partial\xi}\cosh\xi_a\right. \\ &\left.\quad - \frac{2}{Re}\left(\frac{\sin\eta}{\cosh\xi} - \frac{\cosh\xi}{\sin\eta}\right)\right] \end{aligned} \tag{12}$$

$$\begin{aligned} &\frac{\partial}{\partial\xi}\left(\frac{1}{\cosh\xi\sin\eta}\frac{\partial\psi}{\partial\xi}\right) + \frac{\partial}{\partial\eta}\left(\frac{1}{\cosh\xi\sin\eta}\frac{\partial\psi}{\partial\eta}\right) \\ &= (\cosh^2\xi - \sin^2\eta)\,\mathrm{sech}^3\xi_a w \end{aligned} \tag{13}$$

$$\frac{\partial}{\partial\xi}\left(\cosh\xi\sin\eta\frac{\partial V}{\partial\xi}\right) + \frac{\partial}{\partial\eta}\left(\cosh\xi\sin\eta\frac{\partial V}{\partial\eta}\right) = 0 \tag{14}$$

$$\begin{aligned} &\frac{(\cosh^2\xi - \sin^2\eta)\cosh\xi\sin\eta}{\cosh^3\xi_a}\frac{\partial T}{\partial t} + \frac{\mathrm{Pe}}{2}\times \\ &\left[\frac{\partial}{\partial\xi}\left(-\frac{\partial\psi}{\partial\eta}T\right) + \frac{\partial}{\partial\eta}\left(\frac{\partial\psi}{\partial\xi}T\right)\right] - \mathrm{sech}\xi_a\times \\ &\left[\frac{\partial}{\partial\xi}\left(\cosh\xi\sin\eta\frac{\partial T}{\partial\xi}\right) + \frac{\partial}{\partial\eta}\left(\cosh\xi\sin\eta\frac{\partial T}{\partial\eta}\right)\right] = 0 \end{aligned} \tag{15}$$

The initial condition is : at $t = 0$, $T_2 = 1$.
The boundary conditions are:
As $\xi \rightarrow \infty$, u_ξ, $u_\eta = 0$ and $w = 0$.
The interface conditions at $\xi = \xi_a$ are:
$\mathcal{T}_{E,\xi\eta} + (\mu_1/\mu_2)\mathcal{T}_{1,\xi\eta} - \mathcal{T}_{2,\xi\eta} = 0$; $u_{1,\eta} = u_{2,\eta}$; $u_{1,\xi} = u_{2,\xi} = 0$; $\partial V_1/\partial\xi = (\sigma_2/\sigma_1)\partial V_2/\partial\xi$; $V_1 = V_2$; $T_2 = 0$. The shear stress can be written as

$$\begin{aligned} \mathcal{T}_{\xi\eta} &= \left\{\left[\frac{\partial u_\eta}{\partial\xi} + \frac{\sin\eta\cos\eta}{(\cosh^2\xi - \sin^2\eta)}u_\xi\right] + \right. \\ &\left.\left[\frac{\partial u_\xi}{\partial\eta} - \frac{\sinh\xi\cosh\xi}{(\cosh^2\xi - \sin^2\eta)}u_\eta\right]\right\}\frac{1}{(\cosh^2\xi - \sin^2\eta)^{1/2}} \end{aligned}$$

$$\tau_{E,\xi\eta} = (k_1 E_{1,\eta}E_{1,\xi} - k_2 E_{2,\eta}E_{2,\xi}).$$

As a consequence of the assumed oblate spheroidal shape of the drop, the interfacial condition of normal stress balance need not be considered any further.

The Nusselt number based on the drop major diameter is

$$\mathrm{Nu} = \frac{Q\,2a}{A_s(T_s^* - T_b^*)k} = -\frac{2\hat{V}}{A_s a}\frac{1}{T_b}\frac{dT_b}{dt} \tag{16}$$

Here Q is the net rate of heat transport to the drop, $\hat{V}$ is the volume of the drop $\hat{V} = \frac{4}{3}\pi a^2 b$ and A_s is the drop surface area

$$A_s = \int_0^\pi 2\pi\cosh\xi_a\sin\eta\,(\cosh^2\xi_a - \sin^2\eta)^{1/2}\,d\eta.$$

The dimensionless bulk temperature for the dispersed phase can by calculated as

$$\begin{aligned} T_b &= \frac{3}{2}\frac{1}{\sinh\xi_a\cosh^2\xi_a}\times \\ &\int_0^{\xi_a}\int_0^\pi T\cosh\xi\sin\eta\,(\cosh^2\xi - \sin^2\eta)\,d\eta d\xi \end{aligned} \tag{17}$$

SOLUTION TECHNIQUE

The general solution for the electric potential distribution can be obtained by the method of separation of variables as(Smythe, 1968)

$$V(\xi,\eta) = \sum_{n=0}^{\infty}[A_n P_n(\cos\eta) + B_n Q_n(\cos\eta)] \times [A'_n P_n(i\sinh\xi) + B'_n Q_n(i\sinh\xi)] \quad (18)$$

The solutions which satisfy the appropriate boundary conditions are, for the continuous phase:

$$V_1 = \cos\eta\left[C_1\sinh\xi + C_2\left(\sinh\xi\cot^{-1}(\sinh\xi) - 1\right)\right] \quad (19)$$

and for the drop interior

$$V_2 = C_3\cos\eta\sinh\xi \quad (20)$$

where

$$\begin{aligned} C_1 &= E/\cosh\xi_a \\ C_2 &= \left(E(1-R)\tanh\xi_a(1+\sinh^2\xi_a)\right)\left[(R-1)\sinh\xi_a \times (1+\sinh^2\xi_a)\cot^{-1}(\sinh\xi_a) + \sinh^2\xi_a - (1+\sinh^2\xi_a)R\right]^{-1} \quad (21) \\ C_3 &= C_1 + C_2\left(\sinh\xi_a\cot^{-1}(\sinh\xi_a) - 1\right)/\sinh\xi_a \end{aligned}$$

The flow field in the continuous and the dispersed phase and the transient temperature distribution are obtained numerically by a finite difference method. The governing equations (12, 13, and 15) are discretized using central differencing. 61 × 61 node points were used in the drop interior and 101 × 61 node points were used in the surrounding fluid. The electric field is first calculated for each drop deformation. Electric potential is specified in the far field ($\xi = \xi_\infty$). The value of ξ_∞ is chosen such that the distance between any point on the far field boundary and the drop surface is at least $100a$. The governing equations for the stream function and vorticity transport were discretized using hybrid differencing scheme (Patankar, 1980) and were solved in both phases by an iterative procedure. A guess for the surface velocity distribution is used to solve the stream function and vorticity equations in the continuous phase. The continuous phase solution provides the shear stress at the drop interface. Using the calculated interfacial shear stress, the flow field in the drop interior is obtained. This solution provides a new variation for the tangential velocity at the drop surface. The solution procedure is continued until the difference in both the tangential velocity and the shear stress obtained from the continuous and the dispersed phase are less than 1×10^{-6}.

Transient temperature distributions inside the droplet were obtained for Peclet numbers from 5 to 1500. The governing equation is discretized by central differencing technique. Alternating direction implicit (ADI) method with a tridiagonal algorithm was employed to solve the energy equation. The time step was varied from 0.00001 for short times to 0.0001 at large t. For a typical run, the solution of the flow field took less than 30 minutes on a HP 9000/712/80 workstation. The solution to the energy equation required from 10 minutes (high Pe) to 30 minutes (low Pe). To study the effect of grid refinement, results for Pe = 200 and $b/a = 0.7$ were also obtained with doubling the node points in each direction. The change in the Nusselt number variation was found to be less than 1%.

RESULTS AND DISCUSSION

We have studied the effect of drop deformation on the heat transport in a drop suspended in a uniform electric field. A range of drop deformations is considered from $b/a = 2.0$ to $b/a = 0.5$. The transient heat transport is studied in the limit of bulk of the resistance to heat transfer being in the drop. Under this condition, the transport is governed by the drop Peclet number based on the maximum tangential velocity. A range of Peclet numbers from 5 to 1500 is considered. The numerical computations are carried out for a suspended drop ($\rho_1/\rho_2 = 1$). We have used $\mu_1/\mu_2 = 1$ in our computations. Most of the numerical calculations were performed with $R = 0.1$ and $S = 2$. For one of the cases ($Pe = 200$ and $b/a = 0.7$), the calculations were also carried out by changing the viscosity ratio to 0.1 and 5. The heat transfer results remained essentially unchanged. In all the earlier studies on heat transport in a spherical drop suspended in uniform electric field, the assumption of Stokes flow ($Re \ll 1$) has been used. To make fruitful comparisons with the available results, we have used a value of drop Reynolds number as 0.1 in our computations.

The flow field inside and outside the drop is shown in Figures 2(a) and (b) for $b/a = 0.7$ and 1.5, respectively. As expected the streamline distribution is symmetric about the equatorial plane in both the cases. The flow in the lower half moves in a direction opposite to that in the upper half. For an oblate drop, with increasing drop deformation, the center of each circulatory vortex pattern moves towards the equatorial plane. The radial distance of the vortex center from the drop surface is substantially less for a drop with $b/a = 0.7$ compared to that for a liquid sphere. For a prolate drop, the change in the circulatory vortex pattern from that in the sphere is smaller than that observed for an oblate drop. Similar differences in the circulatory motion with a prolate and an oblate drop shape are also reported by Feng and Scott(1996).

The transient Nusselt number variation are shown in

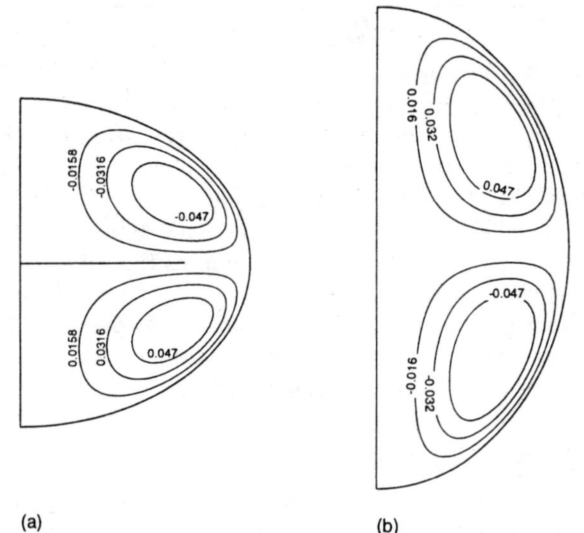

Figure 2: Streamlines in the drop interior. (a) Oblate drop $b/a = 0.7$, (b) Prolate drop $b/a = 1.5$.

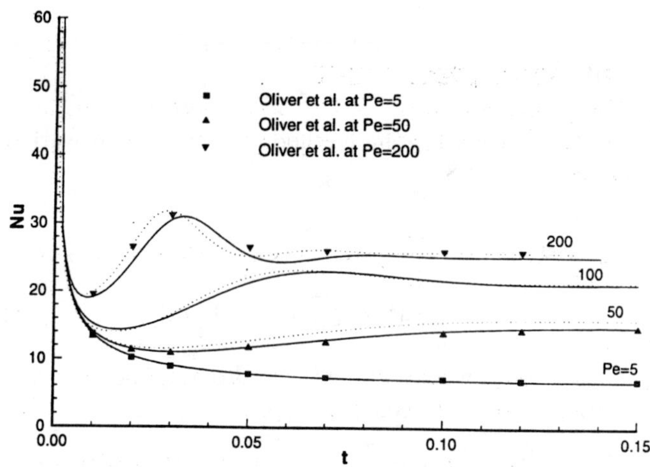

Figure 3: Nusselt number variation for a sphere. Legend: solid lines $b/a = 0.99$, dashed lines $b/a = 1.01$.

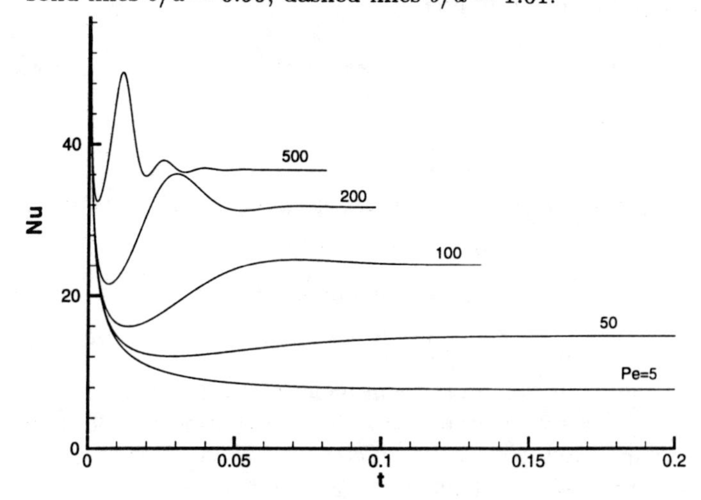

Figure 4: Variation of Nusselt number for an oblate drop $b/a = 0.7$.

Figure 3 for a range of Peclet numbers for a "sphere". The figure shows that the transient Nusselt number variations calculated from our model for $b/a = 1.01$ and $b/a = 0.99$ match well with the result of Oliver et al. (1985) for $b/a = 1$. In the limit of negligible drop deformation, our results for the heat transport are in excellent agreement with those for a sphere. This can be considered as a validation of our computational model. The transient Nusselt number variations for an oblate drop ($b/a = 0.7$) are shown in Figure 4 and those for a prolate drop ($b/a = 1.5$) are shown in Figure 5. For all Peclet numbers, for short times, conduction is the dominant heat transport mechanism as seen from the sharp initial drop in the Nusselt number. This is due to the steep temperature gradients near the drop surface. For low Peclet numbers, conduction remains to be the dominant mechanism at all times. For higher Peclet number, the Nusselt number decreases for short times, but starts increasing thereafter as the cold fluid from the drop interior is brought near the drop surface due to the circulatory motion. The Nusselt number oscillations are more pronounced for high Peclet numbers. This is due to the shorter time scale for the circulatory motion in the drop interior at high Peclet number. The Nusselt number variation eventually attains a steady state. The steady state Nusselt number value increases as the Peclet number is increased. The steady state Nusselt number values become increasingly independent of the Peclet number for very large Peclet numbers. The bulk temperature variation with dimensionless time is shown in Figures 6 and 7 for $b/a = 0.7$ and 1.5, respectively. The dimensionless bulk temperature decreases rapidly for short times. The transient bulk temperature variations show that for higher Peclet numbers the drop bulk temperature decreases more rapidly than that for lower Peclet numbers for all drop deformations. The bulk temperature variations are qualitatively similar in both prolate and oblate drops. However, for large Peclet numbers, the steady state Nusselt number values are significantly higher for an oblate drop than those for a prolate drop.

A comparison of the steady state Nusselt numbers for all drop deformations may be carried out by defining an equivalent Peclet number and an equivalent Nusselt number for a deformed drop as

$$\hat{Pe} = \frac{U\,2r}{\alpha_2} = Pe\left(\frac{b}{a}\right)^{\frac{1}{3}} \tag{22}$$

$$\hat{Nu} = Nu\left(\frac{b}{a}\right)^{\frac{1}{3}} \tag{23}$$

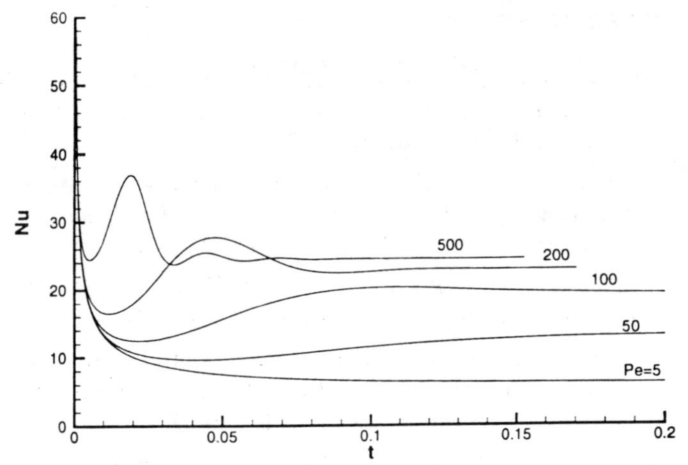

Figure 5: Variation of Nusselt number for a prolate drop with $b/a = 1.5$.

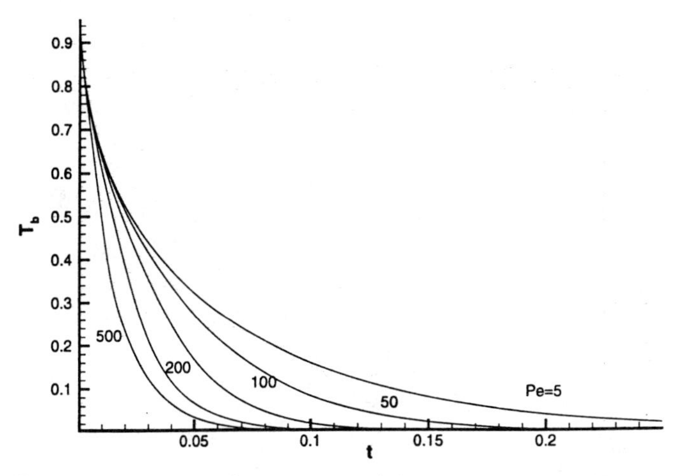

Figure 6: Temporal variation of dimensionless bulk temperature for an oblate drop with $b/a = 0.7$.

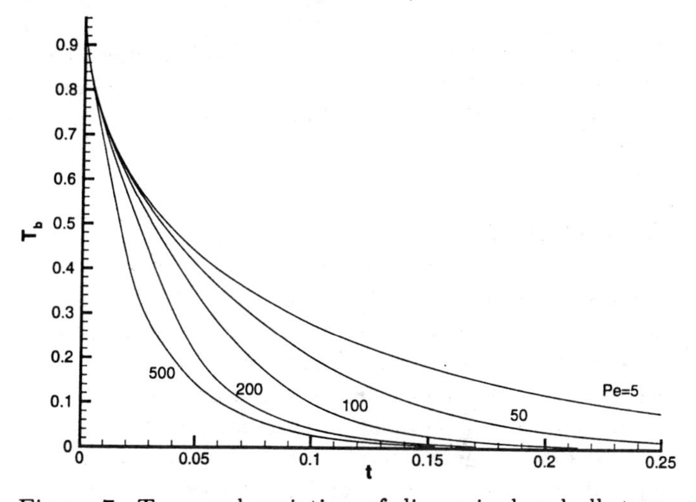

Figure 7: Temporal variation of dimensionless bulk temperature for a prolate drop with $b/a = 1.5$.

where r is the radius of a sphere of equal volume as the deformed drop. $r = a^{2/3}b^{1/3}$. Figure 8 shows the variations of equivalent steady state Nusselt numbers with equivalent Peclet numbers. In the limit of small equivalent Peclet number, the equivalent Nusselt numbers are higher for a deformed drop than those for a liquid sphere. In this range of small $\hat{Pe}$, conduction is the dominant mechanism of heat transport in the drop. Conduction of heat in a deformed drop is much faster than that in a sphere due to the smaller length scale and greater surface area for a deformed drop. In the limit of small b/a, heat conduction length scale for a deformed drop is b (for oblate) or a (for prolate). This is much smaller than a radius of a sphere of identical volume ($r = a^{2/3}b^{1/3}$). Therefore, for small Peclet numbers, the Nusselt numbers for a deformed drop are higher than those for a spherical drop.

It is evident from the flow streamlines shown in Figures 2(a) and (b) that the center of the vortex for a deformed drop is situated near the equator and closer to the drop surface than that for a sphere. At moderate Peclet number, the circulatory vortex in a deformed drop is less effective in bringing the cold fluid from the drop interior to the drop surface. Hence, at moderate Peclet numbers, it is observed that the Nusselt numbers for a deformed drop are lower than those for a sphere.

At very large Peclet numbers, convection is the most dominant transport mechanism and the temperature gradients along each stream line diminish quickly. Therefore, after the initial transients, the heat transport is mainly in the direction perpendicular to the streamlines. For a given Peclet number, the resistance to the heat transport perpendicular to the stream lines will be inversely proportional to the distance between the drop surface and the center of the vortex. This distance is smaller in the case of a deformed drop than that for a sphere. This results in higher Nusselt numbers for deformed drops at high Peclet numbers.

Although the variation of equivalent Nusselt number with equivalent Peclet number is qualitatively similar for oblate and prolate drops, the Nusselt numbers for prolate drops are significantly lower than that for oblate drops. This result can be attributed to the difference in the position of the vortex center in both the cases. Examination of Figures 2(a) and (b) show that the center of each vortex moves away from $\eta = \pi/4$ position as a drop deforms from a spherical shape. However, for a prolate drop, the movement of the vortex center is much smaller than that for an oblate drop. As a result of the difference in the flow field, the Nusselt numbers obtained for a prolate drop are lower than those for an oblate drop.

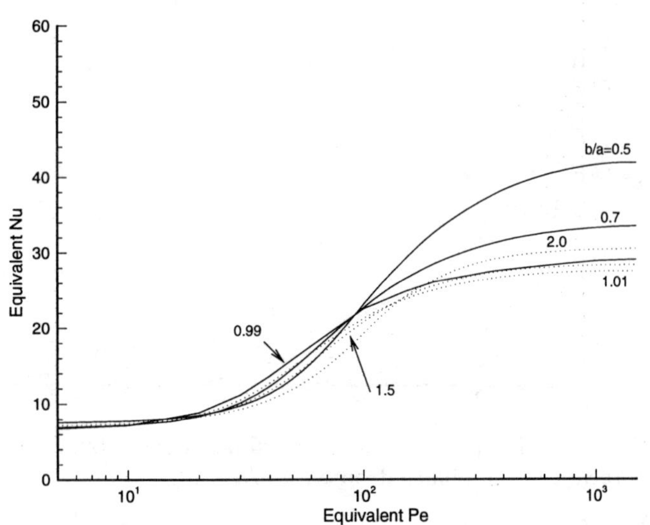

Figure 8: Variation of equivalent Nusselt number ($\hat{Nu}$) with equivalent Peclet number ($\hat{Pe}$) for deformed drops. Legend: solid lines - oblate drops, dotted lines - prolate drops.

CONCLUSIONS

We have studied the transient heat transport in a deformed drop suspended in a uniform electric field. The results for the transient temperature distribution and the Nusselt number are obtained in the limit of bulk of the resistance to the heat transport being in the drop. Results show that for all drop deformations, the steady state equivalent Nusselt numbers become increasingly independent of equivalent Peclet number for large Peclet numbers. At very low and very high Peclet numbers, the steady state Nusselt number is higher for a deformed drop than that for a sphere. However, at intermediate Peclet number, the Nusselt number for a sphere may be higher than that for a deformed drop. The maximum steady state Nusselt number increases as the drop deformation is increased. The maximum steady state Nusselt numbers for oblate drops are higher than those for prolate drops. This indicates that the enhancement of direct contact heat/mass transfer by application of electric field is more effective for a combination of liquids for the continuous phase and the drop phase that leads to an oblate deformation of the drop under the application of an electric field.

ACKNOWLEDGMENTS

The support for this work by the National Science Foundation under grant no. CTS-9409140 is thankfully acknowledged. Partial funding for numerical computations was provided by the University of Cincinnati.

REFERENCES

Allan, R. S. and Mason, S. G. 1962. Particle behaviour in shear and electric fields I. Deformation and burst of fluid drops, *Proc. R. Soc.*, **267 A**, 45-61.

Ayyaswamy, P. S. 1995. Direct-contact transfer processes with moving liquid droplets, in *Advances in Heat Transfer*, J. P. Hartnett and T. F. Irvine, editors, **26**, 1-104.

Clift, R., Grace, J. R. and Weber, M. E. 1978. *Bubbles, drops and particles*, Academic Press, Chapter 7.

Feng, J. Q. and Scott, T. C. 1996. A computational analysis of electrohydrodynamics of a leaky dielectric drop in an electric field, *J. Fluid Mech.*, **311**, 289-326.

Happel, J. and Brenner, H. 1965. *Low Reynolds number hydrodynamics*, Noordhoff International Publishing.

Jog, M. A., Ayyaswamy, P. S., and Cohen, I. M. 1996. Evaporation and combustion of a slowly moving liquid fuel droplet: Higher order theory, *J. Fluid Mechanics*, **307**, 135-165.

Jog, M. A. and Hader, M. A., 1997, Transient Heat Transfer to an Oblate Spheroidal Drop Suspended in an Electric Field, *Int. J. Heat and Fluid Flow*, in press.

Melcher, J. R. and Taylor, G. I. 1969. Electrohydrodynamics: a review of the role of interfacial shear stresses, *Annual Reviews of Fluid Mechanics*, **1**, 111-146.

Oliver, D. L. R., Carleson, T. E. and Chung, J. N. 1985. Transient heat transfer to a fluid sphere suspended in an electric field, *Int. J. Heat Mass Transfer*, **28**, 1005-1009.

Oliver, D. L. R. and DeWitt, K. J. 1993. High Peclet number heat transfer from a droplet suspended in an electric field: Interior problem, *Int. J. Heat Mass Transfer*, **36**, 3153-3155.

Patankar, S. V., 1980, *Numerical Heat Transfer and Fluid Flow*, Hemisphere Publishing Co., Washington, DC.

Sadhal, S. S. and Johnson, R. E. 1986. On the deformation of drops and bubbles with varying interfacial tension, *Chem. Engg. Comm.*, **46**, 97-109.

Smythe, W. R., 1968, *Static and Dynamics Electricity*, 3rd ed., McGraw-Hill, New York.

Taylor, G. I. 1966. Studies in Electrohydrodynamics, I. The circulation produced in a drop by an electric field, *Proc. R. Soc.*, **291 A**, 159-166.

Torza, S., Cox, R. G., and Mason, S. G. 1971. Electrohydrodynamic deformation and burst of liquid drops, *Phil. Trans. R. Soc. London A*, **269**, 295-319.

Vizika, O. and Saville, D. A. 1992. The electrohydrodynamic deformation of drops suspended in liquids in steady and oscillatory electric fields, *J. Fluid Mechanics*, **239** 1-21.

HTD-Vol. 342, National Heat Transfer Conference
Volume 4
ASME 1997

THE EFFECT OF ACOUSTIC MODULATION ON LIQUID SHEET DISINTEGRATION

I-Ping Chung, Charles A. Cook, and Cary Presser
Chemical Science and Technology Laboratory
National Institute of Standards and Technology
Gaithersburg, MD 20899

John L. Dressler
Fluid Jet Associates
Dayton, Ohio 45458

ABSTRACT

A fast stroboscope photographic technique was used to investigate the mechanism of liquid sheet disintegration by acoustic perturbation and the effect of different fluid viscosities on disintegration. Acoustic modulation shortens the breakup length of a hollow-cone pressure-swirl spray. The mechanism of early breakup from modulation is attributed to waves imposed on the liquid sheet. The wavelength generated by the modulation was found to depend on the acoustic driving frequency. The experimental results indicated an optimum wavelength, for which the liquid breakup length is a minimum, appears in an acoustic modulated conical sheet disintegration. The optimum breakup length is at a driving frequency of 10 kHz, which is one of three resonant frequencies in our piezoelectric driver. Different viscosity fluids, ν = 1, 9.6, 14, and 50 cs, were used to examine the effect of viscosity on conical sheet disintegration. Increasing liquid viscosity hampers the spray development and lengthens the sheet disintegration. Acoustic modulation improves the disintegration in a higher viscous fluid but was less effective than in a lower viscosity fluid. A higher input modulated power enhances disintegration. The breakup length is inversely proportional to the logarithm of input power.

INTRODUCTION

In general, an increase in liquid viscosity has an adverse effect on atomization quality. For example, Wang and Lefebvre (1987) reported that an increase of liquid viscosity produced a larger Sauter mean diameter (SMD) in sprays. SMD is usually used to describe atomization quality. Poorer atomization quality (i.e., larger SMD) results in a lower evaporation rate and consequently poorer combustion performance. In order to maintain the same atomization quality at a fixed liquid pressure in liquid injection systems, extra energy or perturbation is required for higher viscosity liquids. One proposed method to improve the atomization of viscous liquids is to perturb the liquid by an acoustically modulated device. Previous work by Takahashi et al. (1995), Zhao et al. (1996), Cook et al. (1996), and Chung et al. (1997) demonstrated that the acoustic modulation of liquids affects atomization quality and the combustion process. However, the physical mechanism that describes the transfer of energy to the liquid during atomization is poorly understood. This paper concentrates on the effect of piezoelectric modulation on liquid sheet disintegration.

Liquid sheet disintegration has been studied extensively in a variety of theoretical and experimental investigations.

Theoretical Analyses: All previous theoretical analyses considered a finite-thickness, two-dimensional inviscid flat sheet due to the complicated hydrodynamic process and continuous attenuation of the liquid sheet thickness for a hollow conical spray. For example, Squire (1953) used linear analysis to solve the equations of wave motion on the surface of a finite-thickness inviscid liquid sheet.

It was concluded that an instability occurs if the Weber number is less than 1 (defined as the ratio of the surface tension and momentum forces) and an optimum wavelength exists where the wave growth rate is a maximum. York et al. (1953) analyzed an identical problem, and concluded that instability and wave formation at the interface between the continuous and discontinuous phases were the major factors in the breakup of a sheet of liquid into droplets. Hagerty and Shea (1955) solved Bernoulli's equation for a two-dimensional inviscid flat sheet. Two basic wave forms were found to be solutions, i.e., sinuous and dilational waves. For the sinuous wave, the minimum unstable wavelength was the same as that obtained by Squire (1953). It was also concluded that the growth rate of a sinuous wave is always greater than that of a dilational wave. Clark and Dombrowski (1972) used a second order analysis to solve the equation of sinuous wave motion on a flat sheet and obtained a solution for liquid breakup length. The theoretical results were compared with the measured values for both flat and conical sheets, and better agreement was found for the conical sheets. The calculated results assumed the existence of an optimum dominant wavelength, as concluded by Squire (1953). Rangel and Sirignano (1991) used both linear and nonlinear analysis to consider the sinuous and dilational instability of an inviscid sheet with finite thickness at different ambient air densities and liquid surface tensions. It was found that at lower gas-to-liquid density ratios the growth rate of the sinuous waves was larger than that of the dilational waves, while at higher density ratios, the dilational waves had a higher growth rate. To summarize, all of the aforementioned theoretical analyses show that wavelength is one major factor in liquid sheet disintegration.

Experimental Studies: York et al. (1953) investigated the performance of hollow-cone swirl nozzles. It was demonstrated that waves or ripples appeared on the surface of the liquid sheet before breakup. It was believed that waves were mainly responsible for the sheet breakup, and that frequency and wavelength of the ripple might be related to droplets size. It was also found that the theoretical infinite flat sheet model was not approximated closely by the experimental results for a conical spray. Dombrowski and Hooper (1962) studied the effect of ambient density on droplet formation in single hole fan sprays. It was found that the waves imparted to the liquid sheet played a major role on disintegration. Crapper et al. (1975) investigated thin fan sprays and did not find a wave of maximum growth rate except at low liquid velocities. It was concluded that the theoretical, two-dimensional, finite-thickness flat sheet analysis was inadequate for a thin fan spray. Further, Crapper and Dombrowski (1984) experimentally demonstrated that a wavelength with an optimum growth rate did not exist in thin fan spray sheets but droplet size might be affected by both disturbance amplitude and frequency. Mansour and Chigier (1990) studied liquid sheet disintegration from a two-dimensional air-assisted nozzle. It was concluded that two modes of sheet breakup existed for air-assist sprays: a mechanical mode due to the effect of liquid pressure inside the nozzle, and an aerodynamic mode due to the effect of air shear.

For an acoustic modulated spray driven by a piezoelectric driver, Takahashi et al. (1995) found that waves appeared on the conical liquid surface when modulation was imparted to the spray. Chung et al. (1997) also found that the piezoelectric driver contains several resonant frequencies where the spray features change significantly. Other than those resonant frequencies, the spray was unaffected by the modulation signals. In this study, the mechanism of sheet disintegration as enhanced by an acoustic modulation was examined for different liquid viscosities. The liquid sheet is a conical shape generated with a hollow-cone, pressure-swirl nozzle. Of interest was also whether or not there exists an optimum dominant modulation wavelength on the conical liquid sheet that amplifies the sheet disintegration. Several parameters were investigated to determine their effect on the conical sheet breakup length. The parameters, besides wavelength (or driving frequency, f), included the input perturbation power (P) and fluid viscosity (ν). The study of these parameters was expected to help increase the understanding of the mechanism for liquid disintegration of a conical sheet.

EXPERIMENTAL TECHNIQUE

The experimental setup is shown schematically in Fig.1. Nitrogen from a regulated at 690 kPa high pressure tank is introduced into a stainless steel liquid vessel to transport liquid through a 5 μm pore sintered filter, metering valve, flowmeter, and then to an acoustically modulated pressure swirl nozzle. Upstream of the atomizer, a pressure gauge is used to measure the fluid line pressure. The nozzle is modulated by a piezoelectric driver which is excited sinusoidally by a function generator at a tunable frequency. The piezoelectric driver consists a pair of piezoelectric transducers. The detailed structure and operation of the piezoelectric driver is described elsewhere (Chung et al., 1997). Before reaching the piezoelectric driver, the sinusoidal signal is amplified by an audio power amplifier and a matching transformer. When a sinusoidal acoustic signal is applied to the piezoelectric driver, a small electro-mechanical displacement of the piezoelectric transducers generates waves that are transmitted through the liquid. The additional waves imposed on the liquid sheet improves disintegration.

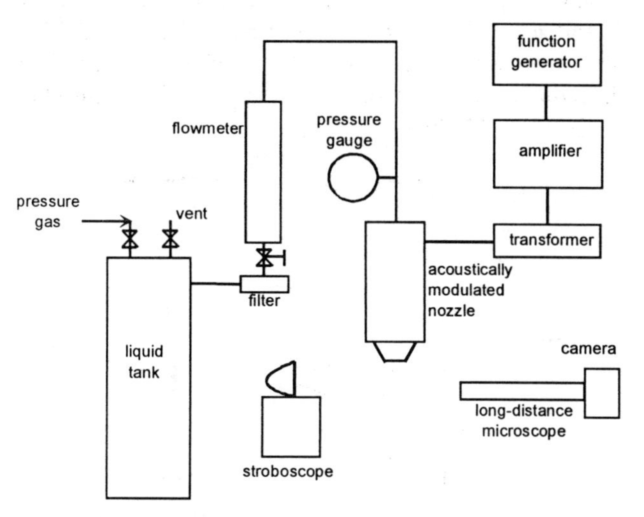

Figure 1 Schematic of experimental setup

A pressure swirl nozzle is used to generate a 60^0 hollow-cone spray. The nozzle orifice diameter (d_o) is 0.025 cm and the designed nominal flow rate is 63 cm^3/min at an injection pressure of 862 kPa. As studied by Chung et al. (1997), the effect of acoustic modulation is particularly prominent at certain resonant frequencies and a part-load injection pressure. Therefore, our experiments were mainly conducted at resonant frequencies and part-load pressure, i.e. 275 kPa for water and 413 kPa for higher viscosity liquids.

A 35 mm camera attached to a long-distance microscope captures the detailed structure of the spray. The magnification of the microscope can be varied by changing the distance between the camera and the objective lens. The maximum magnification factor is 5 for our experimental arrangement. A single-flash stroboscope with a 3 μs pulse duration was used as a back-lighting source to freeze the moving waves on the liquid sheet surface. For each experiment, more than five photographs were taken to record the spray features.

Four different viscosity fluids were used: water, two different volume fractions of glycerol-water mixtures, and silicone oil. The fluid properties measured at 293 K are listed in Table 1.

RESULTS AND DISCUSSION

The observed features are presented in this section to elucidate the process of conical liquid sheet disintegration. In addition, analyses of these results were

liquid	viscosity (cs)	surface tension (kg/s^2)	density (kg/m^3)
water	1.0	0.072	995
glycerol-water ($\approx$ 40/60 by volume)	9.6	0.060	1110
glycerol-water ($\approx$ 50/50 by volume)	14.0	0.059	1160
silicone oil	50.0	0.018	963

Table 1 Fluid properties

carried out to determine the effect of several aforementioned parameters on the liquid sheet breakup length.

A. Observed Spray Features

1. Disintegration Modes

Fraser (1957) described three modes of sheet disintegration as *rim, perforated,* and *wave-sheet* disintegration. The rim mode produces large droplets by pulling threads out from the rim of the sheet. Perforated disintegration occurs when liquid contains thin perforated regions on the sheet surface which coalesce into droplets. Wave-sheet disintegration occurs when waves on the liquid surface reach instability and the aerodynamic force disrupts the liquid into droplets. The photographs presented in Fig.2 show that wave-sheet disintegration is the most dominant breakup mechanism for an acoustic modulation of the liquid spray and the rim mode is the secondary mechanism. In order to obtain the best resolution of the spray and to capture the region of liquid breakup, different magnifications were used for different cases. Since the same nozzle was used for each case, the nozzle size can be used as a reference (the width of the nozzle head $\cong$ 19 mm). Plate 2(a) shows a wave generated by an acoustic modulation on a water (ν = 1 cs) sheet surface. Rim disintegration is ambiguous in this case but appears distinctly in Plate 2(b) where the fluid has a higher viscosity of ν = 14 cs (50/50 glycerol-water). The liquid boundary, because of surface tension, continually curves inward but the acoustic disturbance dislodge threads out from the rim producing large droplets. At the same experimental conditions (i.e., the same fluid at the same injection pressure) but without the acoustic disturbance, Plate 2(c) illustrates the mode of perforated disintegration. Holes appear on the liquid sheet and keep growing until the borders of adjacent holes coalesce and break the liquid sheet into varying size droplets. Rim and perforated disintegration form with high viscosity fluids, for example, at ν = 14 cs in this experimental arrangement, and is consistent with the phenomena reported by Fraser and Eisenklam (1953).

2. Viscosity Effects

As mentioned earlier, an increase in liquid viscosity has an adverse effect on spray atomization. Figure 3 demonstrates the difference in the spray features at various liquid viscosities (ν) for sprays without modulation. In this case, the injection pressure (i.e., Δp = 275 kPa) and nozzle are the same but liquid viscosity increases. The liquid viscosities are 1 (water), 9.6 (40/60 glycerol-water), and 14 cs (50/50 glycerol-water). Again, note that Plate 3(a) has a magnification that is almost three times greater than Plates 3(b) and 3(c). The breakup length for Plate 3(a) is therefore much shorter than for Plates 3(b) and 3(c). A further increase in viscosity to ν = 50 cs resulted in the a string discharge, even though the pressure was pumped to 620 kPa.

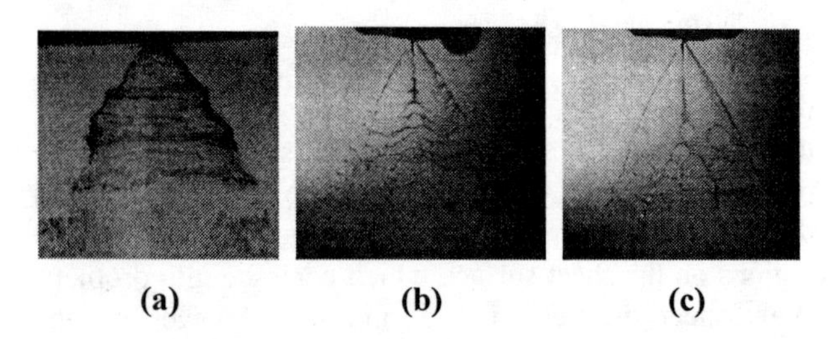

Figure 2 Liquid disintegration modes: (a) wave mode, water (ν = 1 cs) is acoustically modulated at P = 5 W and f = 9.9 kHz; (b) rim mode, glycerol-water mixture (ν = 14 cs) is acoustically modulated at P = 1 W and f = 16.7 kHz; (c) perforated mode, glycerol-water mixture (ν = 14 cs) spray without modulation.

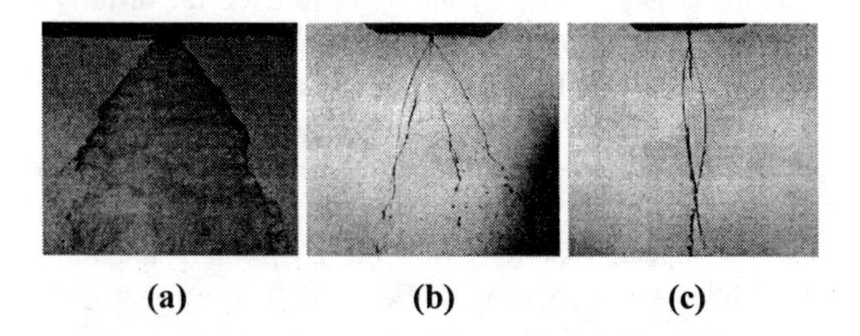

Figure 3 Spray features for different viscosity fluids: (a) ν = 1 cs, (b) ν = 9.6 cs, (c) ν = 14 cs.

3. Spray Development of Higher Viscosity Fluids

Lefebvre (1989) described five stages of a hollow-cone spray at different injection pressures namely, (1) dribble, (2) distorted pencil, (3) onion, (4) tulip, and (5) fully-developed sprays. It is a vivid description for low viscosity fluids, for instance, water (ν = 1 cs and surface tension σ = 0.072 kg/s^2). For a fluid with a higher viscosity of ν = 14 cs and a similar surface tension (σ = 0.059 kg/s^2), the spray structure is similar with increasing injection pressure but a little different shapes, as indicated in Fig. 4. The distinct differences are that perforated disintegration becomes the dominant mode in higher viscosity sprays and no ripples appear on the liquid sheet surface, while ripples usually form on the sheet of lower viscosity sprays (ν = 1 cs) as shown in Fig. 3(a) or as reported by Clark and Dombrowski (1972). In addition, the spray does not become fully developed even at high injection pressures as shown in Fig. 4(d), for an injection pressure of 620 kPa.

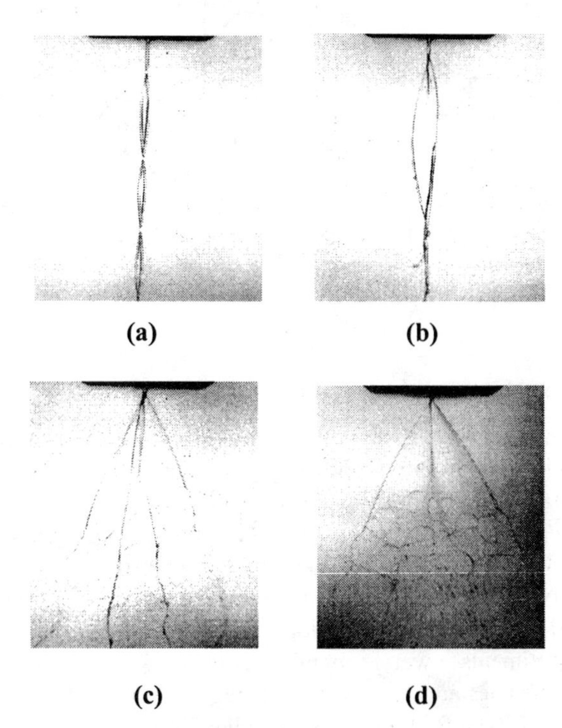

Figure 4 Variation of line pressure for a higher viscosity fluid (ν = 14 cs), the pressures are: (a) 138, (b) 207, (c) 413, and (d) 620 kPa.

4. Acoustic Modulation of Low Viscosity Fluids

The photographs presented in Fig. 5 show a sequence of sprays that are set at the same operating condition (i.e., Δp = 275 kPa, ν = 1 cs, and P = 10 W) but at different driving frequencies (f). Plates 5(a) to 5(c) represent the driving frequencies of 4.5, 9.9, and 18.1 kHz, which are, respectively, three resonant frequencies for this piezoelectric driver. The photographs illustrate that different driving frequencies generate different wavelengths on the liquid surface.

5. Acoustic Modulation of Higher Viscosity Fluids

A sequence of photographs presented in Fig. 6 are for a higher viscosity liquid (ν = 9.6 cs). Because of the high viscosity, the injection pressure was raised to 413 kPa to ensure that the sprays become at least tulip stage as described by Lefebvre (1989). The driving frequencies are similar to the previous case, i.e., 3.4, 9.1, and 16.3 kHz, but the modulated power is higher (P= 50 W). Compared with the lower viscosity case, the wave amplitudes in the higher viscosity case decreased significantly even though the driving power increased fivefold. A further increase in liquid viscosity to $\nu = 50$ cs with a modulation power of 100 W, the discharge becomes a string with dilation waves as shown in Fig. 7.

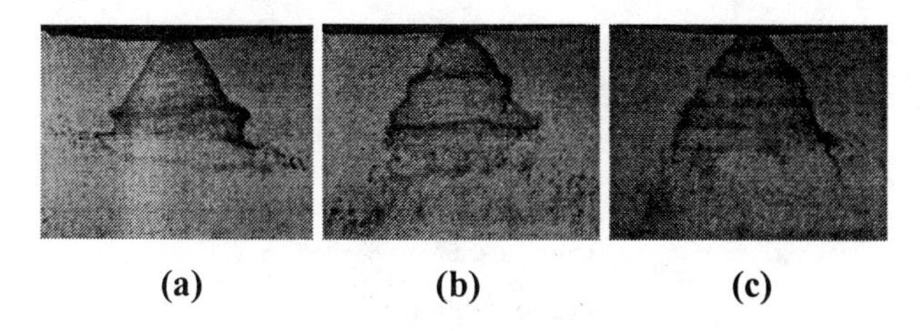

(a) (b) (c)

Figure 5 Variation of the acoustic modulation for water sprays: (a) f = 4.5 kHz, (b) f = 9.9 kHz, (c) f = 18.1 kHz.

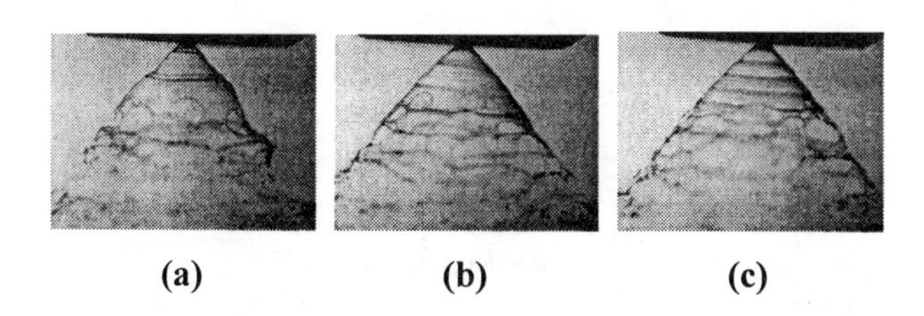

(a) (b) (c)

Figure 6 Variation of the acoustic modulation for a viscous fluid of $\nu = 9.6$ cs at P = 50 W: (a) f = 3.4 kHz, (b) f = 9.1 kHz, (c) f = 16.3 kHz.

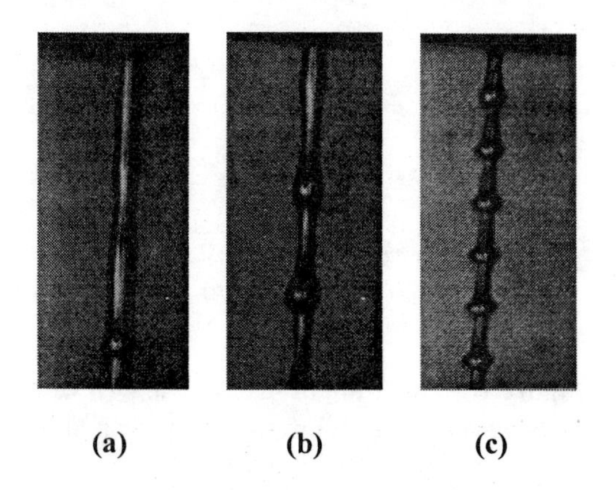

(a) (b) (c)

Figure 7 Variation of the acoustic modulation for a viscous fluid of $\nu = 50$ cs at P = 100 W: (a) f = 3.8 kHz, (b) f = 10 kHz, (c) f = 18.8 kHz.

6. Modulated Power

When the modulated power increases, the disturbance in the liquid increases and generates larger amplitude waves. A higher input modulated power shortens the liquid sheet breakup length as shown in Fig. 8. In this figure, results for three different viscosities are presented at various input modulated powers. Plates 8(a) to 8(c) are for $\nu = 1$ cs (water) at injection pressure $\Delta p = 275$ kPa , 8(d) to 8(f) are for $\nu = 9.6$ cs at $\Delta p = 413$ kPa, and 8(g) to 8(i) are for $\nu = 14$ cs at $\Delta p = 620$ kPa. Note that the photograph magnifications for these three cases are different. The water case has the largest magnification , while $\nu = 14$ cs has the smallest magnification. For the highest viscosity case ($\nu = 14$ cs), the sprays have not become fully developed and the disintegrated droplet sizes are large, but the modulations still improve the droplet breakup.

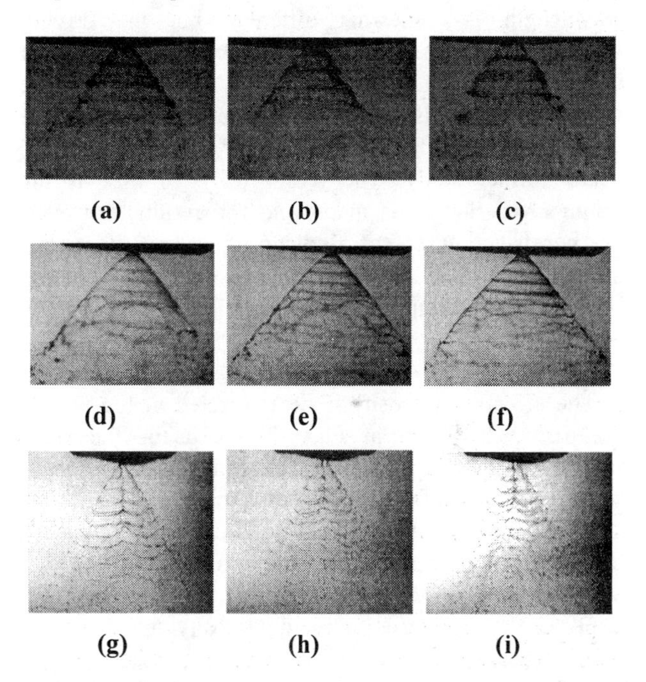

(a) (b) (c)

(d) (e) (f)

(g) (h) (i)

Figure 8 Effect of modulated power on liquid sheet breakup length: (a) to (c) $\nu = 1$ cs and f = 18.1 kHz: (a) P = 10 W, (b) P = 25 W, (c) P = 50 W; (d) to (f): $\nu = 9.6$ cs and f = 16.3 kHz: (d) P = 10 W, (e) P = 50 W, (f) P = 100 W; (g) to (i) $\nu = 14$ cs and f = 16.7 kHz: (g) P = 10 W, (h) P = 50 W, (i) P = 100 W.

B. Parametric Analysis

1. Wavelength Effect

The breakup boundary along the surface of the conical sheet varies at different circumferential locations. In order to measure the breakup length, a specific point on a conical sheet was chosen and the values were averaged from several snapshots of the same spray. The breakup length and wavelength were measured by using a fine scale reticule (40 lines/mm) supplied by the microscope manufacturer. The variance in the wavelength measurements is less than 2%, while the breakup length measurements are more scattered (the largest variation is about 10%). Since the atomization process from a hollow-cone conical spray is chaotic and turbulent, the scatter in the results is reasonable. Figure 9 presents the relationship between wavelength and breakup length for the water spray. The different symbols represent the different input modulated powers. With the reported measurement variance, the results indicate that for each input modulated power a minimum breakup length is always occurs at a wavelength of 1.2 mm, which is at a modulated frequency of 10 kHz. For a higher viscosity case (ν = 9.6 cs), the results presented in Fig. 10 indicate a similar trend to those for ν = 1 cs except that the wavelength is somewhat different, but the driving frequency remains close to 10 kHz. The variation in wavelength generated by modulation is attributed to the change in fluid viscosity. The relationship between the wavelength and the fluid property (mainly viscosity) needs further study. The reason why 10 kHz is the optimal wavelength is unknown. The results, however, are consistent with the findings of several theoretical studies that indicate the existence of an optimal wavelength for liquid sheet disintegration.

2. Viscosity Effect

The effect of viscosity is demonstrated in Fig. 11 by comparing the data in Figs. 9 & 10 for the same modulated power, and normalizing the breakup length with respect to the unmodulated case. For the low viscosity fluid (water, ν = 1 cs), the breakup length for the unmodulated case is 3.4 mm, and it is 5.93 mm for high viscosity fluid (40/60 glycerol-water, ν = 9.6 cs). As expected, an increase in liquid viscosity lengthens the sheet breakup length. For the modulated cases, modulation is less effective for the higher viscosity fluid. For example, at the optimal wavelength case the normalized breakup length for the water spray (ν = 1 cs) is at value of 0.49 (a decrease of 51%) due to modulation, and only 0.67 (a decrease of 37%) for the higher viscosity fluid (ν = 9.6 cs) at the same input modulated power (P = 50 W). It is obvious that a more viscous fluid requires additional energy to ensure disintegration. In order to obtain a quantitative relationship between viscosity and breakup length, a more extensive database is required, and will be the subject of future work.

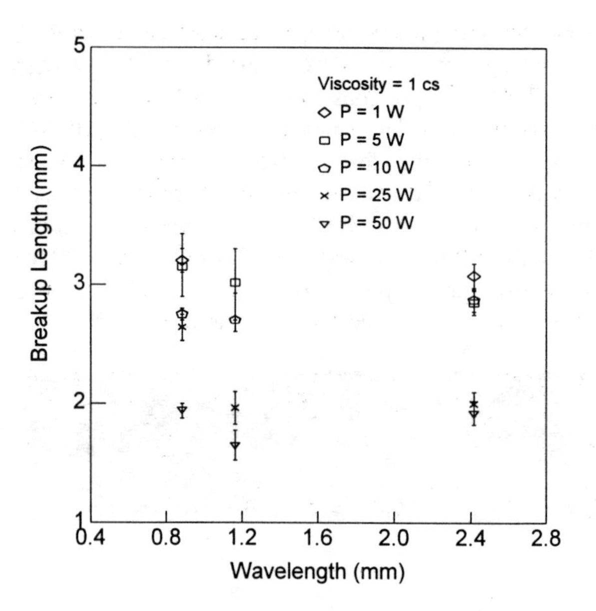

Figure 9 Variation of breakup lengths with wavelength and modulation power for water

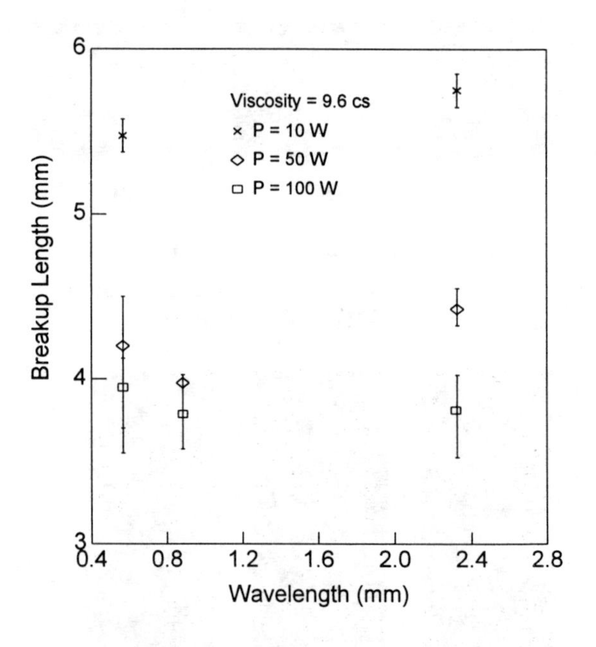

Figure 10 Variation of breakup length with wavelength and modulation power for a higher viscosity fluid

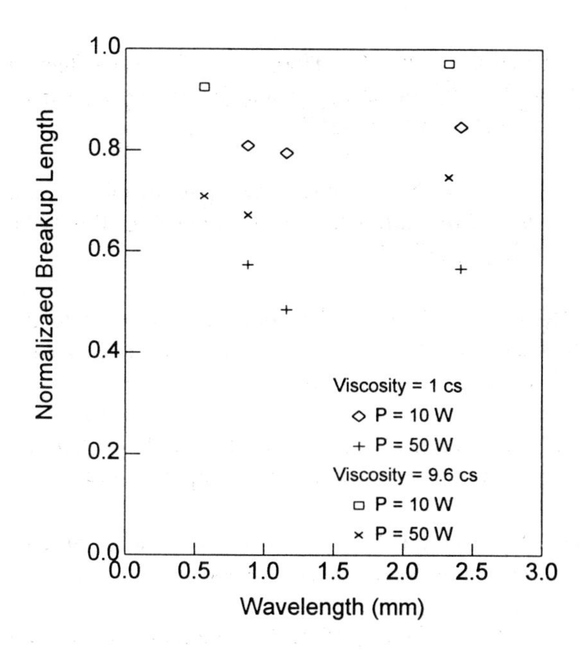

Figure 11 Effect of viscosity on breakup length

3. Dimensionless Parameters

As discussed in previous section, different viscosity liquids result in different sheet breakup lengths. The results shown in Fig. 11 were assumed that the difference of the atomization process mainly comes from the effect of viscosity. However, the liquids used in the experiments contain not only different viscosities but also different surface tensions and densities as shown in Table 1, although the relative magnitudes of the difference in the latter are significantly less than that in the former. In order to include all the effects of different fluid properties, i.e., viscosity, density, and surface tension, the variation of breakup length at different Ohnesorge number (Oh) is calculated and presented in Fig. 12. The Ohnesorge number is defined as the ratio of internal viscous force to interfacial surface tension force. In this figure, the abscissa represented by the wavelength that is normalized by nozzle orifice diameter (d). As expected, the results are similar to Fig. 11 since Ohnesorge number is proportional to viscosity. The liquid with higher Ohnesorge number shows more resistant to atomization.

4. Modulated Power

The relationship between the modulated power and the breakup length is presented in Fig. 13. The breakup length was normalized by the unmodulated case and the abscissa is expressed in terms of the natural logarithm. A linear least-square fit to the data is presented in this figure for a specific fluid viscosity at each wavelength (w.l.). The results show that a higher input modulated power enhances early liquid sheet disintegration and the breakup length is inversely proportional to the logarithm of input power.

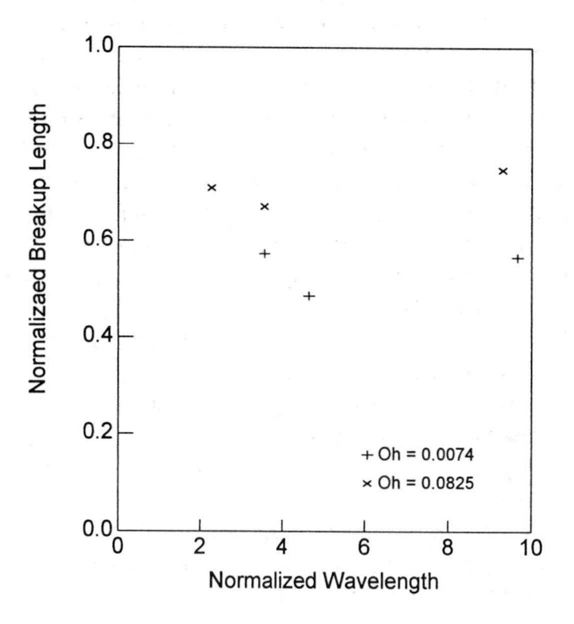

Figure 12 Variation of breakup lengths with Ohnesorge number and wavelength

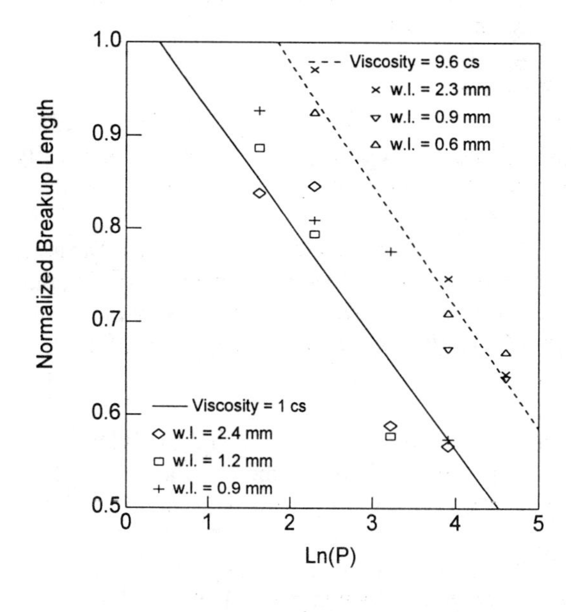

Figure 13 The relationship between modulated power and breakup length

SUMMARY

Acoustic modulation was found to affect the liquid disintegration in a hollow-cone conical spray by shortening the breakup length. The mechanism of early disintegration is attributed to waves imposed on the liquid sheet. Waves on the liquid are generated by the disturbance imparted from the electro-mechanical displacement of piezoelectric transducers. The wavelength transmitted on the liquid sheet depends on the piezoelectric driving frequency and fluid viscosity.

The results indicate that an optimum wavelength is present for acoustic modulated conical sheet disintegration. In the experimental arrangement for this study, the optimum breakup length is at a driving frequency of 10 kHz.

Different viscosity liquids, ν = 1, 9.6, 14, and 50 cs, were used to examine the effect of viscosity on the modulated liquid disintegration process. The improvement in disintegration by modulation is less effective for higher viscosity liquids, but still is significant.

A higher input modulated power enhances early disintegration. The breakup length is inversely proportional to the logarithm of input power.

ACKNOWLEDGEMENTS

One of the authors gratefully acknowledges the support by the National Research Council Postdoctoral Associateship at NIST.

REFERENCES

Chung, I.P., D. Dunn-Rankin, and A. Ganji (1997) Characteristics of A Spray from an Ultrasonically Modulated Nozzle, *Atom. & Sprays* (in press).

Cook, C.A., S.R. Charagundla, C. Presser, J.L. Dressler, and A.K. Gupta (1996), Effect of Acoustic Atomization on Combustion Emissions, *1996 American Flame Research Committee International Symposium*, Baltimore, MD.

Crapper, G.D., N. Dombrowski, and G.A.D. Pyott (1975) Large Amplitude Kelvin-Helmholtz Waves on Thin Liquid Sheets, *Proc. R. Soc. Lond. A.*, vol. 342, pp.209-224.

Crapper, G.D. and N. Dombrowski (1984) A Note on the Effect of Forced Disturbances on the Stability of Thin Liquid Sheets and on the Resulting Drop Size, *Int. J. Multiphase Flow*, vol. 10, pp.731-736.

Dombrowski, N. and P.C. Hooper (1962) The Effect of Ambient Density on Drop Formation in Sprays, *Chem. Eng. Sci.*, vol.17, pp. 291-305.

Fraser, R.P. (1957) Liquid Fuel Atomization, *Sixth Symposium (International) on Combustion*, Reinhold, New York, pp. 687-701.

Fraser, R.P. and P. Eisenklam (1953) Research into the Performance of Atomizers for Liquids, *Imp. Coll. Chem. Eng. Soc. J.*, vol. 7, pp. 52-68.

Hagerty, W.W. and J.F. Shea (1955) A Study of Stability of Plane Fluid Sheets, *J. Appl. Mech.*, pp. 509-514.

Lefebvre, A.H. (1989) Atomization and Sprays, Hemisphere Publishing Co., New York.

Mansour and Chigier (1990) Disintegration of Liquid Sheets, *Phys. Fluids A*, vol. 2 (5), pp. 706-719.

Rangel, R.H. and W.A. Sirignano (1991), The Linear and Nonlinear Shear Instability of a Fluid Sheet, *Phys. Fluids A*, vol. 3, pp. 2392-2400.

Squire, H.B. (1953) Investigation of the Instability of a Moving Liquid Film, *Br. J. Appl. Phys.*, vol.4, pp. 167-169.

Takahashi, F., W.J. Schmoll, and J.L. Dressler (1995) Characteristics of a Velocity-Modulated Pressure-Swirl Atomizing Spray, *J. Prop. & Power*, vol. 11, pp.955-963.

York, J.L., H.E. Stubbs, and M.R. Tek (1953) The Mechanism of Disintegration of Liquid Sheets, *Trans. AMSE*, vol.75, pp.1279-1286.

Wang, X.F. and A.H. Lefebvre (1987) Mean Drop Size from Pressure-Swirl Nozzles, *AIAA J. Propul. Power*, vol. 3, No. 1, pp. 11-18.

Zhao, F.Q., M.C. Lai, A.A. Amer, and J.L. Dressler (1996) Atomization Characteristics of Pressure Modulated Automotive Port Injector Sprays, *Atom. & Sprays*, vol. 6, 4, pp. 461-483.

HTD-Vol. 342, National Heat Transfer Conference
Volume 4
ASME 1997

EFFECT OF MORPHOLOGICAL OSCILLATION AND INTERNAL FLOW ON DROPLET IMPACT, SPREADING, AND SOLIDIFICATION ON A SOLID SURFACE

Jean-Pierre Delplanque

Department of Mechanical and Aerospace Engineering
and
Department of Chemical Engineering and Materials Science
University of California, Irvine
Irvine, California

ABSTRACT

The effect of internal flow and morphology at impact on the spreading behavior of a droplet impinging on a solid surface is investigated. The approach chosen in the present study is numerical simulation. The free surface flow resulting from the droplet impact on a solid surface is modeled using two-dimensional axisymmetric Navier-Stokes equations combined with a volume of fluid fraction transport equation. When solidification is considered, the solidification process is described using a locally 1D, multi-directional algorithm. Morphological deformations of practical interest such as spheroids and superposed semi-spheroids are considered. Internal flow is simulated by specifying the initial velocity field. For instance, the velocity field caused by a Hill's vortex is used to simulate shear-induced internal circulation. Finally, the effect of these phenomena on the droplet solidification behavior is studied. The results show that while the effect of internal circulation is essentially negligible, droplet deformation should be included in the simulation of droplet spreading and solidification.

NOMENCLATURE

D	droplet diameter
e	$=\sqrt{\lvert a^2-b^2\rvert/a^2}$, ellipticity of a spheroid with semi-axes a and b
Eo	$=\gamma(\rho_\ell-\rho_g)D^2/\sigma$, Eötvös number
F	volume of fluid function
$\overline{F_g}$	body forces
$\overline{g}$	gravity acceleration
h_{sf}	latent heat of solidification
k	thermal conductivity
p	pressure
R_0	Droplet radius
R	Radius of the liquid in contact with the substrate
Re	$=\mathcal{V}D/\nu$, Reynolds number
$\mathcal{S}$	local position of solidified layer
Ste	$=k\Delta T/(\alpha\rho h_{sf})$ Stefan number
T	temperature
$\overline{V}$	velocity vector
$\mathcal{V}$	droplet impact velocity
We	$=\rho\mathcal{V}^2D/\sigma$, Weber number

Greek symbols

α	thermal diffusivity
γ	droplet acceleration
θ	contact angle
ϑ	volume fraction open to flow
σ	surface tension coefficient
ν	kinematic viscosity
ξ	$=2R/D$
ρ	density
λ	solidification constant

Subscripts

g	gas
ℓ	liquid
s	solid or substrate

Diacritics

—	vector

INTRODUCTION

The spreading of a droplet after impact on a solid surface is a basic fluid dynamics problem which is at the base of numerous and diverse phenomena such as rain erosion, aircraft icing, agricultural pesticide spraying, and spray-based manufacturing technologies. Experiments and numerical simulations have shown that the possible outcomes of a droplet impact includes bouncing, spreading, coalescence, or splashing (Rein, 1993). Numerical investigations (Trapaga *et al.*, 1992; San Marchi *et al.*, 1993; Liu *et al.*, 1993b; Liu *et al.*, 1993a; Fukai *et al.*, 1993; Liu *et al.*, 1995; Delplanque *et al.*, 1996a; Delplanque and Rangel, 1997; Pasandideh-Fard *et al.*, 1996; Waldvogel and Poulikakos, 1997; Delplanque *et al.*, 1996b) of droplet impact on a solid surface, without or without solidification, typically assume that the droplet is spherical at impact, with no internal circulation. In his extensive review on liquid drop impact, Rein (Renksizbulut and Yuen, 1983) indicates that the effect of drop morphological oscillations on impact behavior has yet to be investigated.

Deviations from the spherical shape can be due to acceleration-induced deformations (Clift *et al.*, 1978), wake shedding induced oscillations (Clift *et al.*, 1978; Rein, 1993), or the droplet formation process (Rein, 1993). In most cases, if the relative velocity is low enough that secondary atomization does not become an issue, the deformed shapes can be modeled using axisymmetric volumes, spheroids or a combination of two semi-spheroids with a common semimajor axis for instance (Clift *et al.*, 1978).

A secondary effect of droplet oscillation is the induced internal flow. Internal circulation may also be induced during the droplet formation process or by shear at the droplet surface from the surrounding fluid. However, the latter type of internal circulation induction mechanism is less likely as it requires the droplet surface to remain pure, free of any potential surfactant (Rein, 1993; Clift *et al.*, 1978).

In droplet-based manufacturing processes (Lavernia and Wu, 1996), it is essential to be able to predict the occurrence of defaults such as voids and cracks because of their importance to the structural integrity and mechanical and thermal behavior of the processed materials. The potential of droplet deformation and internal circulation to be at the source of such defaults cannot be dismissed a priori.

The objective of the present work is therefore to investigate the validity of the assumptions of spherical shape and no internal circulation in droplet impact simulations. Morphological deformations of practical interest such as spheroids and superposed semi-spheroids (Clift *et al.*, 1978) are considered. Internal flow is simulated by specifying the initial velocity field. Namely, the velocity field caused by a Hill's vortex is used to simulate shear-induced internal circulation. Finally, the effect of these phenomena on the droplet solidification behavior is studied.

MODEL DESCRIPTION

The approach chosen in the present study is numerical simulation. The numerical model used was previously developed and is described in detail elsewhere (Kothe and Mjolsness, 1992; Delplanque *et al.*, 1996a). A summary description is provided next for completeness.

The impact of a liquid droplet on a solid surface creates a free surface flow which is assumed to be two-dimensional axisymmetric. The fluid dynamics is described with the Navier Stokes equations:

$$\nabla \cdot \left(\vartheta \overline{\boldsymbol{V}}\right) = 0 \qquad (1)$$

$$\vartheta \frac{\partial \overline{\boldsymbol{V}}}{\partial t} + \nabla \cdot \left(\vartheta\, \overline{\boldsymbol{V}}\, \overline{\boldsymbol{V}}\right) = -\frac{\vartheta}{\rho} \nabla p + \frac{\vartheta}{\rho} \nabla \cdot \tau + \vartheta \overline{\boldsymbol{g}} + \vartheta \overline{\boldsymbol{F}}_b \qquad (2)$$

Here ϑ is a characteristic function used when solidification is considered to describe the growing solid. ϑ=0 in the solid and 1 in the liquid. The evolution of the droplet free surface is predicted using a volume of fluid fraction transport equation:

$$\frac{\partial}{\partial t}(\vartheta F) + \nabla \cdot \left(\vartheta F \overline{\boldsymbol{V}}\right) = 0 \qquad (3)$$

where F is the Volume of Fluid function (F= 1 in the fluid and 0 in the void). These equations are solved using a numerical code developed at Los Alamos National Laboratories, RIPPLE (Kothe and Mjolsness, 1992).

When solidification is considered, the distribution of ϑ must be evaluated. To this end, a multi-directional solidification model (Delplanque *et al.*, 1996a) is used. In this approach, the solid/liquid front is tracked using distributed markers. The displacement of the markers as solidification progresses is evaluated by assuming that the liquid is pure and by approximating the phase-change process with a locally 1D Stefan problem. Growth occurs along the normal to the solid/liquid interface. The local growth rate can then be expressed as:

$$\frac{\partial \mathcal{S}}{\partial t} = \frac{2\lambda^2 \alpha_s}{\mathcal{L}} \qquad (4)$$

where $\mathcal{L}$ is the shortest distance to the substrate and the solidification constant, λ, is obtained from a heat balance at the interface:

$$\lambda = \frac{1}{\sqrt{\pi}} \left\{ \frac{\mathrm{Ste}_s}{\mathrm{erf}[\lambda] \exp(\lambda^2)} - \frac{\mathrm{Ste}_\ell \sqrt{\alpha_\ell / \alpha_s}}{\mathrm{erf}\left[\lambda \sqrt{\alpha_s / \alpha_\ell}\right] \exp\left[\lambda^2 \alpha_s / \alpha_\ell\right]} \right\} \qquad (5)$$

where Ste_s and Ste_ℓ are the solid and liquid Stefan numbers. More information regarding the numerical validation of this model can be found in Delplanque *et al.*(1996a and 1996b). Furthermore, the accuracy of the predictions of this numerical model was recently (Bian *et al.*, 1997) evaluated by comparison with available experimental data both in the fully liquid and the solid-

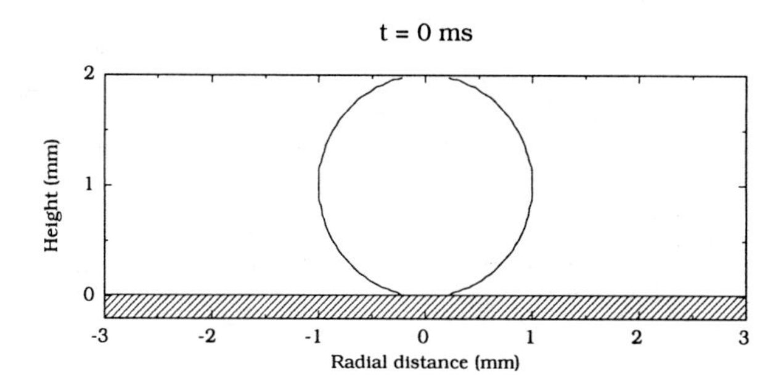

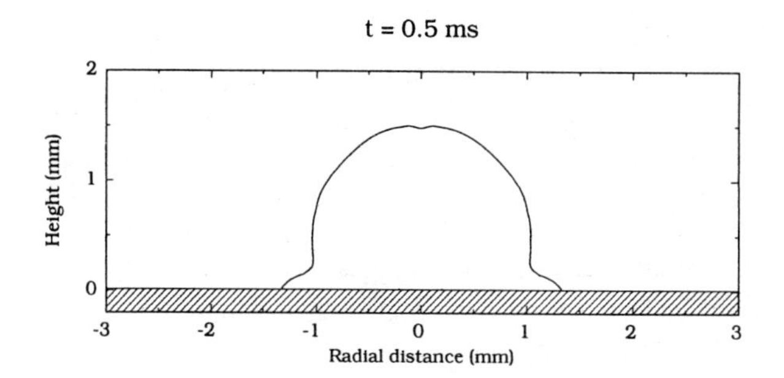

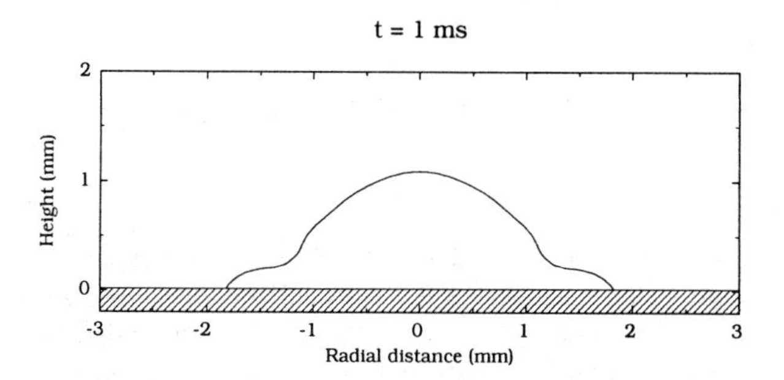

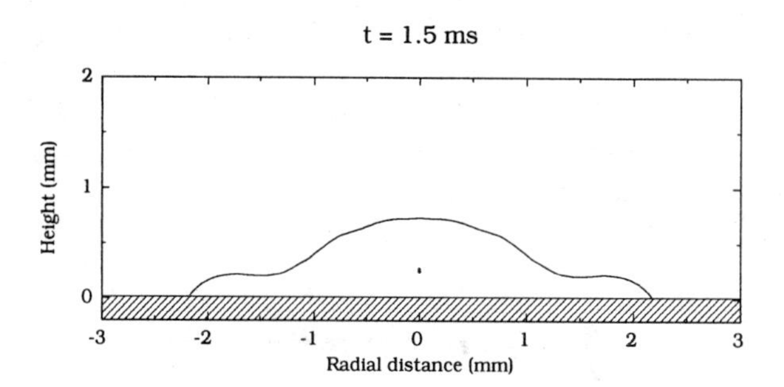

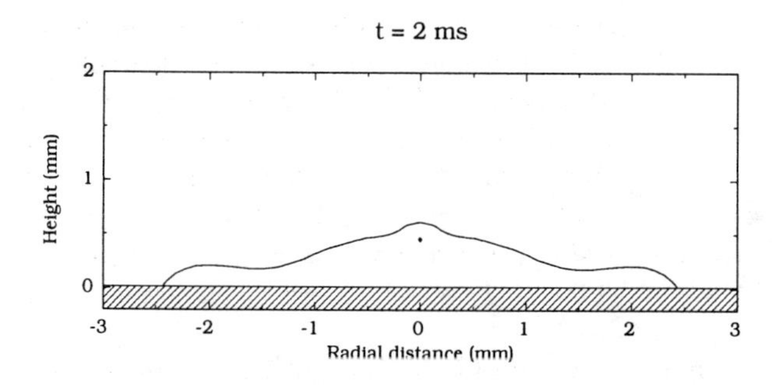

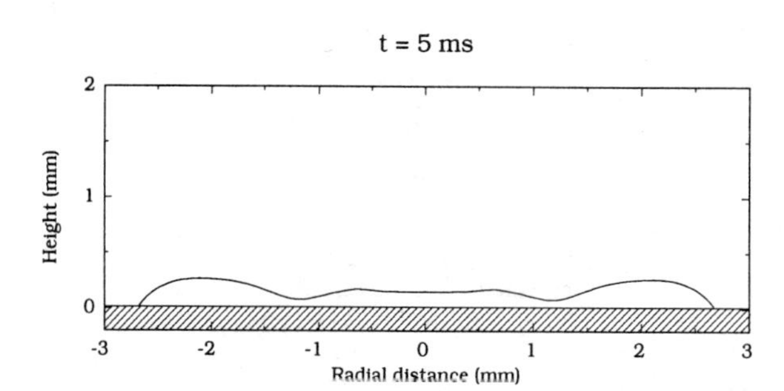

FIGURE 1: Water droplet deformation history. Reference case.

ifying case. In both cases the agreement was found to be quite satisfactory. However, the result quality was found to depend significantly on the ability to estimate or measure the contact angle accurately.

RESULTS AND DISCUSSION

In order to evaluate the influence of deformation and internal circulation on a droplet spreading behavior, a reference case is first defined. A 2 mm (diameter) water droplet impinging on a flat solid substrate at 1 m/s (Re=2112 and We=27) is considered. The contact angle is taken to be constant at 57°. At impact, the droplet deforms and spreads radially up to ξ_{max}=2.7 (t =5 ms) and then recoils. Oscillations are found to stop about 20 ms after impact. The initial spreading phase (0 to 5 ms) will be used as reference for the comparisons (Figure 1) because it is in this initial phase that most of the effects are observed. Furthermore, the ensuing oscillations are, to a large extent, controlled by the physics of the contact line motion which is still the subject of intense research (Anderson and Davis, 1994; de Gennes *et al.*, 1990). The effect of the contact angle value on the droplet spreading behavior is evidenced by the large differences obtained in the predicted droplet edge location history for 57° and 100° (Figure 2). Pasandideh-Fard *et al.* (Pasandideh-Fard *et al.*, 1996) showed for this same case, that the contact angle varied between 40° and 110° during spreading.

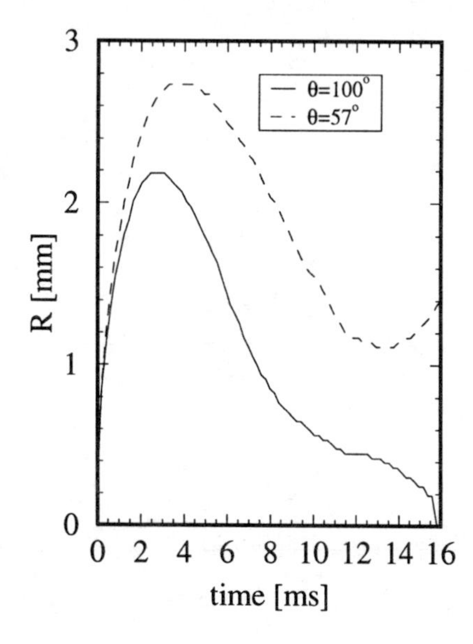

FIGURE 2: Water droplet spreading history. Effect of contact angle.

Influence of Droplet Deformation at Impact

Because of their relevance to actual droplet deformation, both spheroidal and semi-spheroidal morphologies are considered.

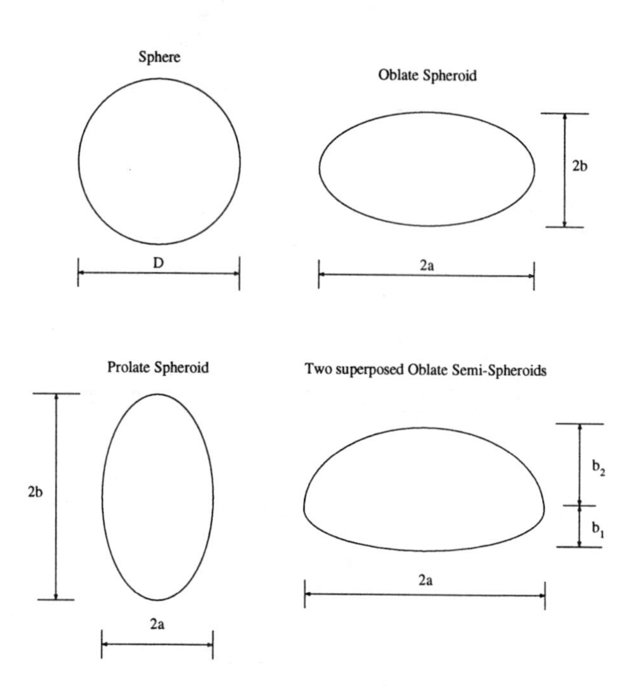

FIGURE 3: Typical droplet morphologies

The simplest type of deformation is the spheroidal geometry (Figure 3). For a given undeformed spherical droplet, diameter D, the value of the semi-axes a and b can be determined as functions of the ellipticity, e, from the following equations:

$$\frac{2b}{D} = \left(1 \mp e^2\right)^{1/3} \quad (6)$$

$$\frac{2a}{D} = \left(1 \mp e^2\right)^{1/6} \quad (7)$$

where the plus sign is used for prolate drops and the minus sign for oblate drops. Such morphologies exist in wake-induced shape oscillations (Clift *et al.*, 1978). The actual shape of the droplet at impact depends on the phase of these oscillation at the time of impact. In order to investigate this deformation regime, the impact of spheroidal drop on a flat substrate is considered. Various values of the ellipticity, from prolate to oblate, are considered to represent the range of possible shapes taken by the drop during an oscillation cycle.

Figure 4 compares the spreading history of an oblate droplet (e = 0.94) to the reference case (dotted lines). The droplet morphology expectedly departs significantly from that of the reference case; the axisymmetric side-jet (t=0.5 ms) is thinner, and the core of the spreading droplet remains flatter, almost linear during the spreading phase. Furthermore, the deformed droplet is found to experience a much shorter spreading/recoil cycle (1.5 ms). Figures 5 and 6 show how the spreading history is quantitatively influenced by the drop ellipticity. The initial

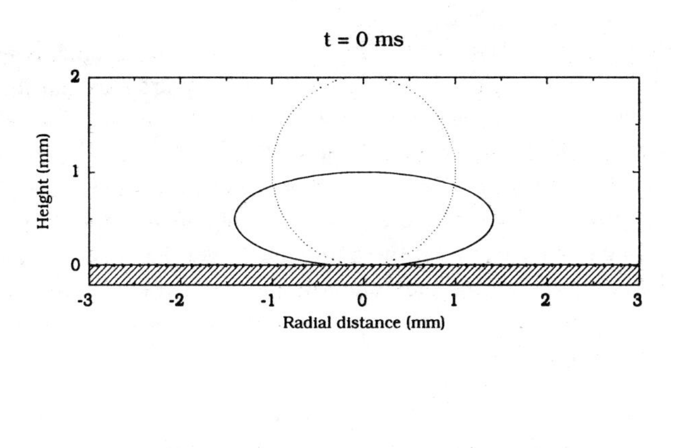

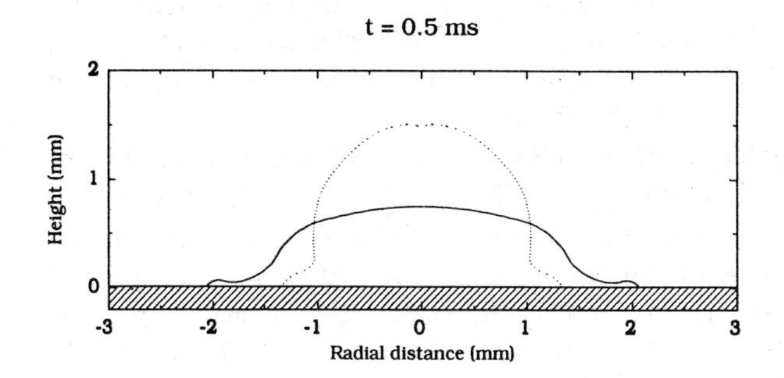

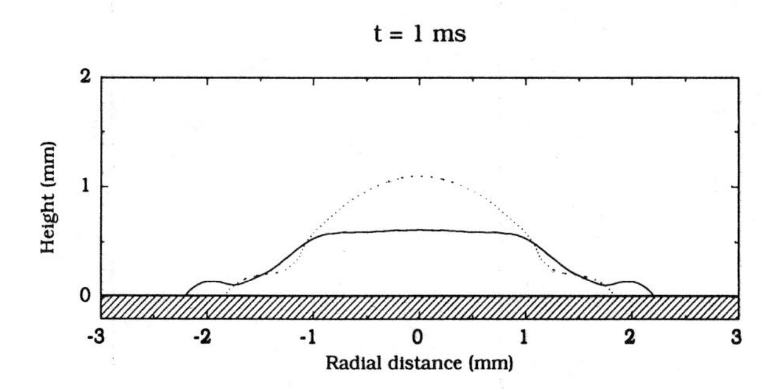

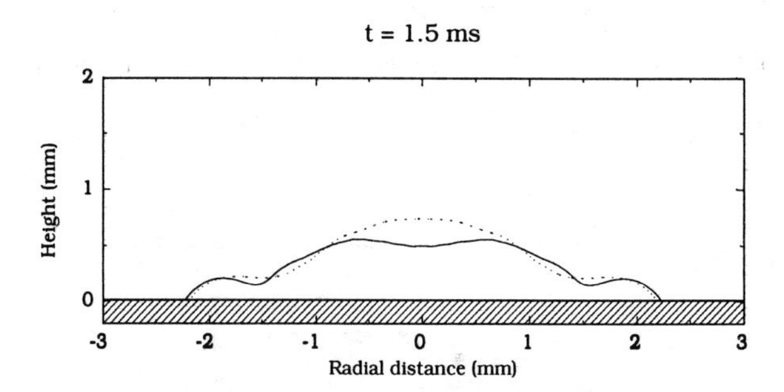

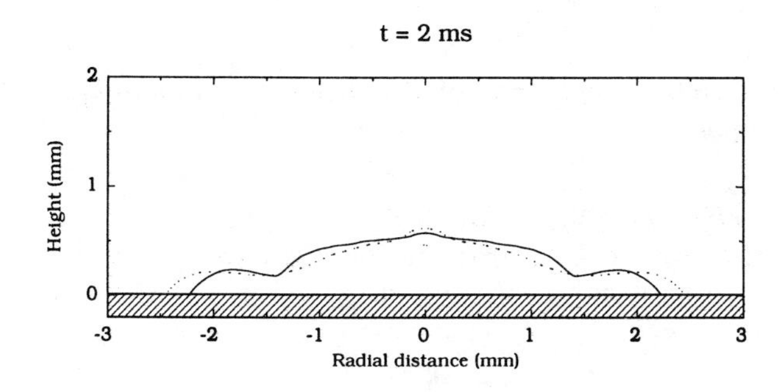

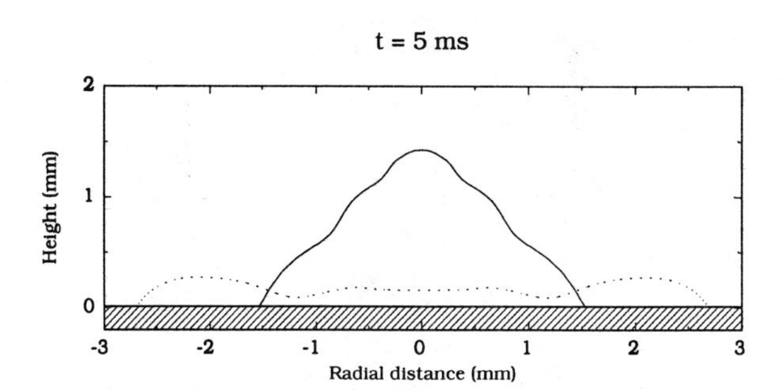

FIGURE 4: Effect of morphology at impact on water droplet deformation history. Oblate drop (e=0.94).

spreading rate is found to increase when the incident droplet morphology is changed from prolate (e=1) to spherical (e=0) to oblate ($e = 0.5$ and 0.94) while the maximum radius reached decreases from 2.9 mm to 2.2 mm. This behavior is attributed to the combined effects of the difference in initial droplet surface energy and the flatter shape of the droplet at the point of impact which, effectively, results in a larger surface area of the substrate being covered right after impact.

Slightly more complex shapes are observed in acceleration-induced deformations. Clift *et al.* (Clift *et al.*, 1978) indicate that the shape assumed by a liquid drop accelerating through air can be approximated by two superposed oblate semi-spheroids (Figure 3). The aspect ratio ($(b_1 + b_2)/2a$) and shape factor ($b_1/(b_1 + b_2)$) are related to the drop Eötvös (or Bond) number (Clift *et al.*, 1978):

$$\frac{b_1 + b_2}{2a} = \frac{1}{1 + 0.18(Eo - 0.4)^{0.8}}, \quad (0.4 < Eo < 8) \quad (8)$$

$$\frac{b_1}{b_1 + b_2} = \frac{0.5}{1 + 0.12(Eo - 0.5)^{0.8}}, \quad (0.5 < Eo < 8) \quad (9)$$

For values of Eo below these ranges, the aspect ratio and shape

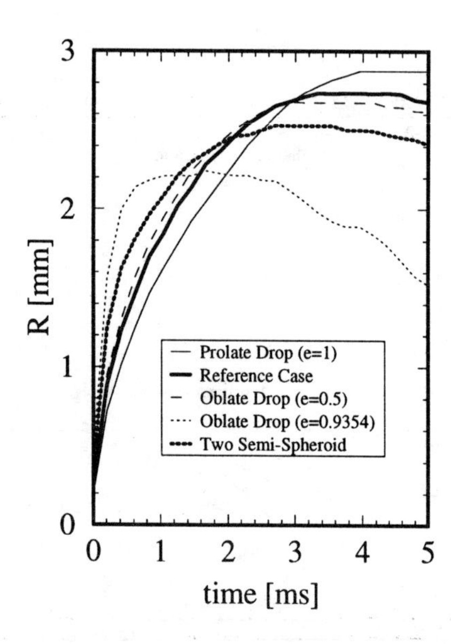

FIGURE 5: Water droplet spreading history. Effect of morphology at impact.

factor are equal to 1 and 1/2 respectively, corresponding to a sphere. For a given Eo and diameter of the equivalent spherical drop D, Equations 8 and 9 are solved together with the requirement that the deformed drop have the same volume than the equivalent spherical drop:

$$\frac{D}{a} = 2\left(\frac{b_1 + b_2}{2a}\right)^{1/3} \quad (10)$$

to yield a, b_1, and b_2. For instance, if the water droplet considered in the reference case is accelerated by drag so that its Eo number is 4 (corresponding roughly to a 20 m/s relative velocity) it will assume such a shape with a=1.15 mm, b_1=0.57 mm, and b_2=0.95 mm. The predicted behavior of this deformed drop is shown in Figure 7. Because of its oblateness, the droplet is found to have an initial spreading rate larger than that of a spherical droplet and the maximum radius reached by the droplet in the initial spreading phase is 8% smaller than in the reference case (Figure 6).

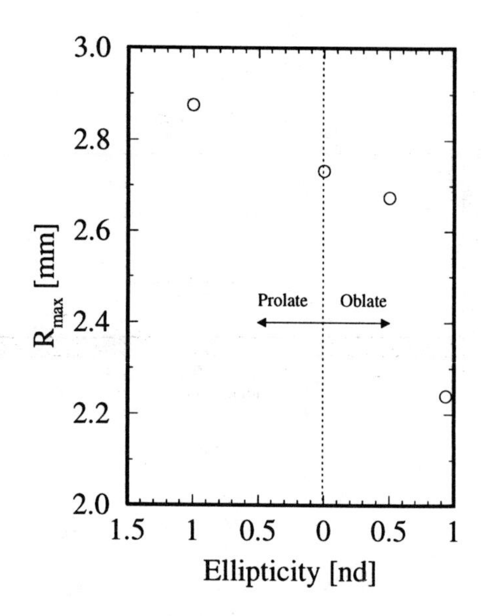

FIGURE 6: Influence of ellipticity on maximum spreading radius.

Influence of Internal Circulation

In order to investigate the potential effect of internal circulation on liquid droplet spreading the classical Hill's spherical vortex is considered.

$$\psi = \frac{1}{2}Ar^2\left\{R_0^2 - (r^2 + y^2)\right\} \quad (11)$$

where A is the vortex strength r is the distance measured form the axis of symmetry, and y the distance measured along the axis of symmetry. It is acknowledged that actual flow patterns, when they exist, are likely to be more complex, but this classical flow allows a simple evaluation and should provide an upper bound of the magnitude of the effects that can be induced by internal circulation since, for instance, surfactants significantly reduce the vortex strength.

Figure 8 compares the deformation after impact of an initially-spherical droplet with internal circulation ($i = 1.0$, where $i = A\mathcal{V}/R_0^2$ is the normalized vortex strength, non-dimensional) to the reference case. This plot shows that the droplet shape is significantly affected by internal circulation. The direct effect of the Hill's vortex on the velocity field is a non-zero radial velocity component at the droplet periphery, which does not exist in the unperturbed field. Consequently, the liquid height along the centerline remains larger than in the reference case (Figure 8) for most of the initial spreading phase (from 1 to 5 ms). For instance, at t=1.5 ms, the liquid height at the centerline is 70% larger than in the reference case. However, the effect of internal circulation on the spreading rate is not as marked. The maximum radius reached does not differ from the reference case value by more than 4% (Figure 9).

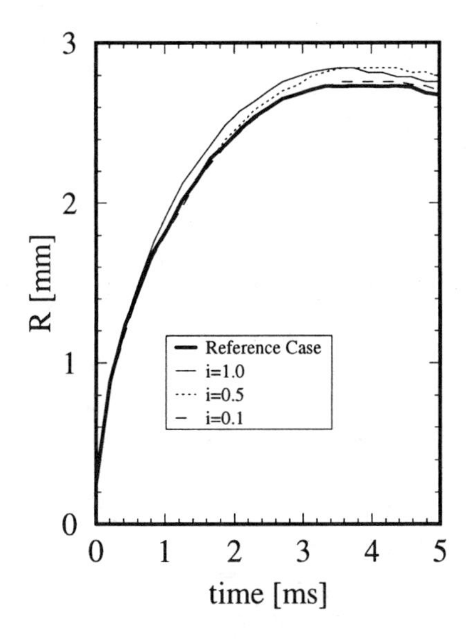

FIGURE 9: Water droplet spreading history. Effect of internal circulation.

Solidification

The case of a spherical liquid copper droplet (4.5 mm, 1531 K, 2.5 m/s) impinging on a cold (300 K) substrate is used as a reference to investigate how the conclusions drawn above are affected by solidification. This case is illustrated in Figure 10. The computations are stopped when there is less than 10% of the liquid left. As expected (Delplanque *et al.*, 1996b), the solidification proceeds by a succession of liquid-jet overflows. The solidification process is completed in less than 3 ms with a solid splat radius of 4.6 mm and height at the centerline of 0.9 mm. This spreading mechanism is propitious to the creation of pores however, in the case considered here, the pores created are smaller or just as large as the mesh size.

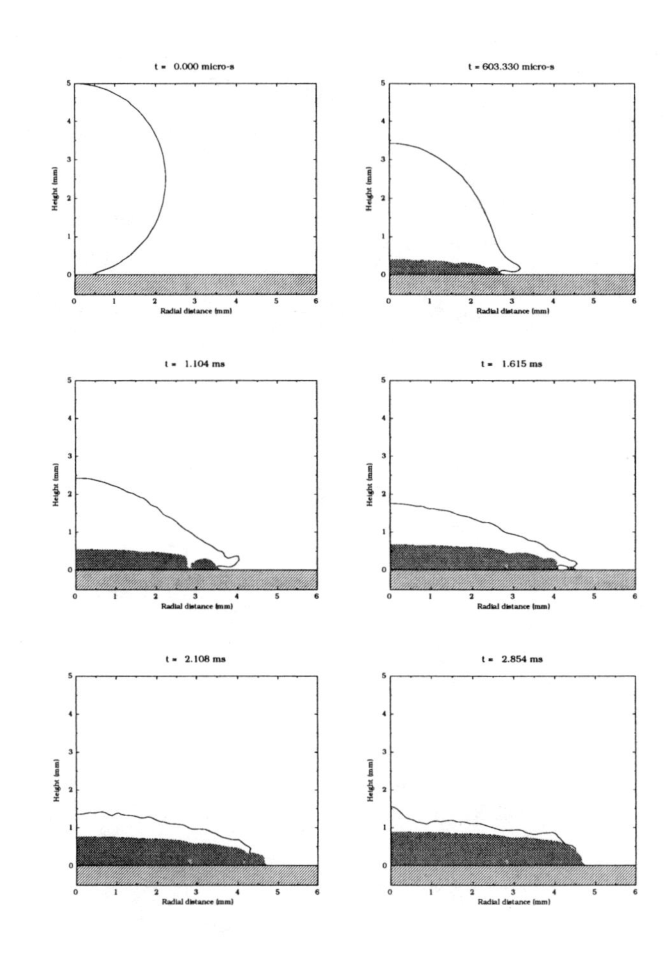

FIGURE 10: Spreading and solidification of a copper droplet. Reference case.

deformation (a=2.58 mm, b_1=1.29 mm, and b_2=2.14 mm, corresponding to $Eo = 4$) are considered next. As in the case without solidification, internal circulation and droplet deformation induce a noticeably different droplet morphology history. In both cases the final solid splat radius is slightly smaller than in the reference case, but the difference is too small (between 3 and 4%) to be significant. Therefore, the influence of droplet shape at impact on the final splat size is much less than that observed on the maximum spreading radius in the case without solidification. This is due to the fact that, in the case considered here, the final splat size is controlled by the solidification process.

The solid fraction history is basically unaffected by internal circulation (Figure 11), while droplet deformation causes solidification to start at a significantly faster rate because more of the liquid is initially in contact with the substrate. This results in a reduced solidification time (10% shorter). An interesting consequence of the flow field modifications induced by internal cir-

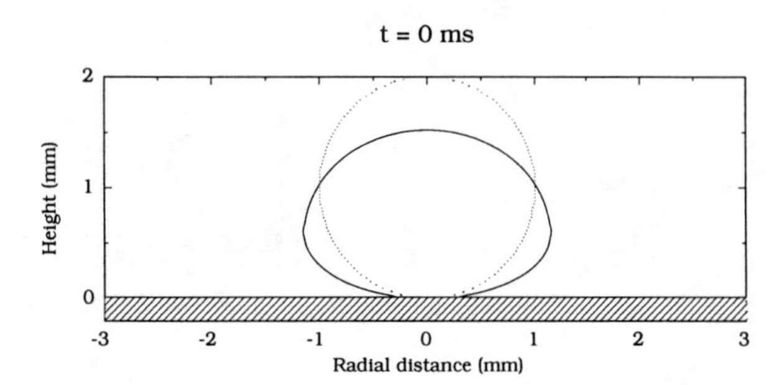

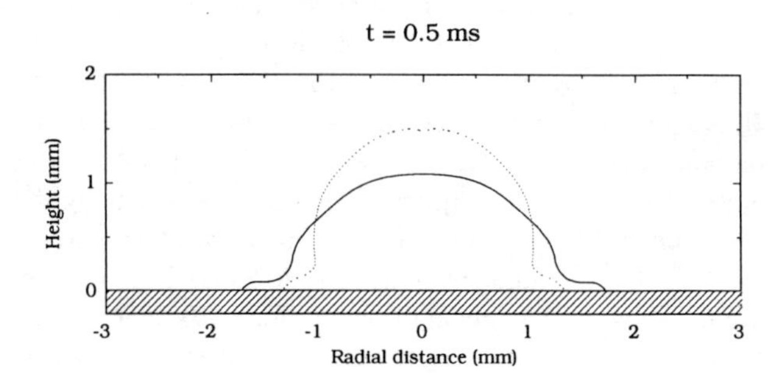

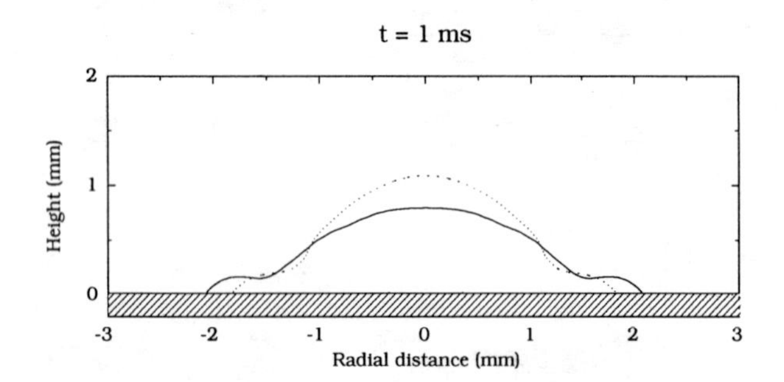

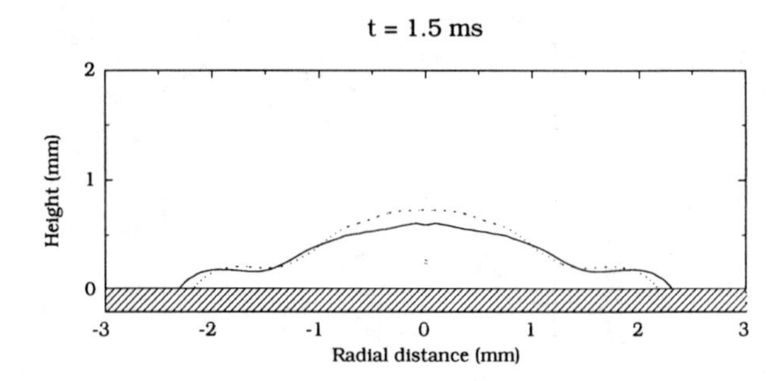

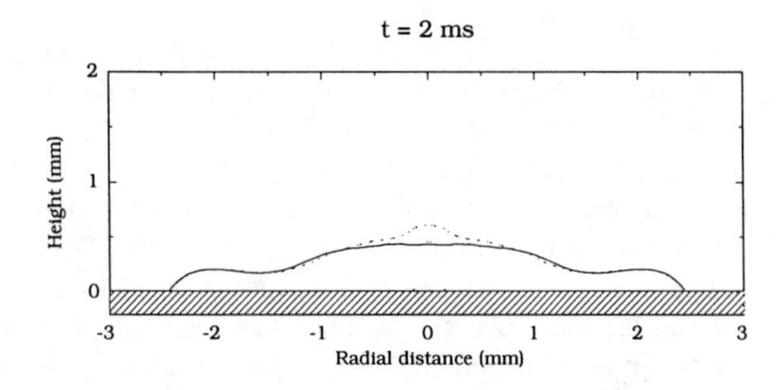

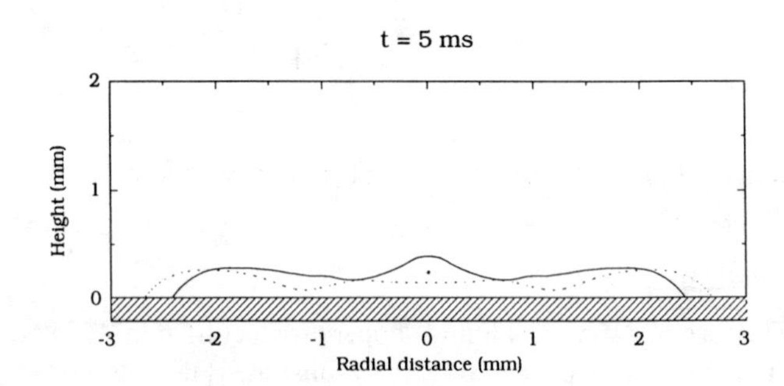

FIGURE 7: Effect of morphology at impact on water droplet deformation history. Two semi-spheroids.

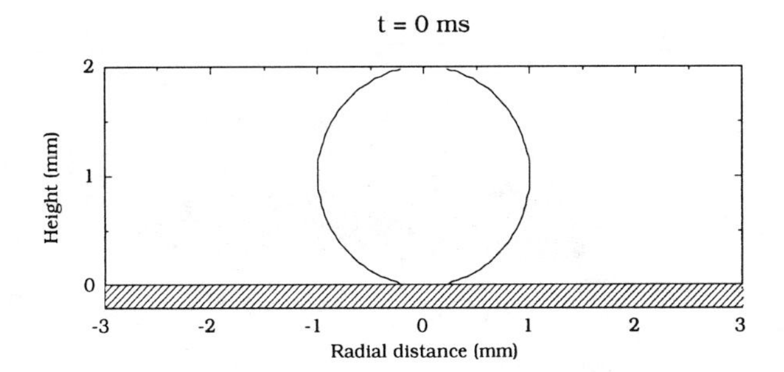

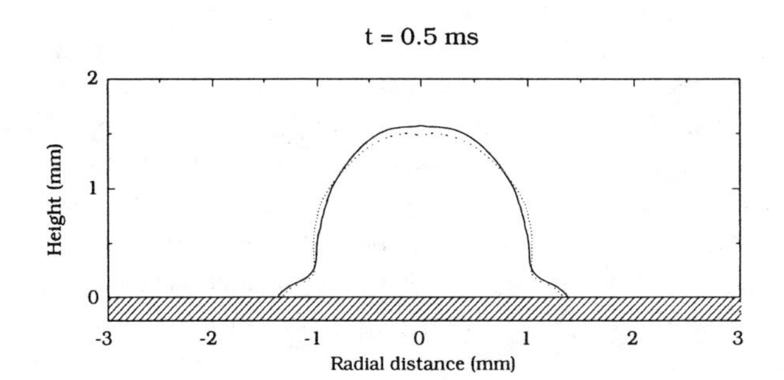

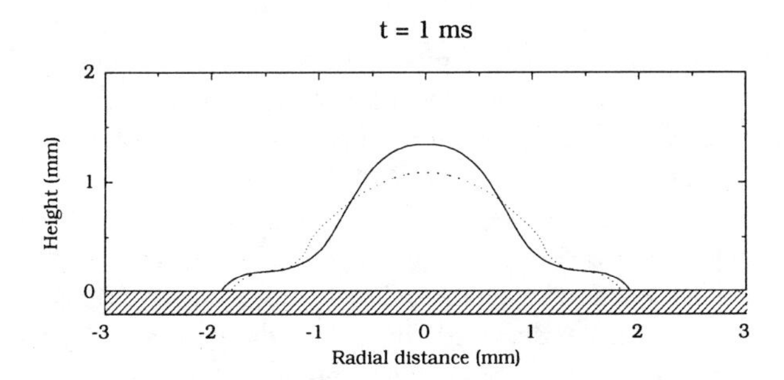

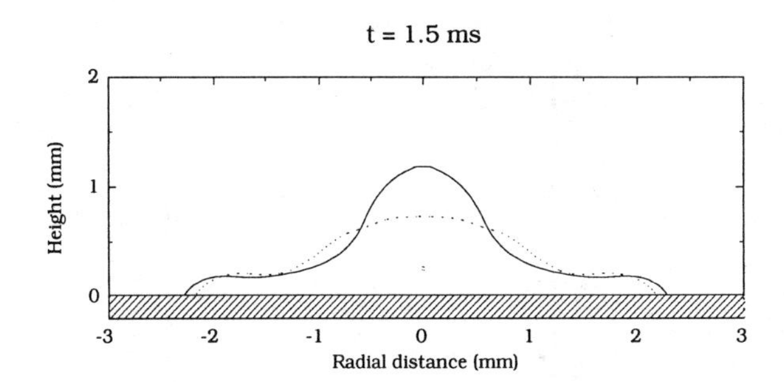

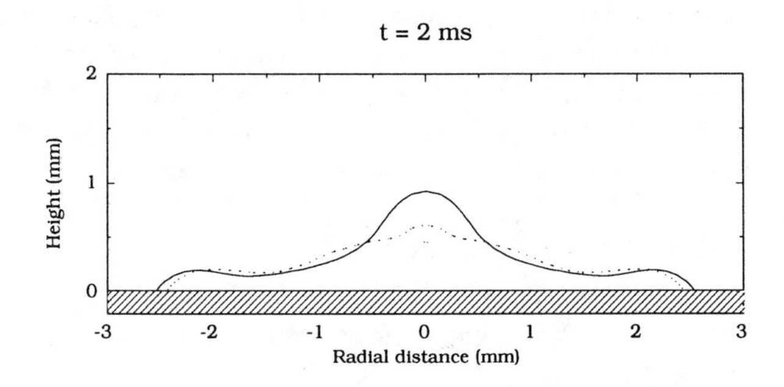

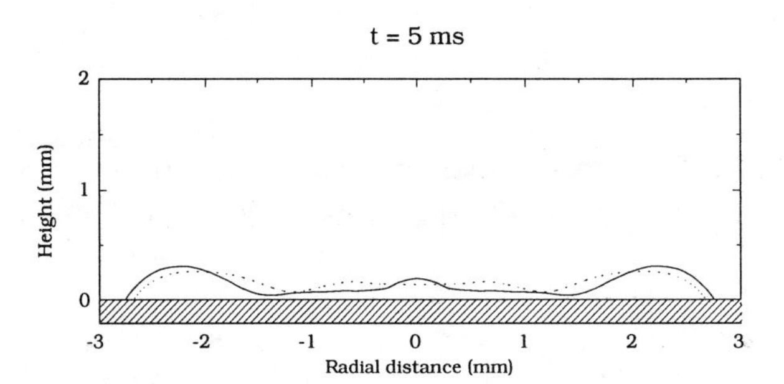

FIGURE 8: Effect of internal circulation on water droplet deformation history.

culation is the increased size of the pores generated by liquid-jet overflow (Figure 12). Liquid jet overflow has been previously reported (Delplanque *et al.*, 1996b) to enhance the non-uniformity of the deposited solid thickness thus yielding a rougher substrate for the subsequent drops which, in turn, promotes porosity formation in the substrate region (Liu *et al.*, 1995). The magnitude of this effect has been shown to depend on the Reynolds and Weber numbers at impact (Delplanque *et al.*, 1996b). The base case considered here shows no noticeable effect of liquid-jet overflow on the final splat morphology (Figure 10). Some voids are formed within the solidified layer, but they are not resolved properly, their size is below or comparable to that of the mesh (Figure 12). With internal circulation however, the final splat morphology reveals the presence of a pore that is better resolved (4 cells high by 2 cells wide, 200 μm × 100 μm). This phenomenon was found to be significant only for the largest vortex strength considered.

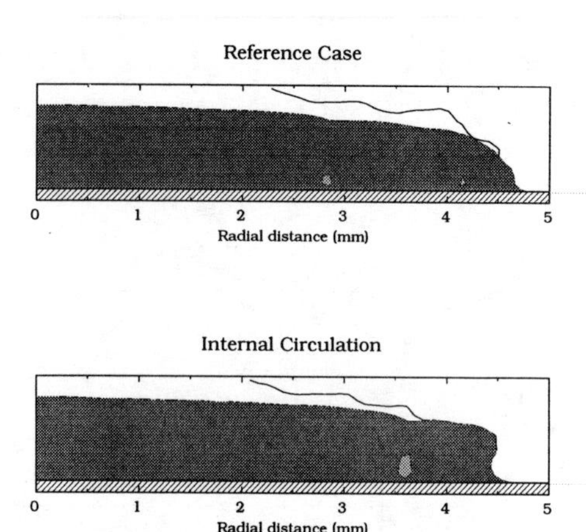

FIGURE 12: Effect of internal circulation and droplet morphology on the final splat morphology.

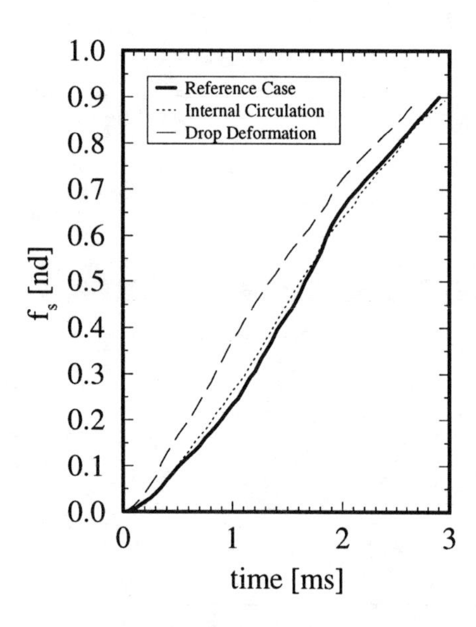

FIGURE 11: Effect of internal circulation and droplet morphology on the solid fraction of a solidifying copper splat.

CONCLUSIONS

The validity of standard assumptions, spherical shape and no internal flow, made in the simulation and modeling of droplet impact and spreading on a solid surface was evaluated.

It was found that axisymmetric deformations such as those created by shape oscillation or droplet acceleration affect the spreading characteristics substantially. Oblate droplets have a larger initial spreading rate and reach a smaller maximum spreading radius than spherical droplets (18% smaller for an oblate droplet with $e = 0.94$ and 7% smaller for an acceleration-deformed droplet with Eo=4). When solidification occur, this effect is reduced if the splat size is solidification controlled. The solidification time decreases for oblate droplets (about 10% for an acceleration-deformed droplet with Eo=4) because such deformations accelerate liquid/substrate contact.

The present study also shows that the effect of internal circulation on droplet spreading and solidification is negligible even when the influence on the internal velocity field of damping phenomena such as shape oscillations or the presence of surfactants is not included. A notable exception to this conclusion is pore formation. The results presented here indicate that for the largest internal vortex strength considered, the formation of pore larger than in the base case is promoted.

ACKNOWLEDGMENT

This work was sponsored in part by the National Science Foundation (DMI 95-28684), a Collaborative UC/Los Alamos Research (CULAR) grant, and Sulzer Metco Inc. The support of the University of California, Irvine through a Faculty Career Development award (1996-97) is gratefully acknowledged.

REFERENCES

Anderson, D. M., and Davis, S. H., 1994, "Local fluid and heat flow near contact lines," *Journal of Fluid Mechanics*, Vol. 268, pp. 231–265.

Bian, X., Delplanque, J.-P., and Rangel, R. H., 1997, "Droplet deposition: comparison of model and numerical-simulation predictions with experimental results.,", To be presented at the 1997 National Heat Transfer Conference, Baltimore, MD.

Clift, R., Grace, J. R., and Weber, M. E., 1978, *Bubbles, drops and particles*, Academic Press.

de Gennes, P. G., Hua, X., and Levinson, P., 1990, "Dynamics of wetting: local contact angles," *Journal of Fluid Mechanics*, Vol. 212, pp. 55–63.

Delplanque, J.-P., Lavernia, E. J., and Rangel, R. H., 1996a, "Multi-directional solidification model for the description of micro-pore formation in spray deposition processes," *Numerical Heat Transfer, Part A: Applications*, Vol. 30, pp. 1–18.

Delplanque, J.-P., Lavernia, E. J., and Rangel, R. H., 1996b, "Simulation of micro-pore formation in spray deposition processes,", HTD-Vol. 336/FED-Vol. 240. Presented at the 1996 ASME IMECE, Second International Symposium on Multiphase Flows and Heat Transfer in Materials Processing, Atlanta, Georgia.

Delplanque, J.-P., and Rangel, R., 1997, "An improved model for droplet solidification on a flat surface," *Journal of Materials Science*, Vol. 32, No. 6, pp. 1519–1530.

Fukai, J., Zhao, Z., Poulikakos, D., Megaridis, C. M., and Miyatake, O., 1993, "Modeling of the deformation of a liquid droplet impinging on a flat surface," *Physics of Fluids*, Vol. 5, No. 11, pp. 2588–2599.

Kothe, D. B., and Mjolsness, R. C., 1992, "RIPPLE – a new model for incompressible flows with free surfaces.," *AIAA Journal*, Vol. 30, No. 11, pp. 2694–2700.

Lavernia, E., and Wu, Y., 1996, *Spray Atomization and Deposition*, John Wiley & Sons, Inc., New York, N. Y.

Liu, H., Lavernia, E. J., and Rangel, R. H., 1993a, "Numerical simulation of impingement of molten Ti, Ni, and W droplet on a flat substrate," *Journal of Thermal Spray Technology*, Vol. 2, No. 4, pp. 369–378.

Liu, H., Lavernia, E. J., and Rangel, R. H., 1993b, "Numerical simulation of substrate impact and freezing of droplets in plasma spray processes," *Journal of Physics D: Applied Physics*, Vol. 26, pp. 1900–1908.

Liu, H., Lavernia, E. J., and Rangel, R. H., 1995, "Modeling of molten droplet impingement on a non-flat surface," *Acta Metallurgica et Materialia*, Vol. 43, No. 5, pp. 2053–2072.

Pasandideh-Fard, M., Qiao, Y. M., Chandra, S., and Mostaghimi, J., 1996, "Capillary effects during droplet impact on a solid surface," *Physics of Fluids*, Vol. 8, No. 3, pp. 650–658.

Rein, M., 1993, "Phenomena of liquid drop impac on solid and liquid surface," *Fluid Dynamics Research*, Vol. 12, pp. 61–93.

Renksizbulut, M., and Yuen, M. C., 1983, "Numerical study of droplet evaporation in a high-temperature stream," *Journal Heat Transfer*, Vol. 105, pp. 389–397.

San Marchi, C., Liu, H., Lavernia, E. J., and Rangel, R. H., 1993, "Numerical analysis of the deformation and solidification of a single droplet impinging onto a flat substrate," *Journal of Materials Science*, Vol. 28, pp. 3313–3321.

Trapaga, G., Matthys, E. F., Valencia, J. J., and Szekely, J., 1992, "Fluid flow, heat transfer, and solidification of molten metal droplets impinging on substrates — comparison of numerical and experimental results," *Metallurgical Transactions B-Process Metallurgy*, Vol. 23B, pp. 701–718.

Waldvogel, J. M., and Poulikakos, D., 1997, "Solidification phenomenain picoliter size solder droplet deposition on a composite substrate," *International Journal of Heat and Mass Transfer*, Vol. 40, No. 2, pp. 295–209.

HTD-Vol. 342, National Heat Transfer Conference
Volume 4
ASME 1997

MODELING OF TWO-PHASE (AIR-WATER) FLOW THROUGH A JET BUBBLE COLUMN

D. Mitra-Majumdar and B. Farouk
Department of Mechanical Engineering and Mechanics
Drexel University
Philadelphia, Pennsylvania

Y. T. Shah
College of Engineering
Drexel University
Philadelphia, Pennsylvania

ABSTRACT

Cocurrent upward air-water turbulent flow through a jet bubble column has been studied numerically using a mathematical model based on the fundamental governing equations of fluid motion. The predicted axial variations of the gas holdup are compared with experimental measurements. The results show that in all the cases, the use of a constant bubble diameter in the model fails to capture the trend of the experimental data. The predicted velocity fields of the two phases and the radial velocity profiles for the liquid phase are also presented in the paper. Large recirculation of the liquid phase is observed, while the gas phase shows no recirculation for the cases studied. These results are consistent with the experimental observations.

NOMENCLATURE

$\mathbf{B}''$	body forces
C_d	drag coefficient
C_f	interphase friction coefficient
d	bubble diameter
F	interfacial force
g	acceleration due to gravity
G_b	production of turbulent kinetic energy by body forces
H_{cyl}	height of the cylindrical portion of the jet bubble column
H_{cone}	height of the conical portion of the jet bubble column
D_{out}	diameter of the outlet of the jet bubble column
D_{in}	diameter of the inlet of the jet bubble column
k	turbulent kinetic energy
p	pressure
P	shear production of the turbulent kinetic energy
R	volume fraction
Re_l	local Reynolds number of flow
u	axial component of velocity
v	radial component of velocity
$\mathbf{V}_r$	relative velocity vector
U	superficial velocity
V	velocity vector

Subscripts

d	dispersed
eff	effective
G	gas phase
i	phase index
L	liquid phase

Superscript

int	interfacial

Greek Symbols

ε	dissipation rate of turbulent kinetic energy
Γ	momentum exchange coefficient
μ	dynamic viscosity
ρ	density

INTRODUCTION

A bubble column is a vertical, tubular vessel into which a mixture of liquid, solid and gas is generally injected concurrently at the bottom. Multiphase bubble column reactors are extensively used in the petrochemical and biochemical industries and in recent years, for more complex reaction systems. A jet bubble column, on the other hand, has an inlet conical section or cone attached to a cylindrical upper section and gas and liquid are fed concurrently into the bottom of the column. Attractiveness of these columns in the process industries is due to the fact that the jet is highly turbulent in these columns, effectively dispersing the gas phase throughout the column (Nishikawa et al. , 1976). Jet bubble columns are observed to be more power efficient than aerated mixing vessels and also vertical/horizontal pipes containing two-phase gas-liquid flows.

A majority of the past modeling of the bubble columns are based on empirical correlations based on experimental evidence (Deckwer, 1992; Fan, 1989; Smith et al., 1986; Joshi, 1983). Even though these correlations fit the data well, they are restricted to very narrow ranges of operating parameters. It has become increasingly more important to develop numerical models to predict the flow characteristics in such reactors. Moreover, the models, based on fundamental governing equations of fluid motion, provide a better insight to the overall flow structure and can be used more easily for scale-up studies. In recent times with the advent of powerful computers and with the better understanding of the multiphase flows, use of computational models based on the fundamental governing equations to predict the flow characteristics in bubble columns and other multiphase reactors is gaining momentum (Alajbegovic et al., 1996; Mitra-Majumdar et al., 1995; Lahey et al., 1993; Jakobssen et al., 1992; Turkoglu and Farouk, 1991; Torvik, 1990). These researchers have shown that such models can adequately predict the hydrodynamics and mixing in multiphase flow reactors. In the present study, we use a hydrodynamic model, for the prediction of the flow characteristics in a jet bubble column. Use of such models for the study of flow in a jet bubble column has not been reported in the literature.

A schematic of the problem of interest is shown in Figure 1. The height of the cone portion of the jet bubble column (H_{cone}) is 0.9 m, the height of the cylindrical portion of the column is column (H_{cyl}) is 0.63 m, the diameter of the inlet of the column (Din) is 0.025 m and the outlet diameter is 0.3 m. These dimensions are chosen such that direct comparison of the predictions can be made with experimental data (Borole et al., 1993). A large increase of the flow area is encountered by the flow. Gas and liquid are introduced at the bottom of the column. Air and water are used in this study. In the experimental study, water is allowed to escape through a hole in the side of the column, while the air escapes from the top of the column. In order to simulate this, a three-dimensional domain would have to be employed and the computations would be highly CPU intensive. In the numerical simulations an axisymmetric flow field is considered. The gas and liquid are allowed to escape together from the top of the column and a zero gradient boundary condition has been applied to the exit of the column (See Figure 1). The axisymmetric assumption will not affect the flow field in the cone area. The study attempts to numerically predict the flow field and phase distribution in the column. The objectives of the study are the following:

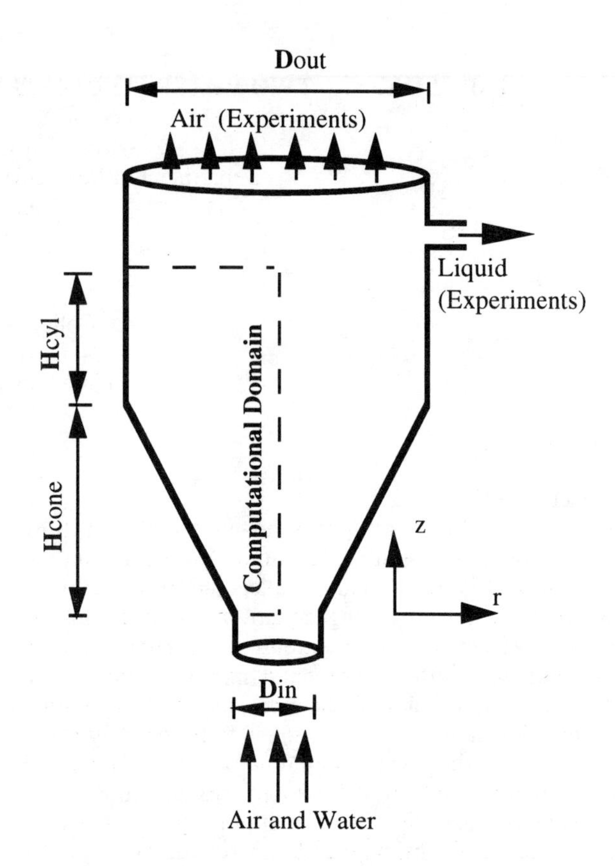

Figure 1. Schematic of the jet bubble column.

i) investigate the applicability of a computational fluid dynamic model, based on the fundamental governing equations of fluid motion to analyze gas-liquid flows through a jet bubble column.
ii) validate the models by comparing the model predictions with experimental data.
iii) improve the model to predict the experimentally observed trends.

Three different flow conditions were used for the model validation. The results of the axial variation of the gas volume fraction and the flow field are presented in the paper.

MATHEMATICAL MODEL

The mathematical formulation used in the proposed study is based on the Multi-fluid model approach. Gas and liquid phases are considered to be two distinct phases, interpenetrating into and interacting with each other. A k-ε turbulence model was used to include the effects of turbulence for the multiphase flows considered. Mass conservation equations, momentum equations and equations for the turbulence parameters, together with the equations for the inter-phase momentum transfer, were solved numerically to determine the flow structure in the jet bubble column. The Reynolds averaged field variables for each phase were weighted by the volume fraction of that phase.

The following assumptions were made in the derivations of the governing equations

a) The pressure in all phases are the same within a computational cell.

b) The flow is axisymmetric.

c) The effect of the gas phase turbulence on the flow field is negligible, due to its low density. The governing equations for the flows considered are presented below (Rosner, 1986).

Mass Conservation Equation

In the absence of phase change, the mass balance for phase 'i' with volume fraction R_i (where $R_L + R_G = 1.0$) yields the following :

$$\frac{\partial}{\partial t}(R_i\rho_i) + \mathrm{div}(R_i\rho_i\mathbf{V}_i) = 0 \qquad (1)$$

Momentum Equations

The momentum equation for each phase can be written as

$$\frac{\partial}{\partial t}(R_i\rho_i\mathbf{V}_i) + \mathrm{div}(R_i\rho_i\mathbf{V}_i\mathbf{V}_i) = \mathrm{div}(\pi) + R_i(\mathbf{B}'') + \mathbf{F}_i^{int} \qquad (2)$$

where $\mathbf{B}''$ is the body force term and $\mathbf{F}_i^{int}$ is the inter-phase momentum transfer term. The div(π) term in the above equation is the contact stress operator and consists of the pressure gradient and the viscous stress terms, Rosner (1986).

Equations for the Turbulence Parameters

Reynolds averaged forms of the above equations (1-2) are considered in the present study. Gas phase turbulence has been ignored in the present study duc to thcir small dcnsity. A k-ε model has been employed to include the effects of the turbulence for the liquid phase only. Following the work of Sato et al. (1981) the effect of the presence of the bubbles on the liquid phase turbulence structure is included in the model. This has been described in details in Mitra-Majumdar et al. (1995). The transport equation for the turbulent kinetic energy for a phase can be given by

$$\frac{\partial}{\partial t}(R_i\rho_i k_i) + \nabla.(R_i\rho_i\mathbf{V}_i k_i) - \nabla.((\mu_i + \frac{\mu_{t,i}}{\sigma_k})\nabla k_i) = R_i(P_i + G_{bi} - \rho_i\varepsilon_i) \qquad (3)$$

where P is the shear production of the turbulent kinetic energy and is given by

$$P_i = \mu_{eff,i}\,\nabla\mathbf{V}(\nabla\mathbf{V} + (\nabla\mathbf{V})^T) - \frac{2}{3}\nabla.\mathbf{V}(\mu_{eff,i}\,\nabla.\mathbf{V} + \rho_i k_i) \qquad (4)$$

and G_{bi} is the production rate of the turbulence energy by the body forces and is given by

$$G_{bi} = -\frac{\mu_{eff,i}}{\sigma_t\rho_i}\,g.\nabla\rho_i \qquad (5)$$

In the present case G_{bi} is zero, since the density of the phases are taken to be constant.

The transport equation for the dissipation rate of the turbulent kinetic energy (ε) is given by

$$\frac{\partial}{\partial t}(\rho_i\varepsilon_i) + \nabla.(\rho_i\mathbf{V}_i\varepsilon_i) - \nabla.((\mu_i + \frac{\mu_{t,i}}{\sigma_\varepsilon})\nabla\varepsilon_i) = C_1\frac{\varepsilon_i}{k_i}(P_i + C_3\max(G_{bi},0)) - C_2\rho_i\frac{\varepsilon_i^2}{k_i} \qquad (6)$$

where P_i and G_{bi} are defined earlier. In the k-ε model the turbulent viscosity was calculated as

$$\mu_{t,i} = C_\mu\rho_i\frac{k_i^2}{\varepsilon_i} \qquad (7)$$

In the model described above, standard values of the turbulence model constants have been used (Spalding, 1980; Clarke and Wilkes, 1989).

Interphase Momentum Exchange

The major component of the forces due to interactions between the phases is due to the drag caused by the slip between the phases. The inter-phase drag force, $\mathbf{F_{dr}}$ per unit volume between two phase can be written as,

$$\mathbf{F_{dr}} = C_f.\,\mathbf{V}_r \qquad (8)$$

where C_f is the inter-phase friction coefficient and $\mathbf{U}_r$ is the slip between the phases. C_f can be written as (Ishii and Zuber, 1979),

$$C_f = (0.75\ C_d \left| V_r \right| R_d \rho_c)/d \qquad (9)$$

where,

d = bubble diameter

R_d = dispersed phase volume fraction

C_d = local drag coefficient

Multi-phase flows through a reactor may show a large variation of phase concentrations over its flow domain. Also, there is a large variation in the local phase velocities. A constant drag coefficient is thus not suitable for the prediction of these flows. The local drag coefficient is influenced by factors which include the phase concentrations in a zone, the local slip velocity and the Reynolds number based on the bubble diameter. The local drag coefficient, C_d, for the bubble-liquid interaction is calculated in the model as a function of the local Reynolds number, Re_1, based on the slip velocity and the bubble diameter, Clift et al., (1978). As seen in equation (9), the bubble size plays a strong role in the interfacial drag. In our past studies, the bubble diameter was held constant for the entire flow domain. However, for the jet bubble column, use of a constant bubble size in the model was found to be inadequate for the prediction of the experimental trends. This issue is discussed in details in the Results section. Other modes of interactions between the phases, such as Magnus effect and the virtual mass effect, are also included in the model. These have been described in details in Mitra-Majumdar et al. (1995).

Boundary Conditions

Due to the symmetrical geometry, only one half of the flow domain has been used for the simulation of the flows (See Figure 1). The boundary conditions for the dependent variables used in the calculations are described below:

(1) At the inlet of the jet bubble column, the mass inflow rate of the phases are specified. Both the inlet velocities and the volume fractions have to be specified for the calculations. Inlet values of k and ε were also specified (Spalding, 1980).

(2) Along the walls, the velocities satisfy the no-slip boundary conditions.

$$v_G = u_G = v_L = u_L = 0 \qquad (10)$$

(3) Along the axis, the symmetry boundary conditions apply.

$$\frac{\partial u_G}{\partial r} = \frac{\partial u_L}{\partial r} = \frac{\partial k}{\partial r} = \frac{\partial \varepsilon}{\partial r} = 0 \qquad (11)$$

$$\frac{\partial R_G}{\partial r} = \frac{\partial R_L}{\partial r} = 0 \qquad (12)$$

$$v_G = u_G = v_L = u_L = 0 \qquad (13)$$

(4) At the top surface of the computational domain, i.e. at the outlet of the column, fixed pressure boundary conditions apply. The outlet pressure is set to atmospheric. The gradients of the dependent variables are set to be zero at the outlet. Flow disturbances due to bubble departure at the open surface has been neglected.

SOLUTION TECHNIQUE

For the numerical solution, the conservation equations are discretized using a finite volume technique. Fully transient form of the formulation has been considered. A fully implicit differencing scheme is applied for the transient formulation of the governing equations. Time steps of the order of hundredths of a second are used. The advection terms are discretized using the hybrid diferencing scheme and the diffusion terms are discretized using a second order accurate central differencing scheme. A non-staggered grid version of the IPSA (Inter-Phase-Slip-Algorithm, Spalding 1980) has been used to solve for the variables in model, where the convection coefficients are calculated using the Rhie-Chow interpolation scheme (1982).

The computations are initiated with initial guesses for the flow variables and the calculations are terminated when the solutions become steady in a "computational sense". The overall mass source residual was used as a measure of the convergence It is defined as the sum of the absolute values of the net mass fluxes in every computational cell. For a given timestep, the iterations are continued until the overall mass source residual becomes less than 0.001. The CPU time requirement (in an IBM Model 360 workstation) of these calculations are rather large due to the very small time steps used in the calculations. Small time steps in the range of 0.01 - 0.02 seconds were used in the calculations. For all the results presented in this paper, a grid size of 60 x 40 (z x r) was used. Grid independency studies were carried out. Sufficient grid stretching was done in order to resolve the boundary layer completely.

RESULTS

To study the flow characteristics in the jet bubble column, three different experimental flow conditions were chosen. These experimental conditions, Borole et al., (1993), encompass a broad range of flow conditions. They are summarized in Table 1. The superficial velocities, defined as the flowrate per unit cross-section of the cylinder area, are shown in Table 1.

Table 1. Flow conditions used in the numerical simulation of air-water flow through a jet bubble column, Borole et al. (1993).

Case Number	Bubble Diameters*	Gas Superficial Velocity (U_G), m/s	Liquid Superficial Velocity (U_L), m/s
IA	A	0.03	0.006
IB	B	0.06	0.006
IC	C	0.09	0.006
IIA	A	0.03	0.006
IIB	B	0.06	0.006
IIC	C	0.09	0.006
IIIA	A	0.03	0.006
IIIB	B	0.06	0.006
IIIC	C	0.09	0.006

* Discussed in details in following sections

In all cases, for the numerical simulations, the inlet volume fraction of the gas phase was assigned a value of 0.6 and the liquid volume fraction was 0.4. These values were chosen based on our earlier experience in modeling gas-liquid flows through a bubble column. The inlet velocities of both phases were then calculated from the knowledge of the phase superficial velocities and these volume fractions. This is necessary since both the inlet velocities and volume fractions of the two phase need to be specified in the model boundary conditions.

It was found that the use of a constant bubble diameter for the calculations of the interfacial forces was not adequate for the prediction of the characteristics of these flows. This is due to the fact that the bubble sizes in gas-liquid flows through a jet bubble column do not remain constant over the flow domain. The gas jet comes in through the inlet of the column as big bubbles with high velocity. In the later portion of the column, the gas bubbles faces increasing resistance to its motion and slows down due to the combined effect of the liquid phase resistance and the increasing flow area. The large bubbles of the gas break into large numbers of small bubbles. This phenomenon of bubble breakup in the later portion of the jet bubble column has been noticed in experimental studies [Wisecarver, 1996]. This results in effective dispersion of the gas in the flow field and higher mixing. In order to include this physics into the present model, variable bubble size distribution was attempted (equation 9). In the present study, three different bubble size distributions were used in the calculations of the interfacial forces in the model.

Cases A :	$d = 0.005$ m	(constant bubble diameter)
Cases B :	$d = 0.010$ m	$0 < H < 0.2$ m
	$d = 0.008$ m	0.2 m $< H < 0.7$ m
	$d = 0.005$ m	0.7 m $< H$
Cases C:	$d = 0.012$ m	$0 < H < 0.2$ m
	$d = 0.010$ m	0.2 m $< H < 0.4$ m
	$d = 0.007$ m	0.4 m $< H < 0.6$ m
	$d = 0.005$ m	

In cases IA, IIA and IIIA, the bubble diameter was held constant over the whole flow domain. In cases IB, IIB and IIIB, the bubble diameter near the inlet was kept large (0.1 m) and it was decreased to 5 mm in the later portion of the column. In cases IC, IIC and IIIC, the bubble diameter was decreased even more drastically from 0.12 m at inlet to 0.005 towards the end of the cone. Thus, cases IC, IIC and IIIC represents the phenomenon of strongest bubble breakup. As will be shown, the use of a variable bubble diameter in the model brings about a large change in the predictions.

In order to validate the model predictions, the predicted values of the axial variation of the radial averaged gas volume fractions were compared with the experimental data, Borole et al. (1993). The results for the axial variation of the radially averaged gas volume fraction for cases IA, IB and IC are shown in Figure 2. The experimental data shows that

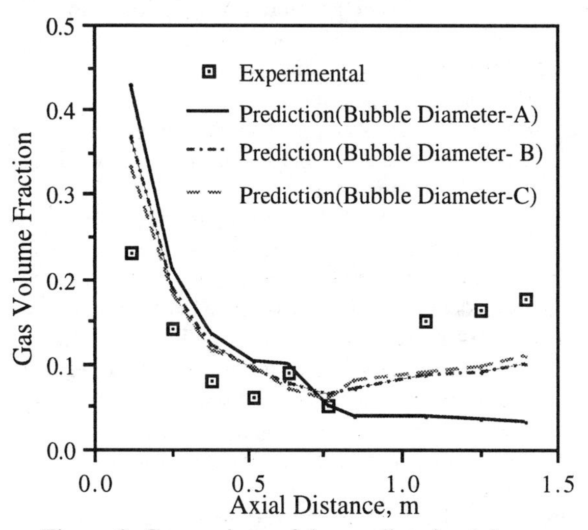

Figure 2. Comparison of the predicted axial variation of the gas volume fraction with the experimental data, Case I.

the gas volume fraction decreases in the initial portion of the column due to the acceleration of the bubbles aided by buoyancy. Then the gas phase volume fraction increases in the later portion of the cone and in the cylindrical portion of the jet bubble column. One of the reasons for this increase could be attributed to the fact that the area increase in the flow domain is large and this causes a deceleration of the phases. However, as can be seen in the predictions, the rise in the gas volume fraction is mainly due to the fact that the bubble breakup happens in the later portion of the conical portion of the jet bubble column and the bigger bubbles break into large number of smaller bubbles. This breakup causes an increase in the effective interfacial area between the gas and liquid phase for the calculation of the drag force. Moreover, it is well known that in gas-liquid flows, smaller bubbles move at lesser speed than the larger bubbles. These effects causes the gas phase to slow down and an increase in the gas volume fraction in the later portion of the jet bubble column. As can be seen from the figure, use of a constant bubble diameter fails to predict this trend. Use of both sets of variable diameter distributions in the model predict this trend. Case IC gives a better fit to the experimental data than Case IB. This indicates that there is indeed a strong phenomenon of bubble breakup in the column. In the initial portion of the column, however, the predicted values of the gas volume fraction is higher than the experimental data. This is observed with all the three bubble diameter distributions used in the model. The large acceleration of the gas phase in this portion of the column, as observed in the experimental data, is not captured by the model. Moreover, the data shows that at the end of the column, the gas volume fraction is still increasing. The model predictions shows a plateau in this region. Probable reason for this could be attributed to the fact that in the experimental study, the gas phase escapes through the top of the column and the liquid phase is allowed to escape only through a hole at the side of the column (See Figure 1). No gas-liquid disengaging section was present in the experimental setup. In the numerical simulations the gas and the liquid are allowed to bubble out through the top of the computational domain and a zero gradient boundary condition was applied to the top of the computational domain. This could lead to the discrepancies between the predicted values of the gas volume fractions and the experimental data. In order to resolve this discrepancy, a three-dimensional model would have to be applied and would be highly CPU-intensive.

The results for the axial variation of the radially averaged gas volume fraction for Cases IIA, IIB and IIC are shown in Figure 3 and that for Cases IIIA, IIB and IIIC are shown in Figure 4. Again the gas volume fraction first decreases in the initial portion of the column and then increases. The comparisons of the model predictions similar to that observed for Case I. Again, the use of a constant bubble diameter was not adequate to capture this trend. The use of variable diameter in the model gave a better fit to the experimental data than that seen with the use of a constant bubble diameter in the model. As shown in table 1, the liquid flowrate is the same in all the three experimental conditions studied in the paper, while the gas flowrate is different. Case IIIA, IIB and IIC have the highest gas flowrate amongst these cases. It was observed experimentally that as the gas flow rate is increased, the volume fraction of the gas phase in the column increases.

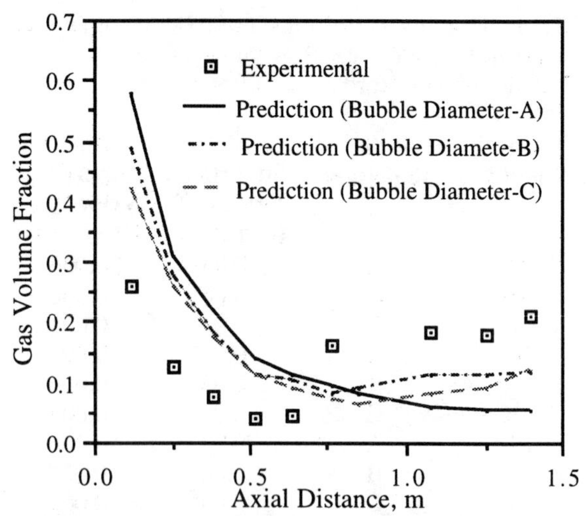

Figure 3. Comparison of the predicted axial variation of the gas volume fraction with the experimental data, Case II.

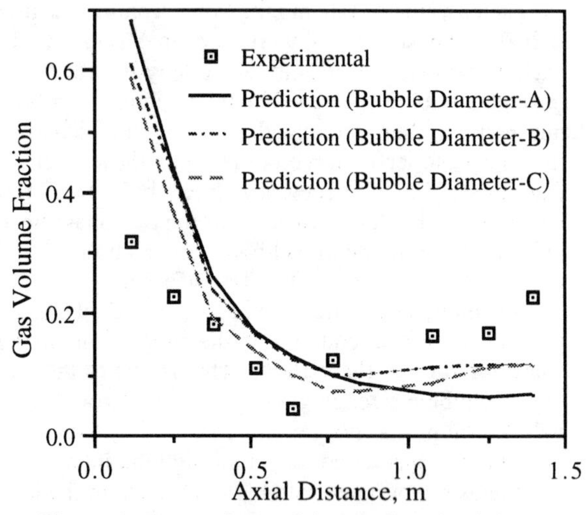

Figure 4. Comparison of the predicted axial variation of the gas volume fraction with the experimental data, Case III.

Comparison of the Figures 2-4 shows that this is predicted by the model. Comparison of the model predictions with other parameters of the flow (such as the velocities of the liquid phase) would help in the further development of the model.

All the results for the flow field and the velocities of the phases presented next are from the simulations using the variable bubble diameter distribution Case C. The velocity vectors for the gas phase and the liquid phase for Case IC are shown in Figure 5. In the initial portion of the column, i.e. near the inlet of the cone, the phases move solely upwards. This is probably due to the fact that the cone inlet diameter is small and the phases, principally the gas phase, come in at large velocity. At a later portion of the column, a recirculation zone appears for the liquid

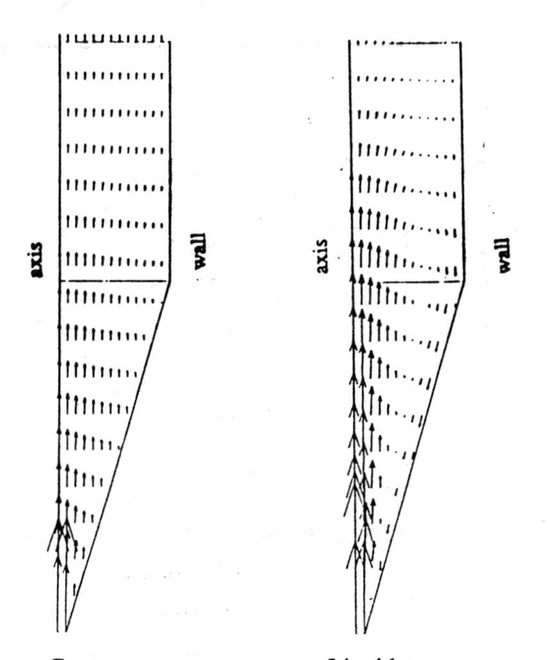

Gas (Scale : 1 cm = 6.4 m/s) Liquid (Scale: 1 cm = 2.4m/s)

Figure 5. Velocity vectors for Case I (Bubble Diameter- Case C).

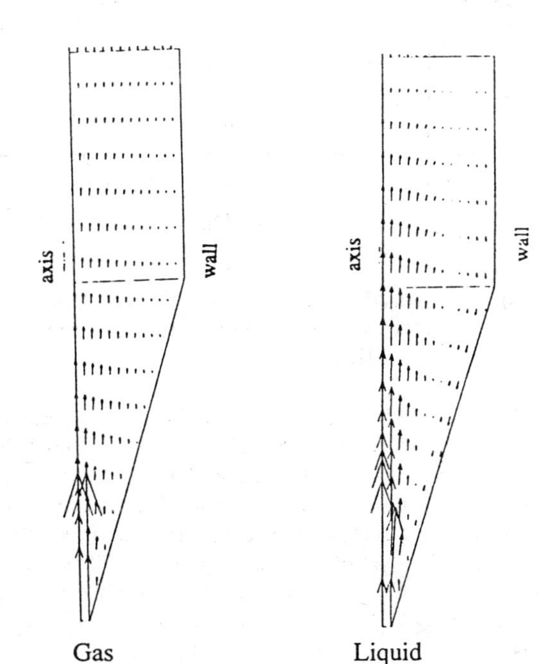

Gas (Scale : 1 cm = 10.0 m/s) Liquid (Scale: 1 cm =3.3 m/s)

Figure 6. Velocity vectors for Case II (Bubble Diameter- Case C).

phase. This is due to effect of the buoyancy. Coupled with the large density variations between the phases, the effect of gravity is to accelerate the gas phase. Moreover the gas phase concentrates towards the center of the column. This causes a large variation of the velocity of both the gas and the liquid phase from the axis of the column to the wall of the column. This leads to the appearance of the recirculation zone for the liquid phase. Experimental observations also showed that the motion of the phases are upwards near the column inlet and recirculation appears at a later portion of the column. The velocity vectors for the gas and liquid phases for Case IIC is shown in Figure 6 and that for the Case IIIC is shown in Figure 7. The trend of the velocity fields are the same in all the three cases. However, the velocity of the phases increases as the gas flowrate is increased.

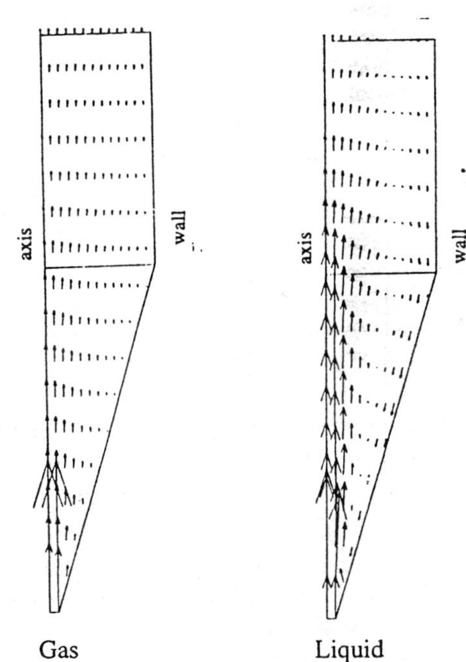

Gas (Scale : 1 cm = 14 m/s) Liquid (Scale: 1 cm = 4.7 m/s)

Figure 7. Velocity vectors for Case III (Bubble Diameter- Case C).

The radial variation of the axial component of the velocity of the phases at 0.5 m from the inlet of the cone and at 0.75 m from the inlet of the cone, for Case IC, are presented in Figure 8(a) and 8(b) respectively. As described in the earlier section, the gas phase velocity is solely upwards while the liquid phase shows recirculation. The velocity of both the phases, however, decreases from the column center towards the wall. The magnitude of the velocities are

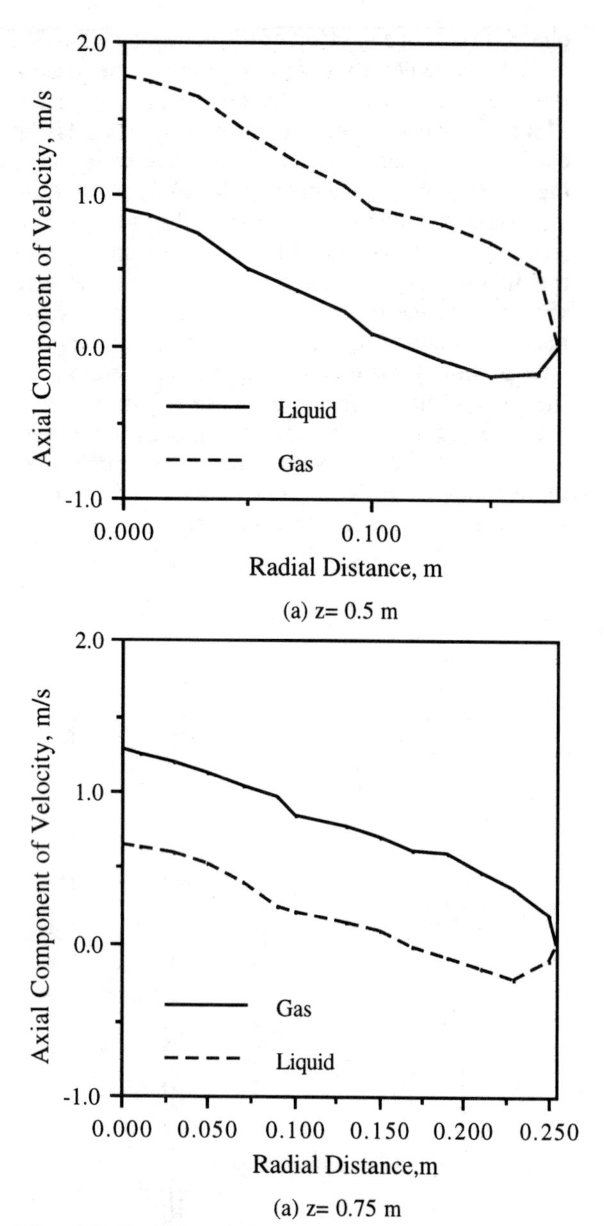

(a) z= 0.5 m

(a) z= 0.75 m

Figure 8. Radial variation of the velocity of the phases at two different axial locations, Case IC.

smaller at 0.75 m as compared to that at 0.5 m. This is due to the increased area of cross-section of the column and due to the greater effect of the drag between the phases. The effect of the bubble breakup, which increases the interfacial drag, also plays a major role in the slowing down of the motion of the phases. This, in turn, increases the void fraction. Also, the zone of recirculation is larger at the higher elevations. However, magnitude of the radial variation of the velocity is decreased. The radial variation of the axial component of the velocity of the phases at 0.5 m from the inlet of the cone and at 0.75 m from the inlet of the cone, for Case IIC, are presented in Figure 9(a) and 9(b) respectively and

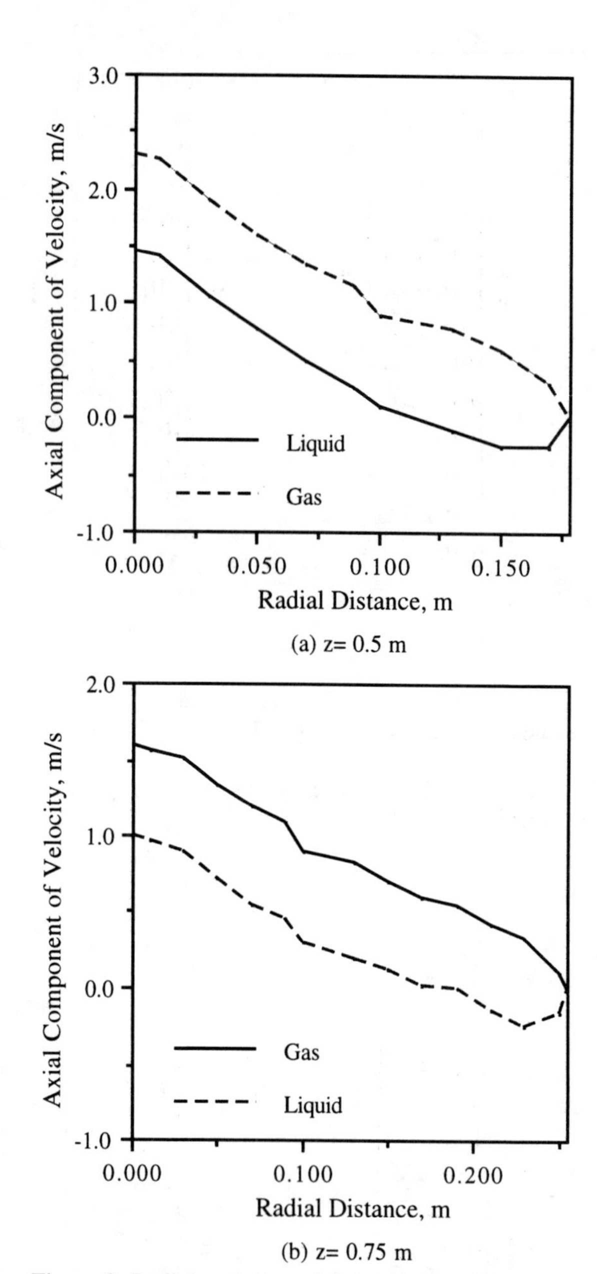

(a) z= 0.5 m

(b) z= 0.75 m

Figure 9. Radial variation of the velocity of the phases at two different axial locations, Case IIC.

those for Case IIIC, are presented in Figure 10(a) and 10(b) respectively. Similar trend of the radial variation of the velocity of the phases, as observed in Case I, is seen in these cases also. The velocity of the phases, however, increase as the gas flow rate is increased. Comparison of Figures 8 -10, it can be seen that at the same axial distance from the inlet, the magnitude of the recirculation is stronger at a higher gas flow rates. This is due to the fact that at higher flowrates the gas phase tends to concentrate more towards the axis of the column and the motion of the gas phase in the radial direction is reduced. This leads

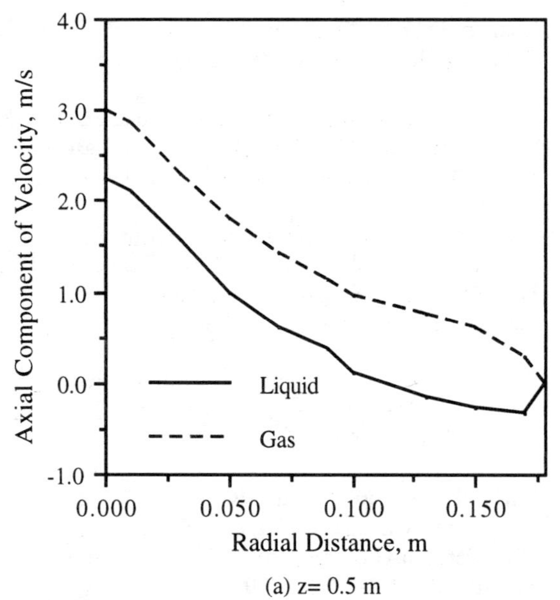

(a) z= 0.5 m

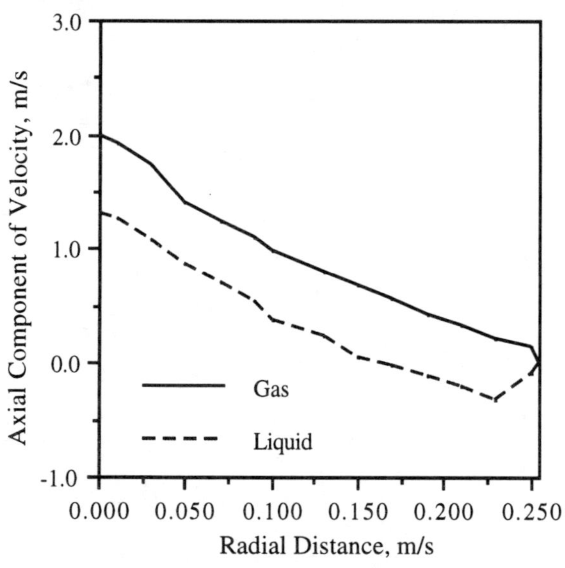

(b) z= 0.75 m

Figure 10 Radial variation of the velocity of the phases at two different axial locations, Case IIIC.

to larger radial variation of the velocities of the phases and larger recirculation of the liquid phase.

Overall, the present model predicts the experimental data and observations well. Incorporation of the phenomenon of bubble breakup in the simulation of these flows was found to be necessary for a good fit between the model predictions and the experimental data.

CONCLUSIONS

A computational fluid dynamic (CFD) model based on the fundamental governing principles of fluid motion has been applied to study the air-water flow through a jet bubble column. The experimental data for the axial variation of the volume fraction of the gas phase were used for the validation of the model. The void fraction first decreases in the cone, due to the acceleration of the gas phase, aided by buoyancy. Then there is an increase in the void fraction in the latter portion of the cone and in the cylindrical portion of the jet bubble column. This trend is predicted by the model, after the incorporation of the variable bubble diameter in the calculations of the interfacial momentum transfer. The gas phase shows no recirculation. The motion of the liquid phase is solely upwards in the initial portion of the column. A large recirculation zone appears for the liquid phase in the latter portion of the column. This trend is consistent with the experimental observations. The model predictions would be compared with the experimental data for experimental parameters other than the gas volume fraction, as and when they become available.

REFERENCES

Alajbegovic, A., Kurul, N., Podowski, M. Z., Drew, D. A. and Lahey, R. T., Jr., "A CFD Analysis of Multidimensional Phenomenon in Two-Phase Flow Using a Two-Fluid Model", Proceedings of the 1996 National Heat Transfer Conference, 387-394, (1996)

Borole, A., Joshi, J. B. and Wisecarver, K., "Hydrodynamics and Mixing in a Two-Phase Jet Bubble Column", Chem. Eng. Comm., Vol. 126, 189-199, (1993)

Clarke, D.S. and Wilkes, N.S., 1989, "The Calculation of Turbulent Flows in Complex Geometries Using a Differential Stress Model", AERE-R 13428, (1989)

Clift, J. , Grace, J. R. and Weber, M. E., "Bubbles, Particles and Drops", Academic Press, (1978)

Deckwer, W. D., "Bubble Column Reactors", Wiley Publishing Co., (1992)

Fan, L, "Gas-Liquid-Solid Fluidization Engineering", Butterworths, (1989)

Ishii, M. and Zuber, N., "Drag Coefficient and Relative Velocity in Bubbly, Droplet and Particulate Flows", AIChE J., 25, 843-854, (1979)

Jakobssen, H. A., Svendsen, H. F. and Hjarbo, K. W., "On the Prediction of Local Flow structure in Internal Loop and Bubble Column Reactors Using a Two-Fluid Model", Int. Conf. on Gas-Liquid Solid Reactor

Design, Ohio State Univ., Columbus, Ohio, USA, (1992)

Lahey, R. T., Jr., Bertadano, M. L. and Jones, O. C., Jr., " Phase Distribution in Complex Geometry Conduits", Nuclear Eng. & Des., Vol. 141, 177-201, (1993)

Mitra-Majumdar, D., Farouk, B. and Shah, Y. T., "Transport of Gas-Liquid Flows Through Vertical Columns", Chem. Eng. Comm., Vol. 137, 191-209, (1995)

Nishikawa, M., Yonezawa,Y., Kyama, T., Koyama, K. and Nagata, S., "Studies on Gas Holdup in Gas-Liquid Sparged Vessel", J. Chem. Eng. Japan, Vol 9, 214, (1976)

Rhie, C. M. and Chow, W. L., "Numerical Study of the Turbulent Flow Past an Airfoil with Trailing Edge Separation", AIAA/ASME Third Joint Thermophysics, Fluids, Plasma and Heat Transfer Conference, St. Louis, MO, (1982)

Rosner, D. E., "Transport Processes in Chemically Reacting Flow Field", Butterworths, (1986)

Sato Y., Sadatomi, M. and Sekoguchi, K., "Momentum and Heat Transfer in Two-Phase Bubble Flow", Int. J. Mult. Flow, 7, 167-177, (1981)

Spalding, D.B., "Numerical Computation of Multiphase Fluid Flow and Heat Transfer", EPRI WS-78-143, 2, Electric Power Research Institute, Palo Alto, California, 219-275, (1980)

Torvik, R. and Svendsen, H.F., "Modeling of Slurry Reactors - A Fundamental Approach", Chem. Eng. Science, 45, No. 8, 2325-2332, (1990)

Turkoglu, H. and Farouk, B. , 1991, "Mixing Time and Liquid Circulation Rate in Steelmaking Ladles with Vertical Gas Injection", ISIJ International, Vol. 31, No. 12, 1371-1380, (1991)

HTD-Vol. 342, National Heat Transfer Conference
Volume 4
ASME 1997

BUOYANCY-DRIVEN INTERACTIONS OF BUBBLES IN A SUPERHEATED LIQUID

Asghar Esmaeeli
Department of Mechanical Engineering and Applied Mechanics
The University of Michigan
Ann Arbor, Michigan

Vedat Arpaci
Department of Mechanical Engineering and Applied Mechanics
The University of Michigan
Ann Arbor, Michigan

An-Ti Chai
NASA Lewis Research Center
Cleveland, Ohio

ABSTRACT

Dynamics of two-dimensional buoyant, deformable, and interacting bubbles in a superheated liquid is investigated. The full Navier-Stokes equations and energy equation are solved for the fluid inside and outside the bubbles by using a front tracking/finite difference technique. Simulations of two tandem bubbles with two different initial positions show that the shapes of the bubbles are a strong function of their separation distance. The interactions of two system of twelve bubbles under two different density ratios are studied and the dependency of the results on this parameter is investigated.

INTRODUCTION

Bubbles are essential particles in many industrial as well as natural processes. Heat transfer through boiling, fluid cavitation, bubble driven circulation systems in metal processing, and bubble/free surface interactions in oceans are just a few examples of the many roles that bubbles play in the physical systems.

One of the important problems in bubble dynamics is the prediction of vapor bubble growth and the resulting effects on the fluid flow and heat transfer of the system. Bubbles can be found in subcooled as well as superheated liquid. In the former case (i.e. the cavitation bubble), the motion is not affected by the latent heat flow and is controlled by the liquid inertia. In the superheated liquid, the bubble dynamics is controlled by the latent heat flow. The growth of a vapor bubble in a superheated liquid is controlled by three factors: liquid inertia, surface tension, and heat diffusion. These factors determine what is called "the critical radius" which is the minimum radius that a nucleus should start with in order to grow. There are several stages in the bubble growth. At the first stage, when the radius is very small and subsequently the curvature is high, the growth rate is very slow as a result of the straining effect of the surface tension. Due to large vapor inflow, a substantial cooling occurs in the surrounding liquid. At this stage both inertia and thermal effects limit the growth rate. When the radius grows larger and larger, the inertia becomes less and less important until the radius has grown so large that the growth process is limited only by the rate at which heat can be supplied to the bubble wall.

The earliest analysis of a problem in bubble dynamics with phase change was made by Rayleigh (1917) who solved for the collapse of an empty cavity in a large mass of liquid. Rayleigh's solution was confined to axisymmetric, isothermal, and inviscid flow and he also neglected the surface tension. Later, Plesset & Zwick (1954), and Forster & Zuber (1954, 1955) extended

Rayleigh's analysis to the growth of a vapor bubble in a superheated liquid by including surface tension and coupling the momentum equation with energy equation using Clausius-Clapeyron relation. The experiments of Dergarabedian (1953), Kosky (1968), and Florschuetz *et al.* (1969) showed a good agreement with the results of Plesset & Zwick (1954).

The use of numerical simulations for the study of bubbles (drops) with phase change are relatively new. One of the earliest of such studies is the the potential flow solution of Plesset and Chapman (1971) for the collapse of an isothermal spherical vapor bubble in the neighborhood of a solid wall. A number of authors have tried to solve the boiling bubble problem by coupling the momentum equation with a simplified form of the energy equation, see Theofanous *et al.* (1969) and Prosperreti & Plesset (1978), for example. Dalle Donne and Ferranti (1975) were among the first to solve the complete equations of motion and energy. More recent numerical works include boundary fitted/finite element method of Schunk & Rao (1994), finite volume/moving mesh method of Welch (1995), and front tracking/finite difference technique of Juric & Tryggvason (1995).

Our aim is to investigate the dynamics of boiling bubbles rising due to buoyancy and its effect on the fluid flow and the heat transfer of the system.

FORMULATION AND NUMERICAL METHOD

Consider a domain consisting of a liquid and its vapor undergoing a phase change. The material properties of the phases are different but constant within each phase. The governing equations for this problem are conservation of mass, momentum, and energy equations which are valid in each phase and are coupled together by the jump conditions across the interface. Rather than writing the governing equations separately for each of the fluids, a single equation is used which is valid for the entire flow field and takes the jump in properties across the interface into account. The mathematical modeling of the problem is inspired by Peskin's immersed boundary formulation (1977). The momentum equations in conservative form for such a flow are:

$$\frac{\partial \rho \overline{u}}{\partial t} + \nabla \cdot \rho \overline{u}\,\overline{u} = -\nabla p + \nabla \cdot \mu(\nabla \overline{u} + \nabla \overline{u}^T) \qquad (1)$$
$$-\rho \overline{g} + \sigma \oint \kappa \overline{n} \delta(\overline{x} - \overline{x}^f) ds.$$

In the above equation $\overline{u}$ is the velocity, p is the pressure, ρ is the density, μ is the viscosity, $\overline{g}$ is the gravity, σ is the surface tension coefficient, κ is the curvature, and $\overline{n}$ is the unit vector normal to the interface. $\delta(\overline{x} - \overline{x}^f)$ is a delta function which is zero everywhere except at the interface where $\overline{x} = \overline{x}^f$. ds is the differential arch-length of the interface.

The energy equation in conservative form is:

$$\frac{\partial \rho c T}{\partial t} + \nabla \cdot \rho c \overline{u} T = \nabla \cdot k \nabla T + \qquad (2)$$
$$\oint \overline{h} \rho_f (\overline{v}_i - \overline{u}_f).\overline{n} \delta(\overline{x} - \overline{x}^f) ds,$$

where T is the temperature, c is the heat capacity, and k is the heat conductivity. The last term in the above equation acts as a heat source (sink) which is zero away from the interface and its inclusion results in the conventional energy jump condition across the front. Here, ρ_f and $\overline{u}_f$ are the fluid density and velocity at the interface, and $\overline{v}_i$ is the interface velocity. $\overline{h}$ is a measure of difference in the enthalpies of the liquid and the vapor and is derived using thermodynamics consideration:

$$\overline{h} = h_{fg} + T_{eq}(c_l - c_v). \qquad (3)$$

Here, h_{fg} is the latent heat of evaporation, T_{eq} is the equilibrium temperature (which is also used as a reference temperature), and c_l and c_v are the heat capacities of the liquid and the vapor, respectively.

The mass conservation is:

$$\nabla \cdot \rho \overline{u} = \oint (\rho_l - \rho_v) v_n \delta(\overline{x} - \overline{x}^f) ds, \qquad (4)$$

where, $v_n = \overline{v}_i.\overline{n}$ and the integration of the above equation gives the same mass jump condition as the conventional one.

The introduction of the interface velocity $\overline{v}_i$ adds another unknown to the problem. This new unknown is being taken care of by implementation of the modified Gibbs-Thomson relation at the interface:

$$T_f = T_{eq} + \frac{\sigma \kappa T_{eq}}{\rho_v h_{fg}} - \nu v_n, \qquad (5)$$

where ν is the inverse kinetic mobility. The above equation was first derived by Alexiades & Solomon (1993) and then modified by Juric & Tryggvason (1995) to include the kinetic mobility effect.

The above equations are supplemented by the equations of state for the material properties:

$$\frac{\mathcal{D}\rho}{\mathcal{D}t} = 0; \quad \frac{\mathcal{D}\mu}{\mathcal{D}t} = 0; \quad \frac{\mathcal{D}k}{\mathcal{D}t} = 0; \quad \frac{\mathcal{D}c}{\mathcal{D}t} = 0, \qquad (6)$$

where

$$\frac{\mathcal{D}}{\mathcal{D}t} = \frac{\partial}{\partial t} + \overline{v}_i \cdot \nabla. \qquad (7)$$

In order to solve the above equations, we use a numerical technique similar to the one developed by Juric &

Tryggvason (1995). The momentum and energy equations are discretized over a staggered grid and solved using a modified projection algorithm, which is first order in time and second order in spatial dimensions. Initially (i.e. time n) the density ρ^n, viscosity μ^n, heat conductivity k^n, and heat capacity c^n fields are constructed using the known position of the front. The front is then moved to a new position using the interface velocity at the current time (i.e. $\overline{u}_f{}^n$). The surface tension force is distributed to the grid as a body force using Peskin distribution function (see Peskin 1977) and the density ρ^{n+1} and heat capacity c^{n+1} fields are constructed at the new position. The Navier-Stokes equations are solved in the absence of pressure term and a provisional (i.e. unprojected) velocity field $\overline{u^*}$ is computed. The iterative part is started by guessing the front velocity at the next time step and construction of the mass source at the new position of the front. An elliptic pressure equation is then solved and the provisional velocity is corrected for the pressure to obtain new velocity (i.e. projected velocity) $\overline{u}^{n+1}$. The velocity of the fluid at the front position $\overline{u}_f{}^{n+1}$ is interpolated using a Peskin interpolation function (see Peskin 1977) and the heat source Q^{n+1} is constructed. The energy equation is then solved for T^{n+1} and the Gibbs-Thomson relation is checked for the front points. If this equation is satisfied for all the front points within a predetermined tolerance, the heat conduction coefficient k^{n+1} and viscosity fields μ^{n+1} are computed for the next time step and the calculation proceeds. Otherwise, a new guess is proposed for the front velocity and the computations are repeated.

The individual parameters that control the problem are ρ_l, ρ_v, μ_l, μ_v, k_l, k_v, c_l, c_v, h_{fg}, g, ν, σ, $T_\infty - T_{eq}$, and initial diameter of the bubble, d_0. Nondimensionalization results the following nondimensional parameters: $\lambda = \rho_v/\rho_l$, $\gamma = \mu_v/\mu_l$, $\eta = k_v/k_l$, $\zeta = c_v/c_l$, $Ja = c_l(T_\infty - T_{eq})/h_{fg}$, $Pr = \mu_l c_l/k_l$, $Gr = \rho_l g(\rho_l - \rho_v){d_0}^3/{\mu_l}^2$, $Bo = g(\rho_l - \rho_v){d_0}^2/\sigma$, and $\vartheta = k_l\nu/\rho_l h_{fg} d_0$. Ja is the Jacob number which is ratio of the sensible to the latent heat, Pr is the Prandtl number which is a measure of diffusion of momentum with respect to thermal energy, Gr has a strong resemblance to the Grashof number, Bo is the Bond number and is the ratio of buoyancy to the surface tension, and ϑ is nondimensional inverse kinetic mobility. When we present our results, $l_s = d_0$, $u_s = \sqrt{gd_0}$, and $t_s = l_s/u_s = \sqrt{g/d_0}$ are used as length, velocity, and time scales.

RESULTS

We start our analysis by considering the behavior of a single bubble rising in a liquid. The first frame of figure (1) shows the initial position of the bubble in a rectangular domain. The domain size (nondimensionalized by dividing by bubble diameter) is $x^* = 5$ and $y^* = 10$ and the grid resolution is 64×128. Initially, the vapor and the liquid are superheated and the flow is quiescent. The nondimensional variables are $\lambda = 0.5$, $\gamma = 0.05$, $\eta = 0.025$, $\zeta = 1$, $Ja = 0.1$, $Pr = 0.25$, $Gr = 20$, $Bo = 4000$, and $\vartheta = 0.1$. In the current simulation and the subsequent ones, the domain is periodic in the horizontal direction, wall-bounded (and insulated) at the bottom, and open at the top. At the top boundary, the pressure is fixed to a specific value, $\partial v/\partial y = 0$, and $\partial T/\partial y = 0$, where v is the vertical component of the velocity. We have run simulations with much lower material property ratios, however, here we use somewhat higher ratios to illustrate the basic idea. The sensitivity of the results with the grid resolutions have been tested and it appeared that the grid resolutions used in this study give converged results.

Frames $2-4$ show the velocity field (at every third grid point) and the bubble at equispaced times. The bubble shape is primarily a function of viscous forces, inertial forces, and surface tension. As the bubble size increases, both the front and rear are flattened, with greater flattening occurring at the lower edge (frame 2). Further increase in bubble size results in the indentation at the rear of the bubble (frame 3) and the formation of the skirts (frame 4). As the bubble evaporates, the liquid temperature starts to drop. Due to the convection of the liquid by the vortices on the either side of the bubble, the evaporation rate around the bubble surface is not uniform. Inspection of the temperature field (not plotted in the text) showed a large temperature gradient underneath of the bubble which is exposed to the hot fluid pumped by the vortices from the top.

In the next simulation, we investigate the interactions of two bubbles with an initial tandem configuration. The nondimensional domain size is 5×10 and the resolution is 128×256. All the other nondimensional parameters are the same as the previous run except $Gr = 160$, $Bo = 16000$, and $\vartheta = 0.05$. The initial separation distance of the bubbles centers is $2d_0$. As the bubbles rise, the top one takes an elliptic shape while the bottom one starts to elongate in the vertical direction (frame 2). Subsequent frames shows further indentation in the rear of the top bubble, transformation of the bottom one to a tear drop shape (frame 3) and its further elongation (frame 4). Since the lower bubble is shielded by the top one, it experiences a lower drag and one may speculate that the vertical elongation is as a result of net effect of the forces on the bubble in the horizontal direction which are larger than the drag and other vertical forces. It is seen that the counterrotating vortices inside the bubbles merge together and the hot fluid is now pumped from the front of the top bubble to the rear of the bottom one. Therefore, a higher evaporation rate at the rear of the lower bubble may be expected.

Next, we repeat the above simulation to explore the effect of the separation distance on the dynamics of the flow (Figure 3). Here, the initial separation distance is $4d_0$ and all the other nondimensional numbers are the same as the ones in figure (2). The deformation of the bubbles is now similar to that of the single bubble case studied earlier. Moreover, it is seen that the flow field around each bubble is not very much influenced by the presence of the other one. This is evident from the velocity field where the vortices on the either side of the bubbles have not merged together.

Also investigations about a single bubble or interaction of two of them are quite important in getting insight into the problem, in practical problems it is the collective behavior of systems of large number of bubbles that is of interest. In order to investigate the behavior of the bubbles in larger systems, we have run a few simulations with a relatively larger domain size. Figure (4) shows the interactions of twelve bubbles in a 20×20 domain. The resolution is 256×256 and the nondimensional parameters are the same as the ones in figure (1). The first frame shows the initial positions of the bubbles (which is a weakly perturbed configuration) and the subsequent frames show the velocity field (plotted at every fourth grid point) at equispaced times. It is seen that the bubbles deform similarly. Also, the bubbles at the upper rows have a higher growth rates compared to the bubbles at the lower ones. This is possibly due to the fact the rise of the bubbles in the upper rows reduces the temperature of the liquid, and as a result, the ones in the lower rows will be imposed to the liquid at a lower temperature. As the bubbles rise, the flow field around each bubble is influenced by its neighbors. Once the rear of the bubbles start to indent (frame 3), the vortices start to merge and it is seen that the original uniform flow field evolves to two distinct streams of upgoing flow at the position of the bubbles and a downcoming one in the rest of the domain.

Next, we investigate the interactions of the bubbles in the previous system under a smaller vapor to liquid density ratio. The density ratio is changed by reducing the density of the vapor. With the exception of $\lambda = 0.05$, $Gr = 38$, and $Bo = 7600$, all the other nondimensional numbers remain the same as the previous run. Figure (5) shows the bubbles and the velocity field (plotted at every fourth grid point) at the selected times. The initial positions of the bubbles are the same as the ones in figure (4). Notice that the velocity scale is substantially reduced here for a better visualization. Decreasing the density ratio increases the evaporation rate of the liquid which is evident by comparing the size of the bubbles in the third frames of figure (5) and (4). Contrary to the dimpled elliptic shape of the bubbles in figure (5), here, the bubbles take tear drop shapes. Moreover, inspection of the velocity field shows that the counterrotating vortices inside the bubbles are absent and the velocity field in the liquid is relatively uniform. Figure (6) compares the vectors of heat flux for the last frames of figure (4) and (5) (again, the scale of the vectors for the right frame is considerably reduced for a better visualization). The heat flux is maximum at the rear of the bubbles and relatively uniform in the vertical passages between them.

CONCLUSION

Numerical simulations of boiling bubbles is presented. The results shows the effect of the initial positions of the bubbles and density ratio on the their subsequent interactions. It is seen that bubble dynamics, fluid flow, and heat transfer are a strong function of these two parameters.

BIBLIOGRAPHY

Alexiades, V. and Solomon, A. D., 1993, "Mathematical modeling of melting and freezing processes," Hemisphere, Washington, D. C., pp. 92-94.

Dalle Donne, M. and Ferranti, M. P., 1975, "The growth of vapor bubble in superheated sodium," Int. J. Heat Mass Transfer, Vol. 18, pp. 477-493.

Dergarabedian, P., 1953, "The rate of growth of vapor bubbles in superheated water," J. Appl. Mech., Vol. 20, pp. 537-545.

Florschuetz, L. W., Henry, C. L., and Rashid Khan, A., 1969, "Growth rates of free vapor bubbles in liquids at uniform superheats under normal and zero gravity conditions," Int. J. Heat Mass Transfer, Vol. 12, pp. 1465-1489.

Forster, H. K., and Zuber, N., 1954, "Growth of a vapor bubble in a superheated liquid," J. Appl. Phys., Vol. 25, pp. 474-478.

Forster, H. K., and Zuber, N., 1955, "Dynamics of vapor bubble and boiling heat transfer," A.I.Ch.E. Journal, Vol. 1, pp. 531-535.

Juric, D., and Tryggvason, G., 1995, "A front-tracking method for liquid-vapor phase change,"

ASME-FED, Vol. 234, pp. 141-148.

Kosky, P. G., 1968, "Bubble growth measurement in uniformly superheated liquids," Numer. Engng Sci., Vol. 23, pp. 695-706.

Peskin, C. S., 1977, "Numerical analysis of blood flow in the heart," J. Comput. Phys., Vol. 25, pp. 220-252.

Plesset, M. S. and Chapman, R. B., 1971, "Collapse of an initially spherical vapor cavity in the neighborhood of a solid boundary," J. Fluid Mech., Vol. 47, pp. 283-290.

Plesset, M. S., and Zwick, S. A., 1954, "The growth of vapor bubbles in superheated liquids," J. Appl. Phys., Vol. 25, pp. 493-500.

Prosperetti, A. and Plesset, M., 1978, "Vapor-bubble growth in a superheated liquid," J. Fluid Mech., Vol. 85, pp. 349-368.

Rayleigh, L., 1917, "On the pressure developed in a liquid during the collapse of a spherical cavity," phil. Mag., Vol. 34, pp. 94-98.

Schunk, P. R. and Rao, R. R., 1994, "Finite element analysis of multicomponent two-phase flows with interphase mass and momentum transport," Int. J. Num. Meth. Fluids, Vol. 18, pp. 821-842.

Theofanous, T., Biasi, L., Isbin, H. S., 1969, "A theoretical study on bubble growth in constant and time-dependent pressure fields," Chem. Engng Sci:, Vol. 26, pp. 263-274.

Welch, S. W. J., 1995, "Local simulation of two-phase flows including interface tracking with mass transfer," J. Comp. Phys., Vol. 121, pp. 142-154.

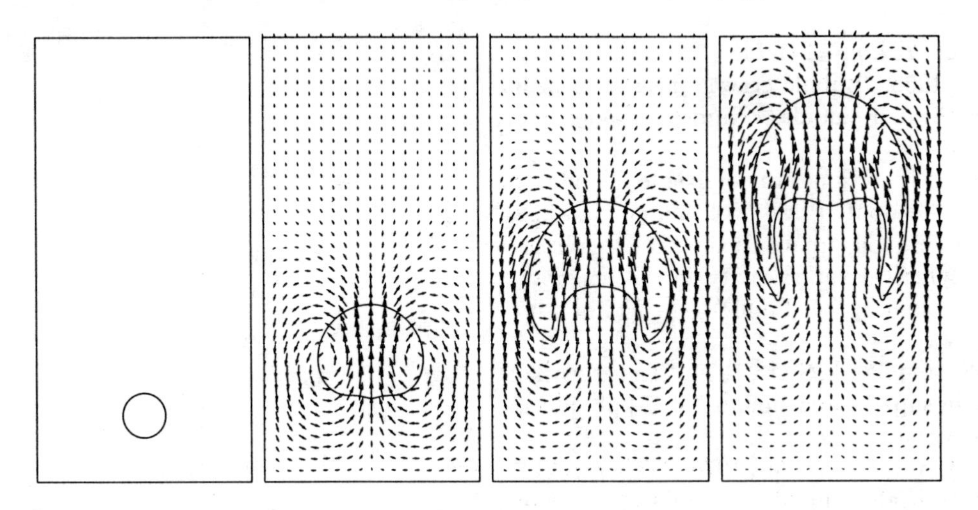

Figure 1: Rise of a single bubble in a superheated liquid. Times are 0, 5.8, 11.9, and 18.5.

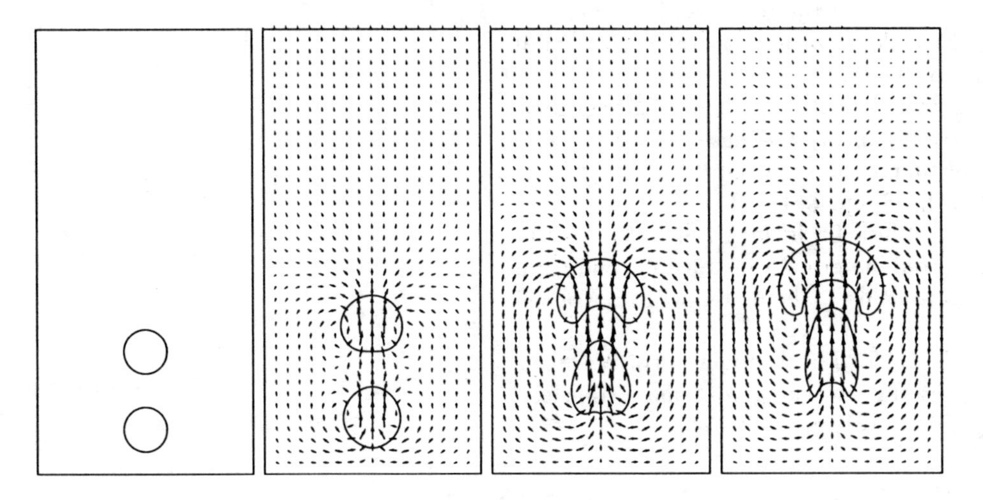

Figure 2: Interactions of two tandem bubbles with small initial separation distance in a superheated liquid. Times are 0, 1.8, 3.6, and 5.4.

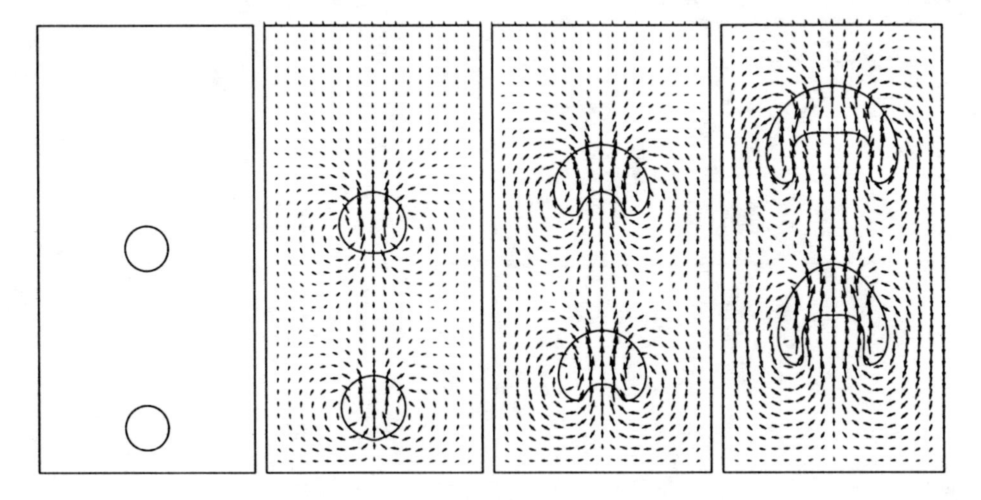

Figure 3: Interactions of two tandem bubbles with large initial separation distance in a superheated liquid. Times are 0, 2.4, 4.4, and 7.7.

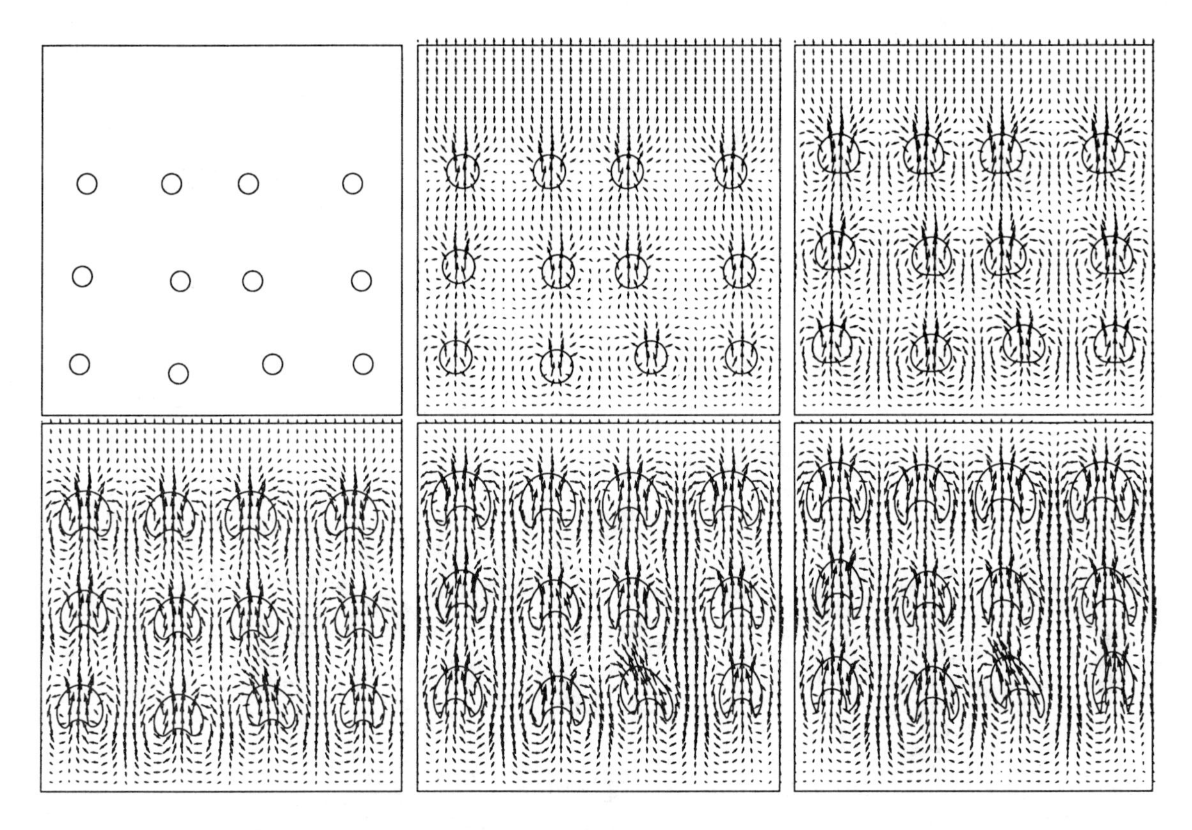

Figure 4: Interactions of twelve bubbles at $\lambda = 0.5$. Times are 0, 2.54, 4.92, 7.3, 9.8, and 11.

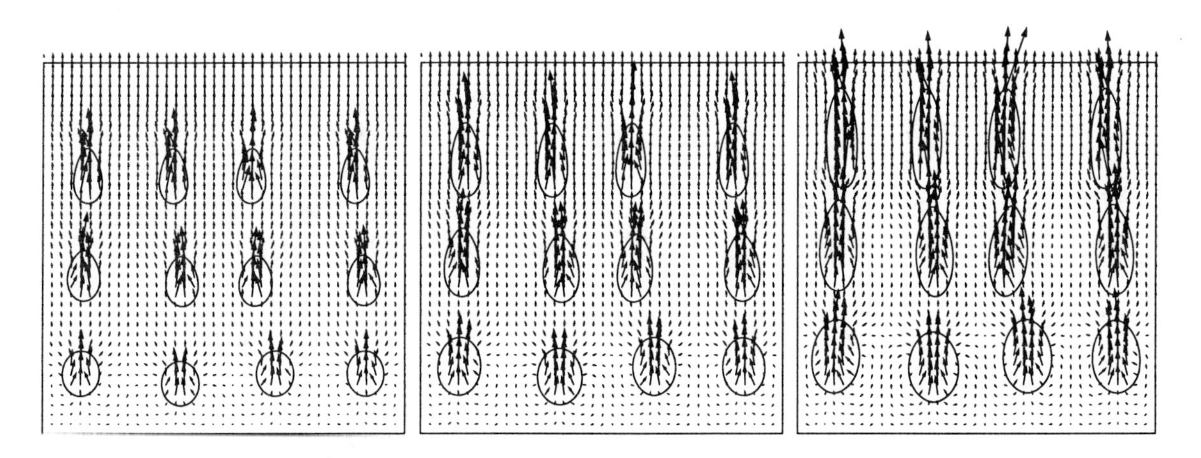

Figure 5: Interactions of twelve bubbles at $\lambda = 0.05$. Times are 0, 1.52, 2.2, and 2.96.

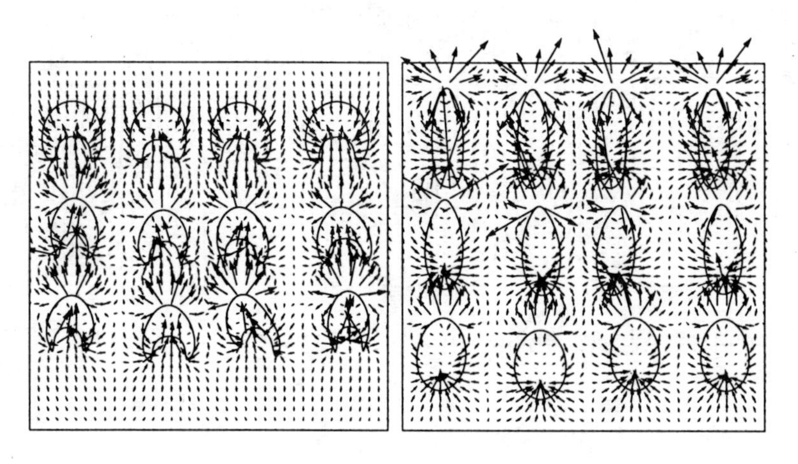

Figure 6: Comparison of the heat flux vector fields of the last frames of the simulations in figure 4 and 5.

HTD-Vol. 342, National Heat Transfer Conference
Volume 4
ASME 1997

THE STUDY OF MARANGONI CONVECTION AROUND VAPOR BUBBLE UNDER MICROGRAVITY

Li Yang **Danling Zeng**
Dept. of Thermal Power Eng., Chongqing Univ.,
Chongqing 630044, People's Republic of China
Tel. (023) 65103442, e-mail: DLZENG@CQU.EDU.CN

ABSTRACT

Transient Marangoni convection under micro and earth gravity conditions around a vapour bubble submerging in superheated liquid-filled rectangular container is simulated with a two dimensional model. A finite difference technique is employed to numerically integrate the unsteady vorticity and heat transport equations. Through simulating the distribution of stream function and temperature, steady state Nusselt Number at the top and bottom walls as a function of Marangoni Number, the response of the maximum stream function and heat transfer to increasing temperature difference of top and bottom walls were obtained.

The results show that Marangoni convection strongly modifies the isotherms at high Marangoni number and Marangoni convection is enhanced by the presence of phase transition at the interface .

NOMENCULATURE

b: height of the region.
c: specific heat
g: gravitational acceleration
h: heat transfer coefficient at liquid-vapor interface
J_m: mass flux
k: liquid thermal conductivity
L: latent heat of evaporation
m: mass
$Ma = -\frac{d\sigma}{dT}\cdot\frac{b\Delta T}{a\eta}$ Marangoni number
$\mathrm{Bi} = \frac{\mathrm{hb}}{\mathrm{k}}$ Biot number
$\mathrm{Pr} = \frac{\nu}{a}$ Prandtl number
R: bubble radius
$Ra = \frac{g\beta b^3\Delta T}{\nu a}$ Rayleigh number
s: width of the region
t: time
T : temperature
u_r: radial velocity on the bubble surface
u_θ: azimuthal velocity on the bubble surface
U: horizontal velocity
V: vertical velocity
x: horizontal coordinate
y: vertical coordinate

Greek symbols

α : thermal diffusivity
β : coefficient of volumetric expansion
η : dynamic viscosity
γ : kinematic viscosity

ρ : density of the liquid
σ : surface tension
$\frac{d\sigma}{dT}$:temperature coefficient of surface tension
ψ : stream function
$\omega = \nabla \times \overline{V}$ vorticity

Subscripts

b: bottom wall
n: polytropic exponent
t : top wall
v: vapor
θ : azimuthal
r: radial

INTRODUCTION

Flows induced by nonuniform surface tension at a liquid-vapor interface were firstly reported by C. G. M. Marangoni in 1871. On a liquid surface ,surface tension strongly depends on the temperature of the fluid. Variations in temperature can influence the transport of heat、 mass and momentum near an interface, especially in small-scale systems. The fluid flow driven by surface tension gradient is termed Marangoni convection which has been investigated by various authors (Gaddis,1971, Cary et al, 1973). Straub(1992) simulated Marangoni convection around gas bubble with adiabatic surface. His study revealed that even under earth gravity condition Marangoni flow cannot be overcome by natural convection in certain configurations. Depending on the boundary conditions, thermocapillary flow can either argument or counteract natural convection.

Up to now theoretical or experimental study on the buoyancy -Marangoni convection around a vapor bubble with phase transition hasnot been reported. In the present paper, a numerical simulation is made for the motion induced by thermocapillary effects at the bubble surface with phase transition around a vapor bubble. The aim of the study is to investigate the heat transfer process affected by thermocapillary convection. and to know the effects of liquid motion induced by thermocapillary flow with phase transition on the thermal process of a vapor bubble .

MODEL DISCRIPTION

The model is illustrated in Fig.1, where a semi-spherical vapor bubble is submerging in a superheated liquid and attaching on the wall. The system is assumed to be symmetry about y axis. Thus the flow is simplified to be two dimensional.The lateral boundaries are adiabatic and nonslip, whereas the top and bottom boundaries are maintained at different temperature . The fluid is assumed

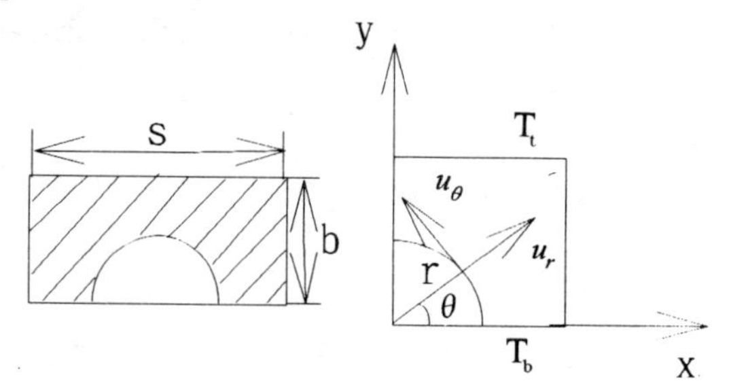

Fig1. Illustration of computational domain

to be initially stationary with linear temperature gradient. Constant physical properties are also assumed.

An expression of the heat flux associated with the mass transition of the fluid through the liquid vapor interface has been derived in Zeng(1994). It reveals that the vapor inside the bubble expands along a polytropic process rather than along the saturation curve. The heat flux includes the latent heat and polytropic heat.The heat transfer coefficient at the liquid vapor interface is given by

$$h = \frac{1}{\Delta T}[J_m L + \frac{m_v c_n}{4\pi R^2} \cdot \frac{dT_v}{dt}] \quad (1)$$

It is considered to be uniform at the whole surface of the bubble.

$$q_v = J_m \cdot L$$

$$q_n = \frac{m_v c_n}{4\pi R^2} \cdot \frac{dT_v}{dt} = \frac{\rho^v R}{3} \cdot \frac{n-k}{n-1} \cdot c^v \frac{dT_v}{dt}$$

FUNDAMENTAL EQUATIONS

The Study of convective heat transfer are reduced to the solution of a nonlinear system of partial differential equations involving the conservation principles of energy、 momentum and mass equations. For the convenience in constructing the numerical algorithm of the present problem, the Navier-Stokes equations is written in the vorticity-stream function form. Using the Boussinesq approximation, the system of equations for convective heat and mass transfer is formulated as follows

$$\frac{\partial \overline{V}}{\partial t}+(\overline{V}\nabla)\overline{V}=-\frac{1}{\rho_o}\nabla P+\nu\nabla^2\overline{V}-g\beta(T-T_o) \quad (2)$$

$$\frac{\partial T}{\partial t}+\overline{V}\nabla T=a\nabla^2 T \quad (3)$$

$$\nabla\cdot\overline{V}=0 \quad \rho-\rho_o=-\rho_o\beta(T-T_o) \quad (4)$$

Navier-Stokes equations in vorticity-streamfunction are

$$\frac{\partial \omega}{\partial t}+\overline{V}\cdot\nabla\omega=-\beta g\frac{\partial T}{\partial x}+\nu\nabla^2\omega \quad (5)$$

$$\omega=-\nabla^2\psi \quad (6)$$

$$\omega=\frac{\partial V}{\partial x}-\frac{\partial U}{\partial y} \quad (7)$$

$$U=\frac{\partial \psi}{\partial y} \qquad =-\frac{\partial \psi}{\partial x}$$

The non-dimensional form of the above equations can be written as follows

$$\frac{\partial \omega^*}{\partial t^*}+\overline{V}^*\cdot\nabla\omega^*=-Ra\cdot\mathrm{Pr}\cdot Ma^{-2}\cdot\frac{\partial T^*}{\partial X^*}+\mathrm{Pr}\cdot Ma^{-1}\nabla^2\omega^* \quad (8)$$

$$\omega^*=-\nabla^2\psi^* \quad (9)$$

$$\frac{\partial T^*}{\partial t^*}+\overline{V}^*\cdot\nabla T^*=Ma^{-1}\cdot\nabla^2 T^* \quad (10)$$

In the process of nondimensionalizing the energy and vorticity transport equation, the following dimensionless parameters are introduced

Prandtl number $\mathrm{Pr}=\frac{\nu}{a}$

Marangoni number $Ma=-\frac{d\sigma}{dT}\cdot\frac{b\Delta T}{a\eta}$

Rayleigh number $Ra=\frac{g\beta b^3\Delta T}{\nu a}$

BOUNDARY CONDITIONS

Thermal boundary conditions

$$X^*=0\ ,\ \frac{\partial T^*}{\partial X^*}=0\ ;\qquad X^*=1\ ,\ \frac{\partial T^*}{\partial X^*}=0 \quad (11)$$

$$Y^*=0\ ,\ T^*=1\ ;\qquad Y^*=1\ ,\ T^*=0 \quad (12)$$

Hydrodynamic boundary conditions

$$X^*=0\ :\qquad \frac{\partial \psi^*}{\partial Y^*}=\frac{\partial^2\psi^*}{\partial X^{*2}}=0$$

$$X^*=1\ :\qquad \frac{\partial \psi^*}{\partial Y^*}=\frac{\partial \psi^*}{\partial X^*}=0$$

$$X^*=0,1\ :\qquad \frac{\partial U^*}{\partial Y^*}=\frac{\partial V^*}{\partial Y^*}=0\ ,\quad \omega^*=\frac{\partial U^*}{\partial X^*} \quad (13)$$

$$Y^*=0,1\ :\qquad \frac{\partial \psi^*}{\partial X^*}=\frac{\partial \psi^*}{\partial Y^*}=0$$

$$\frac{\partial U^*}{\partial X^*}=\frac{\partial V^*}{\partial X^*}=0\ ,\quad \omega^*=-\frac{\partial V^*}{\partial Y^*} \quad (14)$$

At the liquid-vapor interface:

The thermal bounday condition at the bubble surface is

$$k\left(\frac{\partial T}{\partial r}\right)\Big|_{r=R}=h\left(T_r|_{r=R}-T_V\right) \quad (15)$$

The radial velocity at the bubble surface is related to the surface temperature by

$$u_r|_{r=R}=\frac{h}{\rho L}\left(T_V-T_r|_{r=R}\right) \quad (16)$$

Marangoni boundary condition is

$$\eta\left[r\frac{\partial}{\partial r}\left(\frac{u_\theta}{r}\right)\right]\Big|_{r=R}=\frac{1}{R}\frac{\partial T}{\partial\theta}\frac{d\sigma}{dT} \quad (17)$$

NUMERICAL SCHEME

A finite difference technique is employed to numerically integrate the unsteady vorticity and heat transport equations. The alternative direction implicit (ADI) method is adopted which employs the fractional time step concept to generate tridiagonal matrices. The central-difference scheme was used for all space derivatives. For a two-dimensional case in Cartesian coordinates, the circular geometry of the bubble and the rectangular region in Fig.1 are discretized with 40×40 grid using"blocked-off-regions" in connection with regular triangular elements.

The numerical methods are based upon the solution of the above (T,ω,ψ) Boussinesque system.For the solution of it,the values of the vorticity at the boundary are required.The wall vorticities are obtained by Taylor-series expansion of stream function ψ near the boundary points.

$$\omega_i = -\frac{\partial \psi_{i+1}}{\Delta Z} \qquad (18)$$

where the subscripts i and i+1 denote the grid points at the wall and next to the wall, respectively, and ΔZ denotes the grid spacing. Poission equation for the stream function at each time level was solved by successive overrelaxation. The optimum value of the relaxation parameter was determined by

$$\alpha_0 = 2\left(\frac{1-\sqrt{1-\zeta}}{\zeta}\right) \qquad (19)$$

$$\zeta = \left[\frac{\cos\left(\frac{\pi}{M-1}\right)+\beta^2\cos\left(\frac{\pi}{N-1}\right)}{1+\beta^2}\right]^2 \qquad (20)$$

Here, $\beta=\frac{\Delta X}{\Delta Y}$,and M, N denote the grid point number in X , Y direction respectively.

The iteration procedure is terminated when the condition

$$\sum_{i,j}^{M,N}\left|\psi_{i,j}^{k+1}-\psi_{i,j}^{k}\right| \le \varepsilon \qquad (21)$$

is satisfied. where k denotes the number of iteration .

The computation procedure for T,ω,ψ may be written as follows:

1. Assuming the initial values of velocity fields.
2. The temperature T's are determined by solving energy equation.
3. The vorticity equation is solved to evaluate the values of vorticity.
4. The stream function ψ's are obtained by solving Poisson equation.
5. The values of velocity are defined in terms of the stream function as $U=\frac{\partial \psi}{\partial Y}$ and $=-\frac{\partial \psi}{\partial X}$.
6. The stream function ψ's are used to determine the vorticity at the boundary.
7. The whole procedure is then repeated cyclically.

RESULTS

In our study, the bubble is regarded as fixed at a certain location in the liquid, and its radius is 2 mm, the length b and s of the lateral walls are 4 mm. The bottom and top walls are maintained at different temperature T_b and T_t ($T_b > T_t$ or $T_b < T_t$). Water is used as the working fluid in our calculations. The distribution of isotherms under different conditions are given for various Marangoni number and Prandtl number. Firstly, we discussed the results obtained by assuming adiabatic condition at the liquid vapor interface. From Fig.2, the isotherms pattern for Marangoni number of 12,000; 50,000 and 100,000 are drawn ,respectively.

It is seen that when Marangoni number is 12,000 the isotherms is modified slightly near the bubble. As Marangoni number increases to 50,000, the high velocity at the interface induces a large upward flow near the bubble and hence Marangoni convection influences the isotherms dramatically. When the Marangoni number reaches 100,000, the temperature distribution shows a tremendous change in the liquid area . When the Marangoni number or the temperature difference between top and bottom walls is increased, the influence of Marrangoni convection on heat transfer grows.

In the presence of phase transition at liquid vapor interface, the results are different from that obtained by assuming adiabatic condition at the interface. Fig.3 shows the isotherms for a fluid with Marangoni number of 5,000 and 25,000 respectively. The comparison of the two isotherms, Bi=0 and Bi $\neq$ 0 at the interface suggests a significant enhancement in Marangoni convection due to the presence of phase transition.

The displacement of the isotherms in Fig.5 demonstrates the effect Marangoni convection exerts on buoyancy flow. If surface tension driven convection acts in the same direction as the buoyancy convection, Marangoni and buoyancy convection argument each other. Fig.6 illustrates that, even for small Marangoni number , the behaviour is very noticeable.

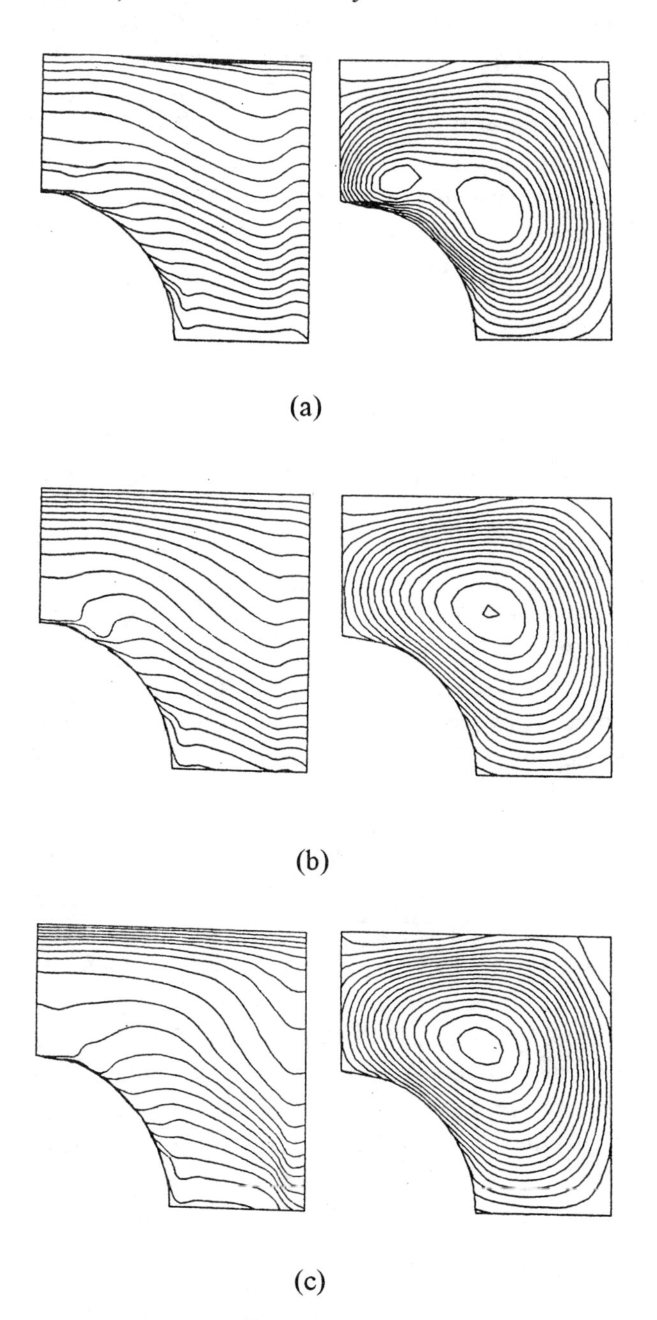

Fig.2 Predicted isotherms(left) and streamlines(right) for various Marangoni number (a)Ma=12000 (b)Ma=50000 (c)Ma=100000 Ra=0; Bi=0; Pr=1.71 (adiabatic condition)

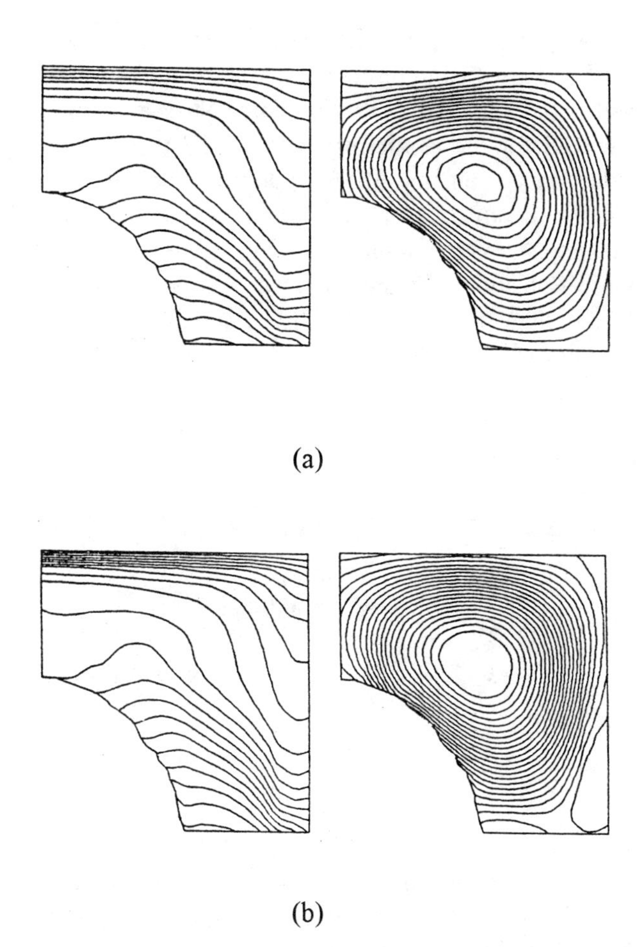

Fig.3 Predicted isotherms(left) and streamlines(right) for various Marangoni number (a) Ma=12000 (b) Ma=50000 ; Ra=0; Bi $\neq$ 0; Pr=1.71 (with phase transition)

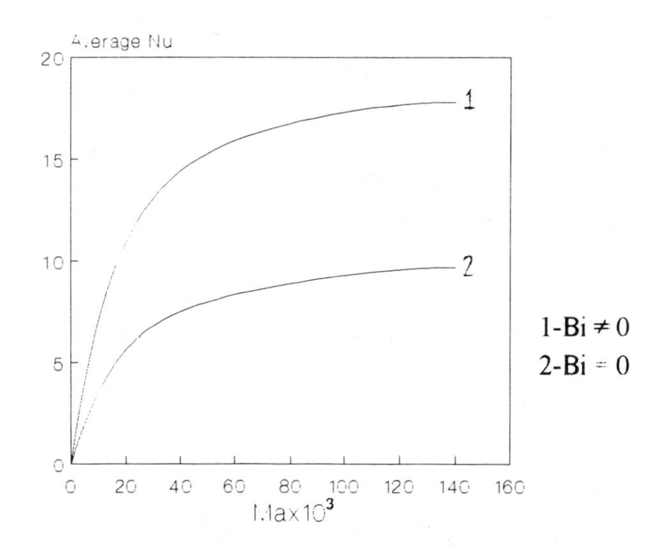

Fig.4 Steady state Nusselt number at the isothermal walls as a function of the Marangoni number

An increase in heat flux at the heated and cooled walls is expressed in terms of Nusselt number at those walls. In Fig.4, the steady state Nusselt number averaged for the isothermal bottom and top walls is depicted as a function of Marangoni number. We find that the steady state Nusselt number increases with growing Marangoni number. Owing to the presence of phase transition, the steady state Nusselt number is higher than that obtained by assuming adiabatic condition at interface under the same Marangoni number.

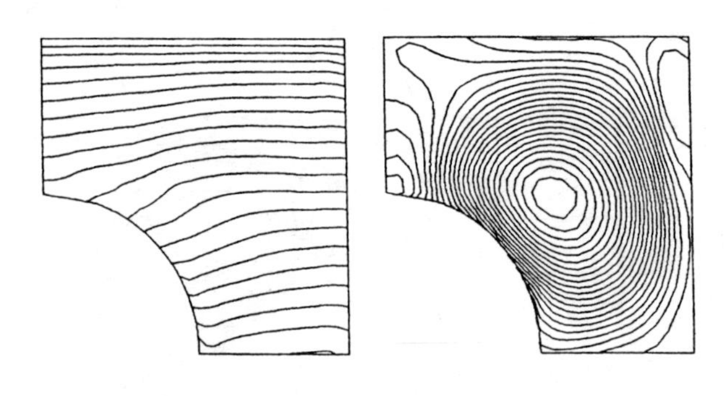

Fig.5 Predicted isotherms and streamlines for Marangoni number Ma=20000; Ra=285000; Bi=0; Pr=1.93 (M-B)

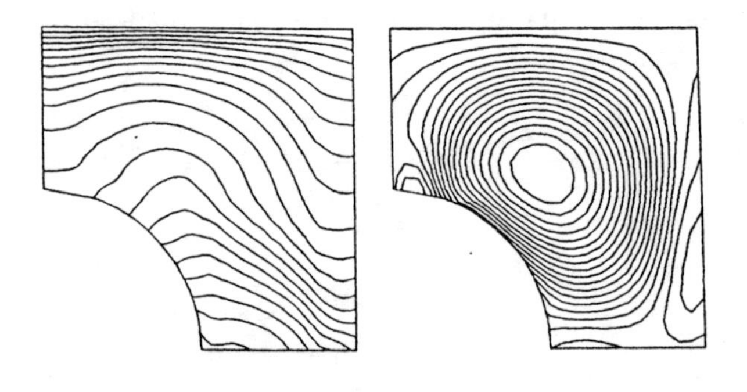

Fig.6 Predicted isotherms and streamlines for Marangoni number Ma=5000; Ra=57850; Bi=0; Pr=2.55 (M+B)

CONCLUSION

The main attention in the present work is paid to the heat transfer in a liquid and the influence of phase transition on thermocapillary convection under microgravity. The analysis has shown that the effects of thermocapillary convection on heat transfer is significant. For small Marangoni number, the diffusive heat transfer is dominant over convective energy transport. But, for high Marangoni number energy transport is dorminated by thermocapillary convection. The temperature field is strongly influenced by Marangoni convection if Marangoni number is large. The accumulation of isotherms at the heated and cooled walls is equivalent to an increase in heat transfer at those walls. Owing to the presence of phase transition at liquid vapor interface, thermocapillary convection is enhanced and the effect of thermocapillary flows on the enhancement of heat transfer is obvious.

REFERENCES

(1) R. B. Baliga, and S. V. Patanba, 1983, "A Control Volume Finite-Element Method of Two-Dimensional Fluid Flow and Heat Transfer", Numerical Heat Transfer, Vol.6, pp. 245-261.

(2) J. D. Cary, B. B. Mikic, 1973, "The Influence of Thermocapillary Flow or Heat Transfer in Film Condensation", Trans. ASME, J. Heat Transfer, ser, c95, 22.

(3) E. S. Gaddis, 1971, "The Effect of Liquid Motion Induced by Phase Change and Thermocapillarity on the Thermal Equalibrium of a Vapor Bubble", Int.J. Heat Mass Transfer. Vol. 15, pp. 2241-2250.

(4) Johannes Straub, Johann Betz, and Rudi Marek, 1992, "Numerical Simulation of Marangoni Convection Around Gas Bubbles in a Liquid Matrix", Proceedings Villth Europen Symposium on Materals and Fluid Sciences in Microgravity, Brussels, Belgium.

(5) Danling Zeng, 1994, "A Mathematical Model of the Growth of a Bubble Based on Non-equilibrium Thermodynamics", Journal of Egineering Thermophysics. Vol.15, No. L, pp. 9-12. (in chinese)

HTD-Vol. 342, National Heat Transfer Conference
Volume 4
ASME 1997

EXPERIMENTAL STUDY OF EFFECT OF VIBRATION ON ICE CONTACT MELTING WITHIN RECTANGULAR ENCLOSURES

Liang Quan, Zongqin Zhang, and Mohammad Faghri
Department of Mechanical Engineering and Applied Mechanics
University of Rhode Island
Kingston, RI 02881

ABSTRACT

Experiments on ice contact melting within a rectangular enclosure under vibrating conditions are performed. Isothermal wall condition is maintained on the test cells with aspect ratios of 0.4, 1.0, and 2.5, respectively. It is shown that melting rates are increased under vibrating conditions and melting enhancement is proportional to the acceleration of vibration. Compared to the stationary experiments, the maximum melting rate enhancement of 170 % is observed. Aspect ratio plays an important role in melting process and the lowest melting rates occur in both stationary and vibrating conditions for aspect ratio of 1.0. The relative melting enhancement by vibration for both high and low aspect ratios are significant. The increase in melting due to vibration is more pronounced for the low Stefan numbers. Preliminary experiments show that horizontal vibration can be more effective than vertical vibration to enhance the melting rate.

NOMENCLATURE

a	maximum acceleration of vibration, $(2\pi f)^2 A$
A	amplitude of vibration
AR	aspect ratio, H / W
Ar	Archimedes number, $(\rho_l - \rho_s)\, g\, H\, W^2 / \rho_l \nu^2$
C	specific heat
f	frequency of vibration
Fo	Fourier number, $\alpha t / H^2$
g	gravitational acceleration
Gr	Grashof number, $g\beta\,(T_w - T_f)\, H^3 / \nu^2$
h_f	latent heat of phase change material
H	height of test cell
k	thermal conductivity
Pr	Prandtl number
Ste	Stefan number, $C_l\,(T_w - T_f) / h_f$
t	time
T	temperature
ΔT	$T_w - T_f$
W	width of test cell

Greek Symbols

α	thermal diffusivity
β	thermal expansion coefficient
δ	thickness of liquid gap between ice and contact surface
ν	kinematic viscosity
ρ	density
ω	angular velocity

Subscripts

f	fusion (melting)
i	initial
l	liquid
s	solid
w	wall

INTRODUCTION

Phase-change is a heat transfer mechanism of fundamental scientific interest. It has a wide range of applications in science and technology, in particular, in material processing, metallurgy, crystal growth, welding, casting, geophysics, thermal storage and energy transport.

There are two kinds of convection during phase change under normal gravity condition. These are the sedimentation of the denser phase and the natural convection in the liquid phase. The sedimentation (or floatation) of the solid phase, which amounts to self generated forced convection, is a direct consequence of the gravitational field. It occurs in melting as well as solidification, when the solid, in bulk or in fragments, does not adhere to the container wall. The sedimentation (or floatation) of the solid phase appears as contact melting problem with the melt layer existing in a thin layer between the heated container wall and the phase-change interface. Since the thermal resistance of this thin layer is very small, very large heat transfer rates can occur. The enhancement of melting rate because of contact melting usually exceeds that due to natural convection (Bejan, 1992). Natural or buoyancy-driven convection is another noteworthy factor in phase-change processes. It results from wall-to-interface temperature difference and can have an appreciable effect on phase change rate and interface shape (Zhang and Bejan, 1989a, b).

An assumption in the analysis of contact melting problem is that the melting region between the solid-liquid phase change materials (PCM) interface and the heated wall is very thin in relation to the characteristic dimensions of the test cell and conduction in this region dominantly controls the heat transfer (i.e., contact melting). Archimedes number, Ar, the measure of floatation or sedimentation, appears as a function of one-fourth power in contact melting which implies relatively weak dependence on the gravitational constant g.

The topic of contact melting has received considerable attentions in the last decade. Sedimentation studies, experimental, analytical or numerical, have been published by many researchers; for example, melting inside containers (Bareiss and Beer, 1984; Sparrow and Myrum, 1985; Moallemi et al., 1986; Hirata et al., 1991; Dong et al., 1991; Asako and Faghri, 1994) and penetration of hot solid bodies into other materials (Emerman and Turcotte, 1983) were among the related investigations. A review of recent work as well as the scale analysis of close-contact melting was presented in a paper by Bejan (1992).

There are little experimental data available for solid-liquid phase change under vibrating conditions and most analytical solutions do not apply when vibration is present. Natural convection under vibrating conditions is very complicated because different convection modes are possible and are sensitive to the container configuration, boundary conditions and the vibration characteristics. Melting processes under the influence of ultrasonic vibration was experimentally investigated by Choi and Hong (1991). Various frequencies in the range of 15 to 55 KHz were applied in the vertical direction through the bottom plate of the rectangular enclosure and constant heat flux boundary condition was maintained on the side wall of the enclosure by a heated plate. The results showed that the overall melting phenomena was significantly affected by the ultrasonic vibrations and the frequencies of the ultrasonic waves.

The objective of this study is to conduct experiments for contact melting of ice under mechanical vibrations to investigate the effects of amplitude, frequency and the direction of vibration on the melting phenomena. The experiments are conducted within rectangular enclosures where the isothermal conditions are maintained at the walls. The aspect ratios of the test cells are 0.4, 1.0, and 2.5, respectively. Stationary melting experiments with different aspect ratios and wall heating conditions were conducted. These results are used as the references to study the phase change with vibration.

EXPERIMENTAL APPROACH

Apparatus

The experimental apparatus consists of a constant-temperature bath, a pump, test cells, phase change material (ice), vibration generators and measurement instruments. Figure 1 shows the schematic of the experimental system. The thermal bath has a rectangular geometry with dimensions 0.33 m x 0.35 m x 1.0 m. It provides a constant-temperature environment for melting experiments. The commercial automotive anti-freeze solution is heated or cooled inside the thermal reservoir and then circulated into the test cell by the pump. This system provides temperatures from -20 °C to 60 °C with a resolution of ±0.2 °C.

Different vibration devices are used to cover a wide range of vibration parameters. The high amplitude, low frequency horizontal and vertical vibrations are provided by a specially designed cam system. The high frequency, small amplitude vertical vibrations are generated by the commercial vibrator (model US-450, Vibco. Inc.). The other device used in the experiments for small amplitude vibration is an electromagnetic vibrator.

Three test cells, with aspect ratios (AR) of 0.4, 1.0 and 2.5, respectively, have been fabricated. The aspect ratio of the test cell is defined as the ratio of height to width. The four walls (top, bottom, and two sides) of test cells are made of 10 mm thick aluminum plates. The aluminum material has the advantages of light weight and high heat conductivity. The cross section areas of three test cells are 52.2 mm x 130.5 mm, 82.5 mm x 82.5 mm, and 130.5 mm x 52.2 mm, respectively. The depth, which is the dimension in the direction normal to the plane of Figure 1 is 152.5 mm. This dimension is large enough to minimize the three-dimensional (or end) effects on the flow and the ice-water interface. While seeking to eliminate these effects, one must compare the 152.5 mm dimension with the average thickness of the liquid region (10-30 mm) not with the width of the enclosed space, which is occupied mostly by ice. The test cells are designed to have the same total volume for the sake of comparison.

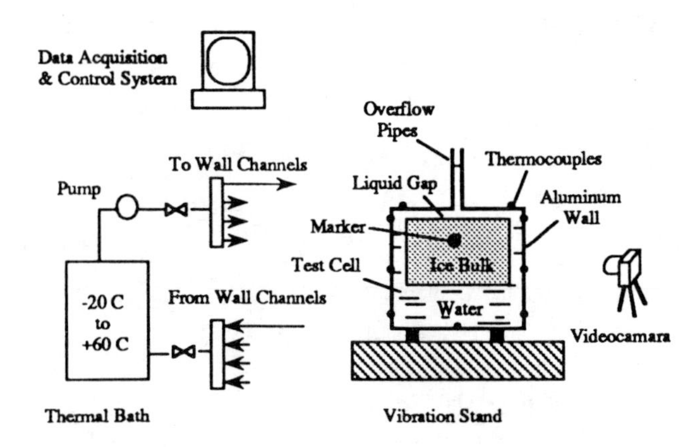

Fig.1 Schematic View of Experimental Setup

A pump is used to circulate the anti-freezing coolant from the thermal reservoir into the channels machined inside the aluminum walls (top, bottom and two sides) to provide the isothermal wall conditions during experiments. These channels are multi-pass, counter-current and carefully designed to assure the uniform temperature distribution along the walls. To facilitate the visual observation of the melting process, the front and back walls are made of two layers 5 mm-thick Plexiglas separated by air gap. This arrangement has a good structural strength as well as thermal insulation characteristics over a wide range of temperatures. The other four heating walls are insulated by Plexiglas to minimize the heat losses. Test cells are firmly mounted on the vibration devices. Two overflow pipes are mounted on the top of the test cell with AR = 1.0. For test cells of AR = 0.4 and 2.5, the overflow pipes are mounted on the front and back walls. During the experiments, the ice water is supplied to test cell through the overflow pipe to provide the contact melting conditions.

Measurements.

The wall temperatures are monitored by thermocouples located at different positions. The temperature distributions along the four walls are measured to assure the uniformity of the isothermal boundary condition. Each thermocouple is made out of chromel-alumel (type K) wire and embedded from the outside, to within 0.5 mm of the inside surface of each wall. WB-T21 terminal panel and QuickLog Data Acquisition and Control software, the products of OMEGA Engineering, Inc., are used in temperature measurement and control. The temperature reading provided by acquisition system is accurate to within ± 0.2°C. Although high velocity constant temperature coolant flow is circulated into the channel, the temperature measured at the contact surface is consistently lower than the bath temperature. The discrepancy is about 5 %. Since the conduction is the dominant heat transfer mechanism, the time-averaged wall temperature at the contact surface was used as the wall temperature.

The instantaneous shape of the ice-water interface is videotaped. The instantaneous shape of the interface is then input into the computer and measured from the transferred computer image. The instantaneous volume of liquid region is calculated numerically by integrating the area defined by the interface described in the preceding paragraph. The accuracy of the liquid volume (or melted fraction) calculation is evaluated for each run by comparing the calculated value with the actual quantity of liquid accumulated at the end of the run. The physical measurement of the end-of-run liquid volume consists of rapidly draining the liquid and measuring the volume. In the melting experiments described in this paper, the differences between the liquid volume calculated from videotape and the volume measured directly are less than 4 %.

The most critical phase of each run is the filling of the enclosure with ice. In order to avoid the formation and trapping of air bubbles, distilled water is used and boiled for 30 minutes before freezing. Only small amount of water is poured into the test cell (about 10 mm deep for each layer) and solidified before the addition of the next layer. Each filling and solidification phase in every run takes approximately two days. The freezing process is performed at the temperature about 10 °C lower than the melting point, so that a compact solid ice crystal structure is obtained. A marker is embedded in the ice during freezing to compare the melting at the top, bottom and sides. Before the start of the experiment, the fully solidified ice is brought to a state of uniform temperature by putting the test cell in the predetermined constant temperature refrigerator for approximately 24 hours. The actual measured initial ice temperatures ranges from - 1.5 °C to - 0.5 °C.

Typical Experimental Run

After the test cell is placed in the constant temperature refrigerator and reached the thermal equilibrium, the test cell is removed from the refrigerator and firmly mounted on the vibration device. Pump is started to circulate the coolant and the vibrator is turned on with a predetermined frequency and amplitude. Bath temperature is constantly monitored and controlled. During the experiment, the wall temperature of test cell is recorded, and the solid-liquid interface is videotaped as the melting continues.

At the end of the run, water will be poured out of the test cell and measured carefully to obtain the solid volume. This result will be used to verify the volume of the solid determined from the computer image.

Uncertainty Analysis

Ideally, all sources of uncertainty should be sampled, and measurements must be repeated a sufficient number of times to assure the reliability of results by using multiple-sample methods. However, repetition is very costly and hence, the data obtained in present experiments is a result of a single-sample estimate. With the guide of an article by Kline (1985), the methodology developed by Kline and McClintock (1953) is used to estimate the uncertainty.

As described before, the difference between the liquid volume fraction measured from the videotape image and that measured from draining test cell at end of experiment is less than 4 %. In the same nominal wall temperature cases, the difference in the actual time-averaged wall temperature varies from 0.1 ~ 0.3 °C. From Figure 6, we can see the slope of melted volume fraction vs. Ste*Fo is weakly dependent on the temperature difference $T_w - T_f$. The contribution of wall temperature fluctuation to the overall uncertainty is small. The initial ice temperatures are controlled within -0.5 ~ -1.5 °C. With consideration of the specific heat and latent heat of ice, the uncertainty associated with initial ice temperature is negligible. Overall the uncertainty of the melted volume fraction is estimated to be 5 %.

The uncertainty of Fo*Ste is relatively easy to analyze. The uncertainty due to the time and length measurement is negligible. If the uncertainty associated with the thermal properties is not considered, the dimensionless time uncertainty is from wall temperature measurement, which depends on the precision of thermocouple. This uncertainty is estimated to be less than 4%.

RESULTS AND DISCUSSION

Some typical experimental images are shown in Figure 2. Figure 2a is the image of AR = 1.0. ΔT = 9.5 °C , t = 16 minute, without vibration. Figure 2b is the image of AR = 0.4, ΔT = 9.6 °C , t = 3.5 minute, with vertical vibration f = 60 Hz, A = 0.23 mm. The image of AR =2.5, ΔT = 9.6 °C , t = 14 minute, with vertical vibration f = 1.1 Hz, A = 37.5 mm is shown in Figure 2c. The ice bulk is in dark shade and water is in light shade. It can be seen from these pictures that the interface of ice-water is distinct and melting pattern is approximately two-dimensional. The shape of the interface is approximately symmetrical with respect to vertical direction except for the case of high AR where the solid ice bulk is inclined to a side wall. This inclination has significant effect on phase change heat transfer and this phenomena will be discussed later.

Table 1 lists all experimental conditions including aspect ratios, vibration parameters, average wall temperatures on the contact surface. It is noted that although the nominal wall temperature is 5, 10 or 15 °C, the actual wall temperature is little lower and varies slightly. The initial ice temperatures are close to melting point. The effects of ice subcooling are small and neglected in this study.

2a

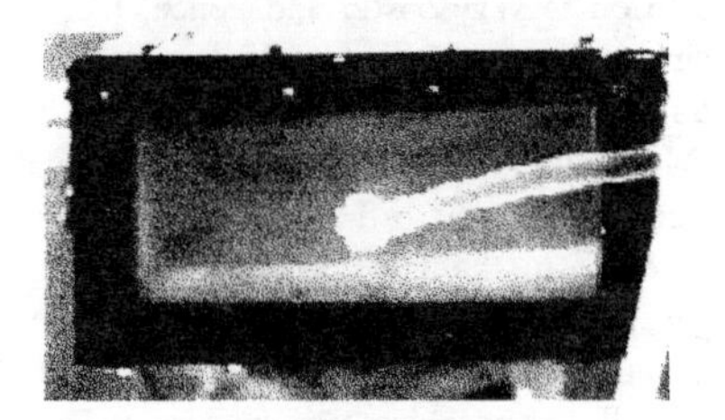

2b

2c

Fig.2 Experimental Images ; a) AR = 1.0; b) AR =0.4; c) AR =2.5

TABLE 1 Time-averaged wall temperatures in various experiments.

No.	AR	Tw	f (Hz)	A (mm)	Direction
1	1.0	9.5	0	0	
2	1.0	9.6	0.55	37.5	vertical
3	1.0	9.5	1.1	37.5	vertical
4	1.0	9.4	1.67	37.5	horizontal
5	1.0	9.3	60	0.23	vertical
6	1.0	9.4	EMV		vertical
7	0.4	9.6	0	0	
8	0.4	9.7	1.1	37.5	vertical
9	0.4	9.6	1.67	37.5	horizontal
10	0.4	9.3	60	0.23	vertical
11	0.4	9.7	EMV		vertical
12	2.5	9.6	0	0	
13	2.5	9.6	1.1	37.5	vertical
14	2.5	9.5	1.67	37.5	horizontal
15	2.5	9.4	60	0.23	vertical
16	2.5	9.6	EMV		vertical
17	1.0	4.8	0	0	
18	1.0	4.7	60	0.23	vertical
19	1.0	14.6	0	0	
20	1.0	14.4	60	0.23	vertical

EMV : Electromagnetic Vibrator

Effects of vibration on ice melting with AR = 1.0 and nominal wall temperature of 10 °C are shown in Figure 3. In order to compare the experimental results with different wall temperature conditions, the Stefan number is included. The dimensionless time, defined as Fo*Ste instead of Fo alone, is used for abscissa. The vertical axis is the melted volume fraction which is defined as the volume ratio of water to test cell. Five sets of vibration experiments data are included in Figure 3. These results are plotted against the stationary ice melting data. The stationary experimental data (T_w = 8.1 °C) of Hirata et al (1991) is also plotted in Figure 3, while the test cell was 50.7 mm x 50.7 mm and Grashof number was calculated as 1.23E+6. Figure 3 shows that the melting rate is related to the maximum acceleration of vibration. For the case of vertical vibration f = 60 Hz ($\pm$1.5Hz) , A = 0.23 mm ($\pm$0.01mm) (a = 3.3 g), an enhancement of the overall melting rate (as much as 115 %) is recorded while compared with the stationary ice melting. It is observed that the ice bulk slightly oscillates in the test cell. This relative motion (or sedimentation) between ice and water is induced by vibration due to their density differences. Ice motion stirs the water in the test cell and thus enhance convective heat transfer rate.

The case of horizontal vibration at f = 1.67 Hz ($\pm$0.03 Hz), A = 37.5 mm ($\pm$ 0.1 mm) (a = 0.42 g) has about 50 % increase of melting rate as well. The relative slide motion of ice along the top surface of test cell is observed. For experiments of A = 37.5 mm and frequency f = 1.1 Hz and 0.55 Hz, respectively (a = 0.18 g and 0.045g), melting rates are increased moderately. The increment for case a = 0.18 g is slightly higher than that of a = 0.045 g. No relative motion of ice is observed in the above cases. It is noted that in contact melting with vertical vibration, acceleration force induced by vibration and buoyancy force are in the same direction.

The direction of this acceleration changes periodically. Compared with the buoyancy force exerted on ice, the periodic downward force, which is proportional to maximum acceleration of vibration a, is not significant to cause the ice to descend from the top contact surface. The vibration melting, with a small electromagnetic vibrating device mounted on the bottom of the cell, has the least effect. The vibration frequency is 120 Hz and the amplitude is too small to be measured. The input power of this electromagnetic device is measured as 10.5 W. The experimental curve collapses with the stationary melting result during the early stage of melting. At later time, the melting curve edges away from the stationary curve when the convection is enhanced due to vibration.

Additional experiment is conducted to estimate the effect of kinetic energy input to test cell under vibration f = 60 Hz, A = 0.23 mm. The test cell (AR = 1.0) is filled with pure water (initial temperature is 9.8 °C) and well insulated. After 4 minutes of vibration, the averaged temperature is measured as 0.30 °C (± 0.05 °C) higher. It can be concluded that the effect of direct kinetic energy input is minor compared with the heat transfer enhancement due to vibration.

There are two kinds of melting processes, i.e., contact melting (which occurs at the top contact surface) and convective melting

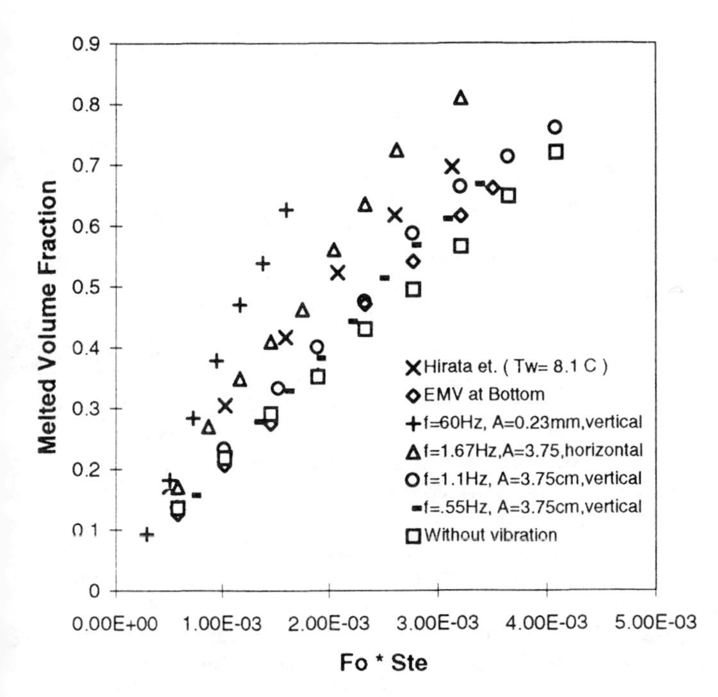

Fig. 3 T_w **near 10 C, AR =1.0, Ste=0.119, Gr=1.8E+6, Ar=2.21E+8**

(includes natural convection and sedimentation induced convection). Contact melting is the dominant mechanism in the present experiment (with and without vibration) as evidenced by the displacements of interface observed from the marker. The effect of vibration on melting is two folds. It promotes the convective melting along the sides (vertical and bottom) of ice bulk. Vibration also increase contact melting on the top as evidenced by the marker position. A preliminary theory is proposed to explain the contact melting enhancement with vertical vibration. For the stationary contact melting, the thickness of the liquid gap is relatively stable. The dominant heat transfer mechanism is conduction across that thin liquid gap. Vibration changes the thickness of the liquid gap between the ice bulk and the top heating surface periodically, therefore it changes the characteristics of the contact melting. When the ice bulk moves closer to the top heating surface, the liquid gap is thinner and heat conduction is enhanced and vice verse. It is assumed that the time-mean thickness of the liquid gap δ during the vibration is approximately the same as that of stationary experiment as shown in our numerical simulations (Shirvanian et al., 1996). The distance away from the contact surface can be specified by a sinusiodal function: $x = d\ (1 + \varepsilon \sin \omega t)$ where $\varepsilon < 1$ and ω is the angular velocity of the ice movement. For a time period of $2\pi/\omega$, the stationary and vibration conduction heat transfer can be expressed as:

$$\int_0^{\frac{2\pi}{\omega}} k\frac{\Delta T}{\delta}\,dt = k\frac{\Delta T}{\delta}\frac{2\pi}{\omega} \qquad (1)$$

and

$$\int_0^{\frac{2\pi}{\omega}} k\frac{\Delta T}{\delta(1+\varepsilon\sin\omega t)}\,dt = k\frac{\Delta T}{\delta\sqrt{1-\varepsilon^2}}\frac{2\pi}{\omega} \qquad (2)$$

respectively.

The relative changes of the conduction with and without vibration can be shown as:

$$\frac{1}{\sqrt{1-\varepsilon^2}} - 1 > 0 \qquad (3)$$

Equation (3) shows that vibration always promotes contact melting. However, for very small ε values which corresponds to small vibration amplitude, the vibration effects on contact melting is negligible.

Effects of vibration on ice melting with AR = 0.4 and nominal wall temperature of 10 °C are shown in Figure 4. Similar melting patterns to the cases of AR = 1 are observed. The case of f = 60 Hz, A = 0.23 mm (a = 3.3 g) increases the melting rate significantly (about 140 %). The other enhancements of melting rates are 75 % (horizontal vibration, f = 1.67 Hz, A = 37.5 mm, a = 0.42 g) and 9 % (vertical f = 1.1 Hz, A = 37.5 mm, a = 0.18 g), respectively

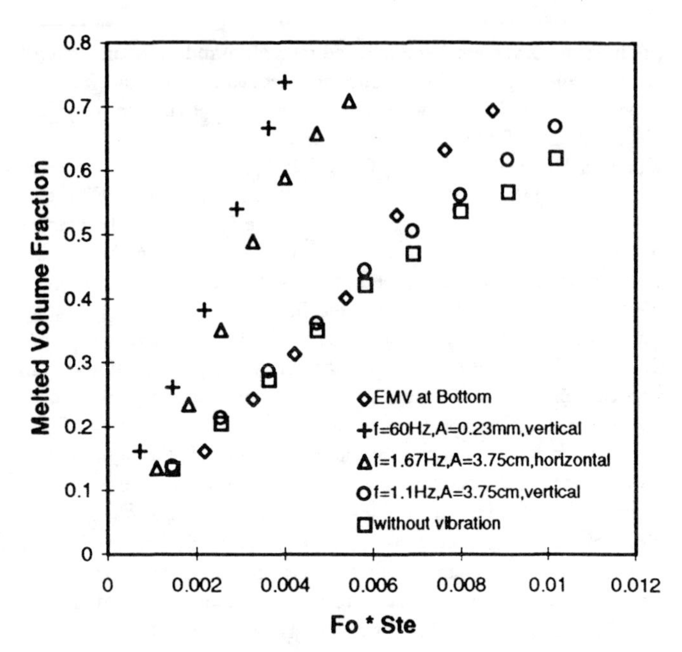

Fig. 4 Tw near 10 C, AR = 0.4, Ste = 0.119, Gr = 4.6E5, Ar=3.51e8

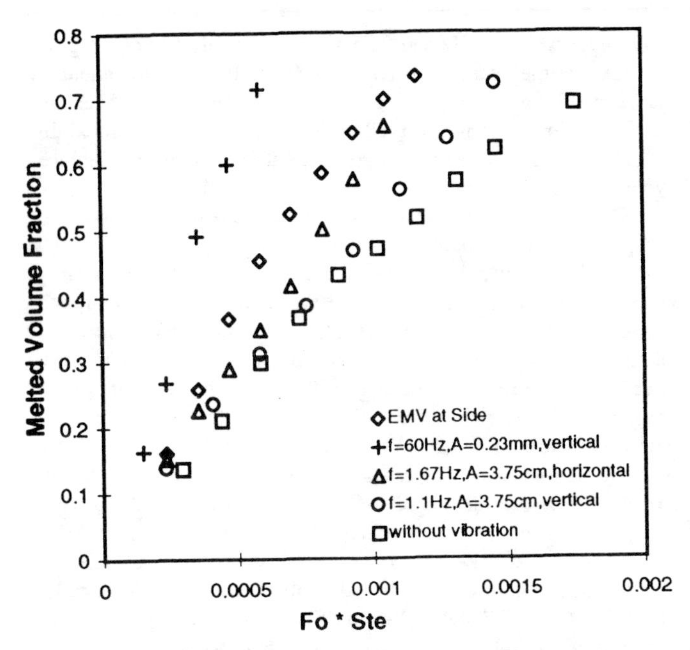

Fig. 5 Tw near 10 C, AR = 2.5, Ste = 0.119, Gr= 7.16E+6, Ar = 1.43E+8

compared with the stationary case. Similar melting rate patterns were observed for experiment using electromagnetic vibrator (f = 120 Hz, input power 10.5 W) mounted at the bottom during the early stages of melting, however, melting rate increases significantly at the later times.

Figure 5 shows the effects of vibration on ice melting with AR = 2.5. The nominal wall temperatures are the same as other AR cases (see Table 1). In most of the experiments, it is observed that ice bulk inclines to the side wall(s), and this results in two (or three) contact melting surfaces (i.e., top and side(s)). The inclination is caused by the asymmetric melting. This phenomenon of inclination was also described by Hirata et al. (1991). Since the starting time and location of solid inclination varies in repeated experiments, there were slight variations between experimental results. The magnitude of the melting rate increment using different vibration device are 170 % (vertical vibration, f = 60 Hz, A = 0.23 mm, a = 3.3 g), 40 % (horizontal vibration, f = 1.67 Hz, A = 37.5 mm, a = 0.42 g), and 14 % (f = 1.1 Hz, A = 37.5 mm, a = 0.18 g), respectively. Also shown in Figure 5 are the melting results with the electromagnetic vibrator (f = 120 Hz, input power 9.5 W) mounted on the bottom and side wall, respectively. It is found that when electromagnetic vibrator is side-mounted, the ice bulk migrates away from the wall mounted with the vibrator and contacts the opposite wall. Contact melting occurs at the entire side surface and top surface of the ice. Therefore, the melting rate is enhanced. Compared with bottom-mounted vibration, melting rate of 50 % increment is measured.

Effects of the wall temperature are showed in Figure 6. The nominal wall temperature are 15, 10 and 5 °C, respectively. The aspect ratio of the test cell is 1.0. For the stationary experiments, the results of melted volume fraction vs. dimesionless time (Fo*Ste) with Tw = 15 °C has the highest value while that with Tw = 5 °C has the lowest. This result is slightly different from the scale analysis from the lubrication theory (Bejan, 1992) where melted volume fraction vs. Fo*Ste is weakly related to $Ste^{-1/4}$. This theory is for pure contact melting and is based on the assumption of constant fluid properties. It is noted that the magnitude of natural convection changes significantly due to the well known water thermal expansion coefficient inversion at 4 °C. The Nusselt number of natural convection is proportional to the $Gr^{1/4}$ for an isothermal wall condition (Bejan, 1984). The natural convection at 5 °C is very weak due to both low thermal expansion coefficient (an order of magnitude smaller) as well as the small temperature difference. In addition, the viscosity of water decreases significantly as the function of temperature between 0 °C to 15 °C. The rate of contact melting is inversely proportional to 1/4 power of viscosity according to the results of scale analysis.

Results of melted volume fraction vs. Fo*Ste under vibrating conditions are different from stationary results. As seen in Figure 6, the case of Tw = 5 °C has the highest melting value while that with Tw = 15 °C has lowest under the same vertical vibration condition (f = 60 Hz, A = 0.23 mm, a = 3.3 g). In stationary experiments, the natural convection is relatively weak at low wall temperature case. Therefore, it may conclude that the effects of vibration on melting are more significant for low temperature heating cases.

The effect of aspect ratio on ice melting is depicted in Figure 7. Three different aspect ratios, i.e., 0.4, 1.0, and 2.5 with nominal wall temperature of 10 °C are evaluated. The abscissa of Figure 7 is the actual melting time. In order to make meaningful comparison, three test cells with various aspect ratios are fabricated with the same volume. It is shown that for stationary experiments, the melting rate for AR = 0.4 and 2.5 are higher than that of AR = 1.0. The higher melting rates for both cases are attributed to the larger

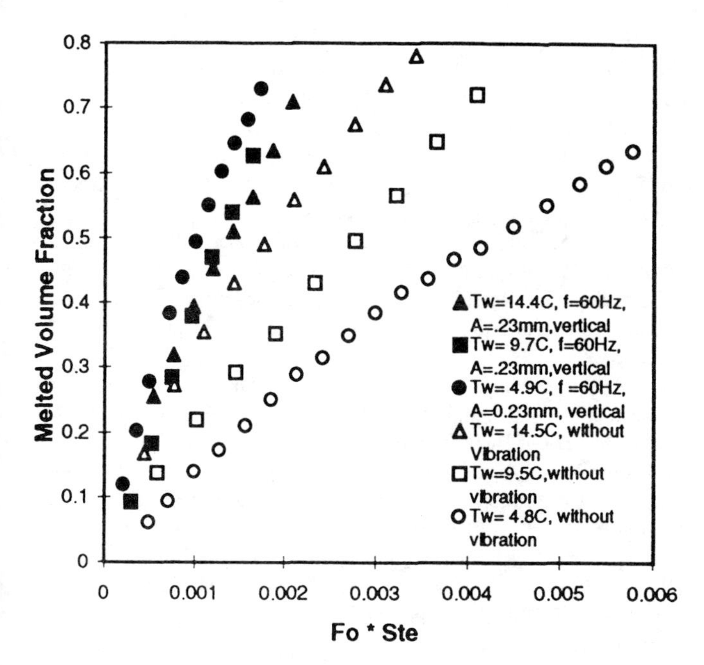

Fig. 6 Ice Melting at AR =1.0

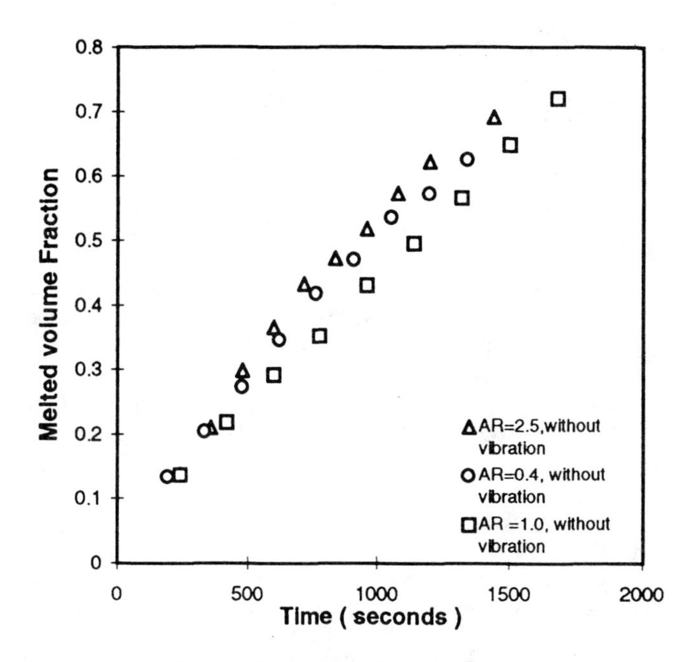

Fig.7 Effects of Aspet Ratio at T_w = 9.5 C

contact surface areas. Although the top contact surface area for aspect ratio 2.5 is smaller, the total contact melting surface area is actually increased by the ice bulk inclination to the side wall(s). It is also noted that test cell AR = 1.0 has the smallest total heating area. For the cases of vertical vibration, A = 37.5 mm, f= 1.1 Hz, melting rate is highest for AR = 2.5 .

Vibration (vertical, f = 60 Hz, A = 0.23 mm, a = 3.3 g) increases melting rate significantly in all aspect ratios as shown in Figures 3, 4 and 5. The relative enhancement in melting for AR = 1.0 is about 115 % while those for aspect ratio 0.4 and 2.5 are about 140 % and 170 %, respectively. The larger relative melting enhancement for AR = 2.5 is caused by the sedimentation driven convection due to the low pressure drag associated with its slender body shape. Slide motion of inclined ice bulk along the side wall also contributes to higher melting rates. In the case of AR = 0.4 where natural convection is the weakest, vibration enhanced convection contributes to the high melting rate. Therefore, it can be concluded that convection induced by vibration has significant effects on melting enhancement. This is evidenced by the relative positions of the marker to the ice-water interfaces. As shown in Fig 3, 4 and 5, the enhancements for AR = 0.4, 1.0 and 1.0 under horizontal vibration with f = 1.67 Hz, A = 37.5 mm, are 40 %, 50 % and 75 %, respectively.

Compared with the stationary problems, phase change process under vibration is much more complicated. Vibration may enhance phase change heat transfer rate in several different ways: 1) sedimentation driven convection is enhanced due to increased interactions between solid and liquid under vibration; 2) vibration promotes the transition of natural convection from laminar flow to turbulent flow; 3) vibration increases the contact melting as described in Equation (3); 4) melting rate is augmented by the slide motion of ice bulk relative to contact surface as seen in cases of AR = 2.5 (vertical vibration) and AR = 0.4 (horizontal vibration), and 5) kinetic energy is input to phase change material through vibration and transformed into heat energy but it is very minor and negligible.

These mechanisms are highly nonlinear and interact with each other. Different vibration parameters, test cell aspect ratios, and heating conditions alternate the magnitude of various mechanisms.

CONCLUSIONS

Experimental studies of ice contact melting within rectangular enclosures under vibrating conditions are conducted with various aspect ratios and wall temperatures conditions. Constant temperature wall condition is applied. It has been demonstrated that in all experiments vibration increases the ice melting rates. The major conclusions are summarized as follows:

1. Melting enhancement is directly related to the maximum acceleration of vibration. In one case (vertical vibration, f = 60 Hz, A = 0.23 mm, a = 3.3 g, aspect ratio = 2.5), melting rate is increased as much as 170 %.
2. Aspect ratio is an important parameter in contact ice melting. For the same ice volume, it is observed that the result of aspect ratio 1.0 corresponds to the lowest melting rate in both stationary and vibrating conditions. The relative melting enhancement by vibration for both high and low aspect ratio cases are very significant.
3. Melting enhancement with horizontal vibration is greater than that with vertical vibration, i.e., slide contact melting is a more effective method to improve phase change heat transfer.
4. Melting enhancement by vibration is more pronounced for the low Stefan number (low wall temperature) cases.
5. Preliminary experiments indicated that the location of the vibrating sources may alternate melting results.

ACKNOWLEDGMENT
This research was supported by the National Science Foundation, Grant No. CTS-9422629.

REFERENCES

Asako, Y., Faghri, M., Charmchi, M., and Bahrami, P. A., 1994, "Numerical Solution for Melting of Unfixed Rectangular Phase-Change Material Under Low-Gravity Environment," Numerical Heat Transfer, Part A, Vol. 25, pp 191-208.

Bareiss, M., and Beer, H., 1984, "An Analytical Solution of the Heat Transfer Process During Melting Inside a Horizontal Tube," *International* Journal of Heat and Mass Transfer, Vol. 27, pp. 739-746.

Bejan, A., 1984, "Convective Heat Transfer" , John Wiley & Sons, New York.

Bejan, A., 1992, "Lubrication by Close-Contact Melting," in: Fundamental Issues in Small Scale Heat Transfer, Bayazitoglu, Y., and Peterson, G. P., eds., ASME HTD-Vol. 227, pp. 61-68.

Choi, K. J. and Hong, J. S.,1991, " Experimental Study of Enhanced Melting Process under Ultrasonic Influnce", AIAA Journal of Thermophysics and Heat Transfer, vol. 5, No. 3, pp. 340-346.

Dong, Z., Chen, Z., Wang, Q., and Ebadian, M.A., 1991," Experimental and Analytical Study of Contact Melting in a Rectangular Cavity", AIAA Journal of Thermophysics and Heat Transfer, Vol. 5, No.3, pp. 347-353.

Emerman, S. H., and Turcotte, D. L., 1983, "Stokes' Problem with Melting," International Journal of Heat and Mass Transfer, Vol. 26, pp. 1625-1630.

Hirata, T., Makino, Y., and Kaneko, Y., 1991, "Analysis of Close-Contact Melting for Octadecane and Ice Inside Isothermally Heated Horizontal Rectangular Capsule," International Journal of Heat and Mass Transfer, Vol. 34, pp. 3097-3106.

Hong, H., and Saito, A., 1992, "The Behavior of Direct Contact Melting in the Unsteady State," Proceedings, 2nd JSME-KSME Thermal Engineering Conference, pp.12-21.

Kline, S.J., 1985, " The Purpose of Uncertainty Analysis ," ASME Journal of Fluids Engineering, Vol. 107, June, pp.153-160.

Kline, S.J., and Mcclintock, F. A., 1953, "Describing Uncertainties in Single-Sample Experiments, " Mechanical Engineering, Vol. 75, Jan., pp. 3-8.

Moallemi, M. K., Webb, B. W., and Viskanta, R., 1986, "An Experimental and Analytical Study of Close-Contact Melting," ASME Journal of Heat Transfer, Vol. 108, pp. 894-899.

Shirvanian, A., Faghri, M., Zhang, Z., and Asako, Y., 1996, "Effects of Vibration on Melting of Unfixed Rectangular Phase-Change Material", to be presented at ASME Annual Meeting, 1997.

Sparrow, E. M., and Myrum, T. A., 1985, "Inclination-Induced Direct-Contact melting in a Circular Tube," ASME Journal of Heat Transfer, Vol. 107, No.3, pp. 533-540.

Zhang, Z. and Bejan, A., 1989a, "Melting in an Enclosure Heated at Constant Rate", International Journal of Heat and Mass Transfer, Vol. 32, No. 6, pp. 1063-1076.

Zhang, Z. and Bejan, A., 1989b, " The Problem of Time-Dependent Natural Convection Melting with Conduction in the Solid ", *International* Journal of Heat and Mass Transfer, Vol. 32, No. 12, pp. 2447-2457.

HTD-Vol. 342, National Heat Transfer Conference
Volume 4
ASME 1997

AN EVALUATION OF SEVERAL HEAT TRANSFER CORRELATIONS FOR TWO-PHASE FLOW WITH DIFFERENT FLOW PATTERNS IN VERTICAL AND HORIZONTAL TUBES

D. Kim, Y. Sofyan, A. J. Ghajar, and R. L. Dougherty
School of Mechanical and Aerospace Engineering
Oklahoma State University
Stillwater, Oklahoma 74078

ABSTRACT

In this study, the general validity of twenty heat transfer correlations obtained from a comprehensive literature review were assessed. These correlations were tested against an extensive set of two-phase flow experimental data available from the literature, for vertical and horizontal tubes and different flow patterns and fluids. A total of 427 data points from four available experimental studies were used for these comparisons. Based on the tabulated and graphical results of the comparisons, appropriate correlations for different flow patterns and tube orientations were recommended.

NOMENCLATURE

A	cross sectional area, ft^2 or m^2
c	specific heat at constant pressure, Btu/(lbm-°F) or kJ/(kg-K)
D	inside diameter of the tube, ft or m
G_t	mass velocity of total flow (= ρV), lbm/(hr-ft^2) or kg/(s-m^2)
h	heat transfer coefficient , Btu/(hr-ft^2-°F) or W/(m^2-K)
k	thermal conductivity, Btu/(hr-ft-°F) or W/(m-K)
L	length of the heated test section, ft or m
$\dot{m}$	mass flow rate, lbm/hr or kg/s
Nu	Nusselt number, dimensionless
P	mean system pressure, psi or Pa
Pa	atmospheric pressure, psi or Pa
ΔP_M	momentum pressure drop, psf or Pa
$\Delta P/\Delta L$	total pressure drop per unit length, lbf/ft^3 or Pa/m
Pr	Prandtl number, dimensionless
Q	volumetric flow rate, ft^3/min or m^3/s
q"	heat flux per unit area, Btu/(hr-ft^2) or W/m^2
Re_M	mixture Reynolds number (= $\rho_L U_M^* D/\mu_L$ in [19]), dimensionless, where $U_M^* = V_L + 1.2(Re_S)^{-0.25} V_S - 12\, Fr_{ED} V_{ED} + 16(Fr_S)^{1.25} V_S$, $Re_S = \rho_L V_S D\,(1-\sqrt{\alpha})/\mu_L$, $V_{ED} = V_{SL} + V_{SG}$, $Fr_{ED} = \alpha D\,(1-\sqrt{\alpha})/V_{ED}^2$, $Fr_S = D(1-\sqrt{\alpha})/V_S^2$, $V_L = V_{SL}/(1-\alpha)$, $V_G = V_{SG}/\alpha$, V_S = slip velocity $= V_G - V_L$
Re_{TP}	two-phase flow Reynolds number, dimensionless $= Re_{SL}/(1-\alpha)$ in [2] $= G_F D/\mu_F$ where G_F = mass flow rate of froth and $\mu_F = (\mu_W+\mu_A)/2$ in [5] $= Re_{SL} + Re_{SG}$ in [6] and [7]
R_L	liquid volume fraction (= $1-\alpha$), dimensionless
T	temperature, °F or °C
V	average velocity in the test section, ft/s or m/s
x	flow quality, dimensionless
X_{TT}	Martinelli parameter [$= \left(\frac{1-x}{x}\right)^{0.9}\left(\frac{\rho_G}{\rho_L}\right)^{0.5}\left(\frac{\mu_L}{\mu_G}\right)^{0.1}$], dimensionless

Abbreviations

A	air or annular flow
B	bubbly flow
B-S	bubbly-slug transitional flow (other combinations with dashes are also transitional flows)
C	churn flow
F	froth flow
H	horizontal
M	mist flow
S	slug flow
V	vertical
W	water

Greek symbols

α	void fraction [= $A_G/(A_G+A_L)$], dimensionless
μ	dynamic viscosity, lbm/(hr-ft) or Pa-s
ρ	density, lbm/ft^3 or kg/m^3
ϕ_g, ϕ_l	Lockhart-Martinelli [43] two-phase gas and liquid multipliers, dimensionless

Subscripts

A	air
B	bulk
CAL	predicted
EXP	experimental
G	gas
L	liquid
MIX	gas-liquid mixture
TP	two-phase
TPF	two-phase frictional
SG	superficial gas
SL	superficial liquid
W	wall

INTRODUCTION

In many industrial applications, such as the flow of natural gas and oil in flowlines and wellbores, the knowledge of two-phase, two-component (liquid and permanent gas) heat transfer is required. When a gas-liquid mixture flows in a pipe, a variety of flow patterns may occur, depending primarily on flow rates, the physical properties of the fluids, and the pipe inclination angle. The main flow patterns that generally exist in vertical upward flow of a gas and liquid in tubes can be classified as bubbly flow, slug flow, froth flow, and annular flow. The main flow patterns that might exist in two-phase gas-liquid flow in horizontal tubes can be classified as stratified flow, slug flow, annular flow, and mist flow. The variety of flow patterns reflects the different ways that the gas and liquid phases are distributed in a pipe. This causes the heat transfer mechanism to be different in the different flow patterns.

Numerous heat transfer correlations and experimental data for forced convective heat transfer during gas-liquid two-phase flow in vertical and horizontal pipes have been published over the past 40 years. In this study, a comprehensive literature search was carried out and a total of 38 two-phase flow heat transfer correlations [1 to 39] were identified. The validity of these correlations and their ranges of applicability have been documented by the original authors, see references 1 to 39. In most cases, the identified heat transfer correlations were derived empirically and were based on a small set of experimental data with a limited range of variables and liquid-gas combinations. In order to assess the general validity of those correlations, they were compared against an extensive set of two-phase flow heat transfer experimental data available from the literature, for vertical and horizontal tubes and different flow patterns and fluids. A total of 427 data points from four available experimental studies [10,40,41,42] were used for these comparisons. The experimental data included different liquid-gas combinations (water-air, glycerin-air, silicone-air), and covered a wide range of variables, including liquid and gas flow rates and properties, flow patterns and pipe sizes.

Table 1 shows twenty of the 38 heat transfer correlations that were identified and tested in this study. The rest of the two-phase flow heat transfer correlations [22 to 39] were not tested since the required information by the correlations were not available through the identified experimental studies. The limitations of the twenty correlations used in this study as proposed by the original authors are tabulated in Table 2. The ranges of the experimental data used to assess the general validity of the correlations listed in Table 1 are provided in Table 3.

RESULTS AND DISCUSSION OF THE COMPARISONS

A summary of the results obtained by comparing the twenty identified two-phase flow heat transfer correlations with the 139 water-air experimental data of Vijay [40], 57 glycerin-air experimental data of Vijay [40], 162 silicone-air experimental data of Rezkallah [41], 48 water-air experimental data of Pletcher [42], and 21 water-air experimental data of King [10] are given in Tables 4 to 7, respectively. These tables give the total number of experimental data points used from each experimental study, the total number of data points for each flow pattern, the number of data points in each flow pattern that were predicted to within ± 30% by the individual heat transfer correlations, and the percent overall mean and r.m.s. deviations for the predictions of each correlation. Note that the magnitudes of mean and r.m.s. deviations in these tables range from 0.08% to 314,035% indicating a wide range of agreement/disagreement of the correlation with the experimental data. The flow patterns for the experimental data were based on the procedures suggested by Govier and Aziz [44], Griffith and Wallis [45], Hewitt and Hall-Taylor [46], Taitel et al. [47], Taitel and Dukler [48], and visual observation as appropriate. For each flow pattern, the tables also highlight the number of data points predicted by the correlation(s) that best satisfied the ± 30% criterion.

The results of comparisons shown in Table 4 indicate that, for bubbly, froth, annular, bubbly-froth, and froth-annular flows, several of the heat transfer correlations did a very good job of predicting the experimental water-air data of Vijay [40] in a vertical tube. However, for slug, slug-annular, and annular-mist flows, only one correlation for each flow pattern showed good predictions. Considering the performance of the correlations for all the flow patterns and keeping in mind the values of the overall mean and r.m.s. deviations, three heat transfer correlations are recommended for this set of experimental data. These are the correlation of Knott et al. [11] for bubbly, froth, bubbly-froth, froth-annular, and annular-mist flows; the correlation of Ravipudi and Godbold [15] for annular, slug-annular, and froth-annular flows; and the correlation of Aggour [1] for bubbly and slug flows. Figures 1 to 3 show how well the recommended correlations for each flow pattern performed with respect to the water-air experimental data of Vijay [40].

Table 1. Heat Transfer Correlations Chosen for This Study

Source	Heat Transfer Correlations
[1]	$h_{TP}/h_L = (1-\alpha)^{-1/3}$ Laminar (L) $Nu_L = 1.615\,(Re_{SL} Pr_L\, D/L)^{1/3}(\mu_B/\mu_W)^{0.14}$ (L) $h_{TP}/h_L = (1-\alpha)^{-0.83}$ Turbulent (T) $Nu_L = 0.0155\, Re_{SL}^{0.83} Pr_L^{0.5}(\mu_B/\mu_W)^{0.33}$ (T)
[2]	$Nu_{TP} = 0.43\,(Re_{TP})^{0.55}(Pr_L)^{1/3}\left(\frac{\mu_B}{\mu_W}\right)^{0.14}\left(\frac{Pa}{P}\right)^{0.17}$
[3]	$Nu_{TP} = 0.060\left(\frac{\rho_L}{\rho_G}\right)^{0.28}\left(\frac{DG_t x}{\mu_L}\right)^{0.87} Pr_L^{0.4}$
[4]	$h_{TP}/h_L = (1-\alpha)^{-1/3}$ (L) $h_{TP}/h_L = (1-\alpha)^{-0.8}$ (T) $Nu_L = 0.0123\, Re_{SL}^{0.9} Pr_L^{0.33}(\mu_B/\mu_W)^{0.14}$
[5]	$Nu_{TP} = 0.029\,(Re_{TP})^{0.87}(Pr_L)^{0.4}$
[6]	$Nu_{TP} = 0.5\left(\frac{\mu_G}{\mu_L}\right)^{1/4}(Re_{TP})^{0.7}(Pr_L)^{1/3}\left(\frac{\mu_B}{\mu_W}\right)^{0.14}$
[7]	$Nu_{TP} = 0.029\,(Re_{TP})^{0.87}(Pr_L)^{1/3}(\mu_B/\mu_W)^{0.14}$ (for water-air) $Nu_{TP} = 2.6\,(Re_{TP})^{0.39}(Pr_L)^{1/3}(\mu_B/\mu_W)^{0.14}$ (for (gas-oil)-air)
[8]	$Nu_{TP} = 1.75\,(R_L)^{-1/2}\left(\frac{\dot{m}_L c_L}{R_L k_L L}\right)^{1/3}\left(\frac{\mu_B}{\mu_W}\right)^{0.14}$
[9]	$Nu_{TP} = 0.26\, Re_{SG}^{0.2}\, Re_{SL}^{0.55}\, Pr_L^{0.4}$
[10]	$\frac{h_{TP}}{h_L} = \frac{R_L^{-0.52}}{1+0.025\, Re_{SG}^{0.5}}\left[\left(\frac{\Delta P}{\Delta L}\right)_{TP}/\left(\frac{\Delta P}{\Delta L}\right)_L\right]^{0.32}$ $Nu_L = 0.023\, Re_{SL}^{0.8} Pr_L^{0.4}$
[11]	$\frac{h_{TP}}{h_L} = \left(1+\frac{V_{SG}}{V_{SL}}\right)^{1/3}$ where h_L is from [21]
[12]	$Nu_{TP} = 125\left(\frac{V_{SG}}{V_{SL}}\right)^{1/8}\left(\frac{\mu_G}{\mu_L}\right)^{0.6}(Re_{SL})^{1/4}(Pr_L)^{1/3}\left(\frac{\mu_B}{\mu_W}\right)^{0.14}$
[13]	$\frac{h_{TP}}{h_L} = 1+0.64\sqrt{\frac{V_{SG}}{V_{SL}}}$ where h_L is from [21]
[14]	$Nu_{TP} = Nu_L\left(\frac{1.2}{R_L^{0.36}}-\frac{0.2}{R_L}\right)$ $Nu_L = 1.615\left[\frac{(Q_G+Q_L)\rho D}{A\mu} Pr_L\, D/L\right]^{1/3}(\mu_B/\mu_W)^{0.14}$
[15]	$Nu_{TP} = 0.56\left(\frac{V_{SG}}{V_{SL}}\right)^{0.3}\left(\frac{\mu_G}{\mu_L}\right)^{0.2}(Re_{SL})^{0.6}(Pr_L)^{1/3}\left(\frac{\mu_B}{\mu_W}\right)^{0.14}$
[16]	$h_{TP}/h_L = (1-\alpha)^{-0.9}$ where h_L is from [21]
[17]	$\frac{h_{TP}}{h_L} = 1+462\, X_{TT}^{-1.27}$ where h_L is from [21]
[18]	$\frac{h_{TP}}{h_L} = \left(1+\frac{V_{SG}}{V_{SL}}\right)^{1/4}$ $Nu_L = 1.86\,(Re_{SL}\, Pr_L\, D/L)^{1/3}(\mu_B/\mu_W)^{0.14}$ (L) $Nu_L = 0.023\, Re_{SL}^{0.8} Pr_L^{0.4}(\mu_B/\mu_W)^{0.14}$ (T)
[19]	$Nu_{TP} = 0.075(Re_M)^{0.6}\frac{Pr_L}{1+0.035(Pr_L-1)}$
[20]	$h_{TP}/h_L = (\Delta P_{TPF}/\Delta P_L)^{0.451}$ $Nu_L - 1.615\,(Re_{SL} Pr_L\, D/L)^{1/3}(\mu_B/\mu_W)^{0.14}$ (L) $Nu_L = 0.0155\, Re_{SL}^{0.83} Pr_L^{0.5}(\mu_B/\mu_W)^{0.33}$ (T)
[21]	$Nu_L = 1.86\,(Re_{SL}\, Pr_L\, D/L)^{1/3}(\mu_B/\mu_W)^{0.14}$ (L) $Nu_L = 0.027\, Re_{SL}^{0.8} Pr_L^{0.33}(\mu_B/\mu_W)^{0.14}$ (T)

Note: α and R_L are taken from the original experimental data for this study. $Re_{SL} < 2000$ implies laminar flow, otherwise turbulent. For Shah [18], replace 2000 by 170. With regard to the eqs. given for [18] above, the laminar two-phase correlation was used along with the appropriate single phase correlation, since [18] recommended a graphical turbulent two-phase correlation.

Table 2. Limitations of the Heat Transfer Correlations Used in This Study (See Nomenclature for Abbreviations)

Source	Fluids	L/D	Orient.	$\dot{m}_G / \dot{m}_L$	V_{SG}/V_{SL}	Re_{SG}	Re_{SL}	Pr_L	Flow Pattern(s)
[1]	A-W, Helium-W, Freon12-W	52.1	V	7.5×10^{-5}-5.72×10^{-2}	0.02-470	13.95-2.09×10^{5}		5.42-6.36	B, S, A, B-S, B-F, S-A, A-M
[2]	W-A	34	V		0.12-4.64	540-2700	16000-112000		B, S, F-A
[3]	Gas-Liquid		H & V						A, M-A
[4]	A-Oil	16	V		0.004-4500		300-66000		B, S, A
[5]	A-W	67	V	45-350		0-4.29×10^{4}	1.4×10^{4}-4.9×10^{4}		F
[6]	A-W A-Glycerin	86	V		0.3-2.5 0.6-4.6		300-14300		B, S
[7]	A-W Gas-Oil-A	14.3	V	244-977 269-513	1-250 0.6-80		>5000 1400-3500		
[8]	Gas-Liquid		H						S
[9]	A-W, A-Poly methylsiloxane, A-Diphenyl oxide	60-80	V			4000-37000	3.5-210	4.1-90	A
[10]	A-W	252	H		1.21-6.94	1570-8.28×10^{4}	22500-11.9×10^{4}		S
[11]	Petroleum oil-Nitrogen gas	118.6	V	1.57×10^{-3}-1.19	0.1-40	6.7-162	126-3920		B
[12]	A-W, A-Ethylene glycol	17.6	V	1.92×10^{-4}-0.1427 0-0.11	0.16-75 0.25-67		5.5×10^{4}-49.5×10^{4} 380-1700	140 @ 37.8°C	B, S, F
[13]	A-W	17	H						B, S, A
[14]	A-85% Glycol, A-1.5% SCMC, A-0.5% Polyox		H				500-1800		S
[15]	A-W, A-Toluene, A-Benzene, A-Methanol		V		1-90	3562-82532	8554-89626		F
[16]	A, W, Oil, etc.; 13 Liquid-Gas combinations	52.1	V		0.01-7030		1.8-1.3×10^{5}	4.2-7000	B, S, C, A, F, B-S, B-F, S-C, S-A, C-A, F-A
[17]	A-W	35	V						B
[18]	A, W, Oil, Nitrogen, Glycol, etc.; 10 combinations		H & V		0.004-4500		7-253000		B, S, F, F-A, M
[19]	A-Liquid	67	V	9.4×10^{-4}-0.059	4-50			4-160	S, A
[20]	A-W, A-Glycerin, Helium-W, Freon12-W	52.1	V		0.005-7670		1.8-130000	5.5-7000	B, S, F, A, M, B-F, S-A, F-A, A-M

Table 3. Ranges of the Experimental Data Used in This Study

Water-Air	$16.71 \leq \dot{m}_L$ (lbm/hr) ≤ 8996	$0.06 \leq V_{SL}$(ft/sec) ≤ 34.80	$231.83 \leq Re_{SL} \leq 126630$
Vertical	$0.058 \leq \dot{m}_G$ (lbm/hr) ≤ 216.82	$0.164 \leq V_{SG}$(ft/sec) ≤ 460.202	$43.42 \leq Re_{SG} \leq 163020$
Data (139 Points)	$0.007 \leq X_{TT} \leq 433.04$	$59.64 \leq T_{MIX}$ (°F) ≤ 83.94	$14.62 \leq P_{MIX}$ (psi) ≤ 74.44
of Vijay [40]	$0.061 \leq \Delta P_{TP}$ (psi) ≤ 17.048	$0.007 \leq \Delta P_{TPF}$ (psi) ≤ 16.74	$0.033 \leq \alpha \leq 0.997$
	$5.503 \leq Pr_L \leq 6.982$	$0.708 \leq Pr_G \leq 0.710$	$11.03 \leq Nu_{TP} \leq 776.12$
	$101.5 \leq h_{TP}$ (Btu/hr-ft^2-°F) ≤ 7042.3	$0.813 \leq \mu_W/\mu_B \leq 0.933$	L/D = 52.1, D = 0.46 in.
Glycerin-Air	$100.5 \leq \dot{m}_L$ (lbm/hr) ≤ 1242.5	$0.31 \leq V_{SL}$(ft/sec) ≤ 3.80	$1.77 \leq Re_{SL} \leq 21.16$
Vertical	$0.085 \leq \dot{m}_G$ (lbm/hr) ≤ 99.302	$0.217 \leq V_{SG}$(ft/sec) ≤ 117.303	$63.22 \leq Re_{SG} \leq 73698$
Data (57 Points)	$0.15 \leq X_{TT} \leq 407.905$	$80.40 \leq T_{MIX}$ (°F) ≤ 82.59	$17.08 \leq P_{MIX}$ (psi) ≤ 62.47
of Vijay [40]	$1.317 \leq \Delta P_{TP}$ (psi) ≤ 20.022	$1.07 \leq \Delta P_{TPF}$ (psi) ≤ 19.771	$0.0521 \leq \alpha \leq 0.9648$
	$6307.04 \leq Pr_L \leq 6962.605$	$0.708 \leq Pr_G \leq 0.709$	$12.78 \leq Nu_{TP} \leq 37.26$
	$54.84 \leq h_{TP}$ (Btu/hr-ft^2-°F) ≤ 159.91	$0.513 \leq \mu_W/\mu_B \leq 0.610$	L/D = 52.1, D = 0.46 in.
Silicone-Air	$17.3 \leq \dot{m}_L$ (lbm/hr) ≤ 196	$0.072 \leq V_{SL}$(ft/sec) ≤ 30.20	$47.0 \leq Re_{SL} \leq 20930$
Vertical	$0.07 \leq \dot{m}_G$ (lbm/hr) ≤ 157.26	$0.17 \leq V_{SG}$(ft/sec) ≤ 363.63	$52.1 \leq Re_{SG} \leq 118160$
Data (162 points)	$72.46 \leq T_W$ (°F) ≤ 113.90	$66.09 \leq T_B$ (°F) ≤ 89.0	$13.9 \leq P_{MIX}$ (psi) ≤ 45.3
of Rezkallah [41]	$0.037 \leq \Delta P_{TP}$ (psi) ≤ 9.767	$0.094 \leq \Delta P_{TPF}$ (psi) ≤ 9.074	$0.011 \leq \alpha \leq 0.996$
	$61.0 \leq Pr_L \leq 76.5$	$0.079 \leq Pr_G \leq 0.710$	$17.3 \leq Nu_{TP} \leq 386.8$
	$29.9 \leq h_{TP}$ (Btu/hr-ft^2-°F) ≤ 683.0	L/D = 52.1, D = 0.46 in.	
Water-Air	$0.069 \leq \dot{m}_L$ (lbm/sec) ≤ 0.3876	$0.03 \leq \dot{m}_G$ (lbm/sec) ≤ 0.2568	$7.84 \leq \Delta P/L$ (lbf/ft^3) ≤ 137.5
Horizontal	$0.22 \leq \Delta P_M/L$ (lbf/ft^3) ≤ 26.35	$0.021 \leq X_{TT} \leq 0.490$	$1.45 \leq \phi_g \leq 3.54$
Data (48 points)	$7.23 \leq \phi_l \leq 68.0$	$73.6 \leq T_W$ (°F) ≤ 107.1	$64.9 \leq T_{MIX}$ (°F) ≤ 99.4
of Pletcher [42]	$7372 \leq q''$ (Btu/hr-ft^2) ≤ 11077	$433 \leq h_{TP}$ (Btu/hr-ft^2-°F) ≤ 1043.8	L/D = 60.0, D = 1.0 in.
Water-Air	$1375 \leq \dot{m}_L$ (lbm/hr) ≤ 6410	$0.82 \leq \dot{m}_G$ (SCFM) ≤ 43.7	$22500 \leq Re_{SL} \leq 119000$
Horizontal	$1570 \leq Re_{SG} \leq 84200$	$0.41 \leq X_{TT} \leq 29.10$	$0.117 \leq R_L \leq 0.746$
Data (21 points)	$136.8 \leq T_{MIX}$ (°F) ≤ 144.85	$184.3 \leq T_W$ (°F) ≤ 211.3	$15.8 \leq P_{MIX}$ (psi) ≤ 55.0
of King [10]	$147.9 \leq \Delta P_{TP}$ (psf) ≤ 3226	$1462 \leq h_{TP}$ (Btu/hr-ft^2-°F) ≤ 4415	$1.08 \leq V_{SG}/V_{SL} \leq 6.94$
	$1.35 \leq h_{TP} / h_L \leq 3.34$	$1.35 \leq \phi_l \leq 8.20$	L/D =252, D = 0.737 in.

From the comparison results shown in Table 5, it can be seen that only a few of the tested heat transfer correlations were capable of predicting with good accuracy the glycerin-air experimental data of Vijay [40] in a vertical tube. Considering the overall performance of the correlations for all the flow patterns, only the correlation of Aggour [1] is recommended for this set of experimental data. The performance of Aggour's [1] correlation in different flow patterns (bubbly, slug, froth, annular, bubbly-slug, and slug-annular) with respect to the glycerin-air experimental data of Vijay [40] is shown in Fig. 4.

For the silicone-air experimental data of Rezkallah [41] in a vertical tube, a few of the correlations predicted the experimental data reasonably well. Again, considering the overall performance of the correlations for all the flow patterns and the values of the mean and r.m.s. deviations, only three of the tested heat transfer correlations are recommended. These are the correlation of Rezkallah and Sims [16] for bubbly, slug, churn, bubbly-slug, bubbly-froth, slug-churn, and churn-annular flows; the correlation of Ravipudi and Godbold [15] for churn, annular, bubbly-slug, slug-churn, churn-annular, and froth-annular flows; and the correlation of Shah [18] for bubbly, froth, bubbly-froth, froth-annular, and annular-mist flows. Figures 5, 6, and 7 show the comparison between the predictions of the three recommended correlations and the silicone-air experimental data of Rezkallah [41].

Table 7 shows the results of comparison for the 48 annular flow water-air experimental data of Pletcher [42] and 21 slug flow water-air experimental data of King [10] in horizontal tubes with the identified heat transfer correlations. For the annular flow data, only the correlation of Shah [18] performed well. Figure 8 compares the performance of this correlation with the experimental data of Pletcher [42]. Also shown in Table 7 are the results of comparison between the heat transfer correlations and the slug flow experimental data of King [10]. The experimental data of King [10] were predicted very well with five of the identified heat transfer correlations. Figure 9 shows how well the correlations of Chu and Jones [2], King [10], Kudrika et al. [12], Martin and Sims [13], and Ravipudi and Godbold [15] predicted the slug flow data of King [10].

Table 4. Comparison of Water-Air Experimental Data (139 Data Points) of Vijay [40] with the Suggested Correlations (See Nomenclature for Abbreviations)

Source	Mean	r.m.s.	Data Points within ±30% for Each Flow Pattern (Total No. of Data Points / Pattern)							
	Dev. (%)	Dev. (%)	B (25)	S (25)	F (25)	A (25)	B-F (7)	S-A (10)	F-A (4)	A-M (18)
[1]	-14.28	56.27	25	25	2	14	4	4	1	
[2]	-44.43	97.11	23	17	23	20	7	5	4	3
[3]	-155.91	541.35			3	8		3		
[4]	-30.85	67.36	4	20		11		2		
[5]	85.25	85.64								
[6]	-218.73	402.73						2		
[7]	-221.14	451.19		7		11		6		
[9]	-155.48	172.42						3		1
[11]	3.76	33.95	25	20	25	19	7	3	4	11
[12]	-71.82	240.25	4	6	6	18	2	2	4	3
[13]	-42.69	89.23	25	22	18	18	6	5	4	2
[14]	5701.	25791.								
[15]	-14.66	86.60		21	16	21	4	7	4	3
[16]	-35.36	80.03	25	22	17	14	7	5	4	4
[17]	-81034.	299137.	2							
[18]	24.86	31.51	25	15	25	3	7	3	4	6
[19]	-135.90	352.79	20	14	25	20	7	2	4	1
[20]	46.26	58.59	21	2	23		6			

Note: Blanks indicate the correlation did not satisfy the ±30% criterion.

Table 5. Comparison of Glycerin-Air Experimental Data (57 Data Points) of Vijay [40] with the Suggested Correlations (See Nomenclature for Abbreviations)

Source	Mean	r.m.s.	Data Points within ±30% for Each Flow Pattern (Total No. of Data Points / Pattern)					
	Dev. (%)	Dev. (%)	B (4)	S (19)	F (17)	A (8)	B-S (4)	S-A (5)
[1]	-13.82	18.44	4	17	15	8	4	4
[2]	-99.03	102.81		2				
[3]	-149.87	285.90		2	4			
[4]	88.51	88.69						
[5]	97.04	97.06						
[6]	-1410.	1844.	1	1				
[7]	-6301.	8960.						
[8]	-624.18	675.32						
[9]	-514.71	567.68						
[11]	-85.93	96.64	3	2			2	
[12]	61.62	61.86						
[13]	-164.31	185.58						
[14]	5994.	18350.						
[15]	66.18	66.69						
[16]	-51.49	54.86	1	1		6		
[17]	-17574.	35334.						
[18]	-50.12	54.00	4	4			3	
[19]	-68.43	140.38		17	2		4	
[20]	26.58	33.12	4	17	9		4	

Note: Blanks indicate the correlation did not satisfy the ±30% criterion.

Table 6. Comparison of Silicone-Air Experimental Data (162 Data Points) of Rezkallah [41] with the Suggested Correlations (See Nomenclature for Abbreviations)

Source	Mean Dev. (%)	r.m.s. Dev. (%)	Data Points within ±30% for Each Flow Pattern (Total No. of Data Points / Pattern)										
			B (26)	S (13)	C (11)	A (25)	F (18)	B-S (7)	B-F (10)	S-C (13)	C-A (12)	F-A (6)	A-M (21)
[1]	-5.57	74.95		3	2	4				1	1		10
[2]	-128.74	193.70	13	3			10		5				
[3]	-127.63	435.79		1	1	7				3	5	3	
[4]	20.42	59.91	24				2	2	8	1			4
[5]	86.69	87.22											
[6]	-426.83	841.52		7				1	1	2			
[7]	-623.48	1326.	1	5	2			4		5			
[8]	-126.83	364.56	4	7	7	6		5		9	7		
[9]	-295.28	366.85		2									
[11]	-4.09	57.41	22	3	1	11	18	3	10	3	3	6	2
[12]	-65.83	130.59	4			5	8	2	5	1	1	6	
[13]	-63.47	149.26	21	3	3	9	11	2	10	3	9		
[14]	12702	46885.	5	3	1			5		7			
[15]	-12.06	85.25	5	4	9	15	10	7	1	12	11	6	1
[16]	-20.02	52.55	26	8	9	14	10	6	10	12	10		6
[17]	-91540.	295080.	14										
[18]	9.28	42.96	25	3	1	11	18	3	10	6	4	6	10
[19]	-528.50	984.09											
[20]	41.38	67.08	10			4	14	3	7	1	1	6	

Note: Blanks indicate the correlation did not satisfy the ±30% criterion.

Table 7. Comparison of 48 Water-Air Experimental Data Points of Pletcher [42] and 21 Water-Air Experimental Data Points of King [10] with the Suggested Correlations (See Nomenclature for Abbreviations)

Source	Annular Flow [42]			Slug Flow [10]		
	Mean Dev. (%)	r.m.s. Dev. (%)	No. of ±30% Data Points	Mean Dev. (%)	r.m.s. Dev. (%)	No. of ±30% Data Points
[1]	-233.85	314.86		-57.46	66.21	3
[2]	insufficient exp. information provided			0.08	16.33	20
[3]	99.93	99.93		-2166.	3448.	
[4]	-232.01	297.41	5	-45.74	54.06	6
[5]	99.97	99.97		68.63	69.03	
[6]	-402.71	434.73		-89.46	95.87	
[7]	-480.49	538.78		-65.33	77.95	5
[8]	insufficient exp. information provided			56.06	59.25	2
[9]	-122.58	141.38	6	-121.91	127.86	
[10]	insufficient exp. information provided			4.77	12.14	21
[11]	-80.79	101.76	6	21.44	26.03	12
[12]	-52.30	59.92	11	-4.30	27.61	18
[13]	-246.76	278.68		8.79	18.90	19
[14]	-616.92	1201.	6	91.82	91.85	
[15]	-193.51	212.15		15.72	18.39	19
[16]	-333.49	405.60		-46.47	57.37	7
[17]	-256486.	314035.		-8791.	13529.	
[18]	-13.92	31.98	33	37.42	39.65	7
[19]	-186.16	198.71		-34.09	163.50	9
[20]	4.34	37.11	26	-44.	53.21	7

Note: Blanks indicate the correlation did not satisfy the ±30% criterion.

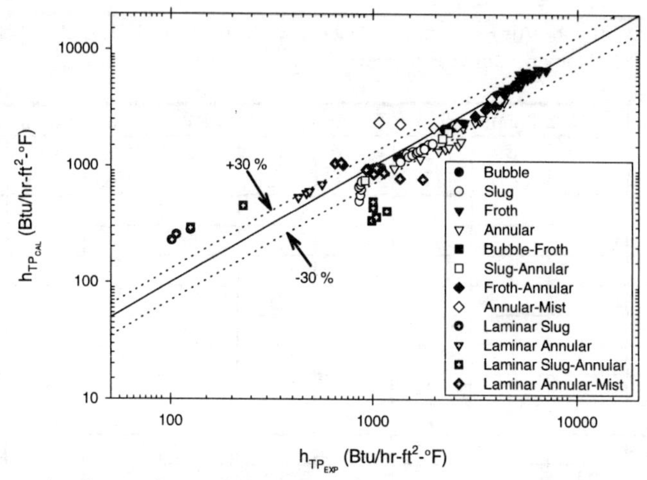

Figure 1. Comparison of Knott et al. [11] Correlation with Vijay's [40] Water-Air Experimental Data

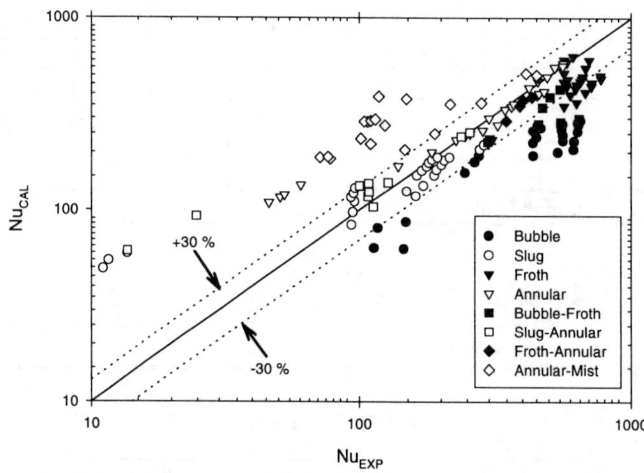

Figure 2. Comparison of Ravipudi & Godbold [15] Correlation with Vijay's [40] Water-Air Experimental Data

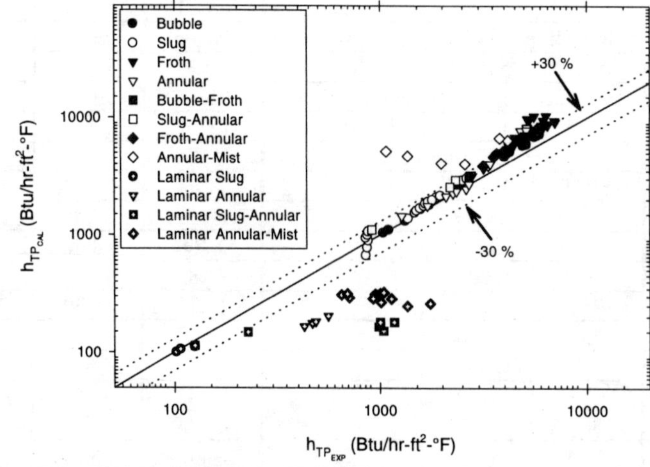

Figure 3. Comparison of Aggour [1] Correlation with Vijay's [40] Water-Air Experimental Data

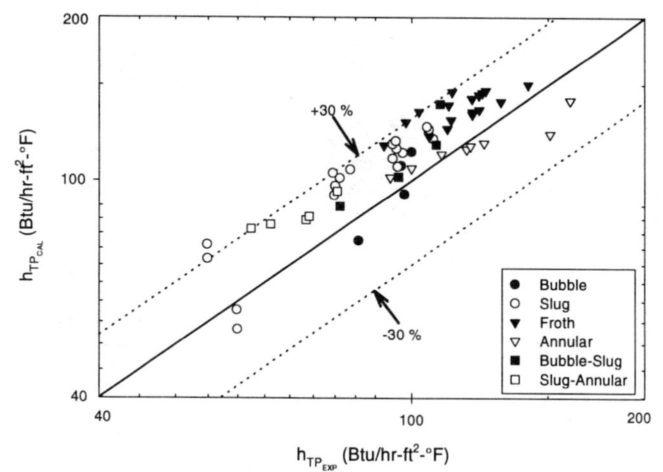

Figure 4. Comparison of Aggour [1] Correlation with Vijay's [40] Glycerin-Air Experimental Data

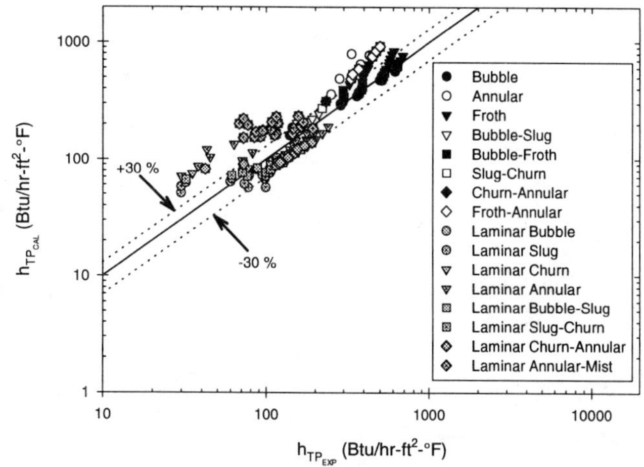

Figure 5. Comparison of Rezkallah & Sims [16] Correlation with Rezkallah's [41] Silicone-Air Experimental Data

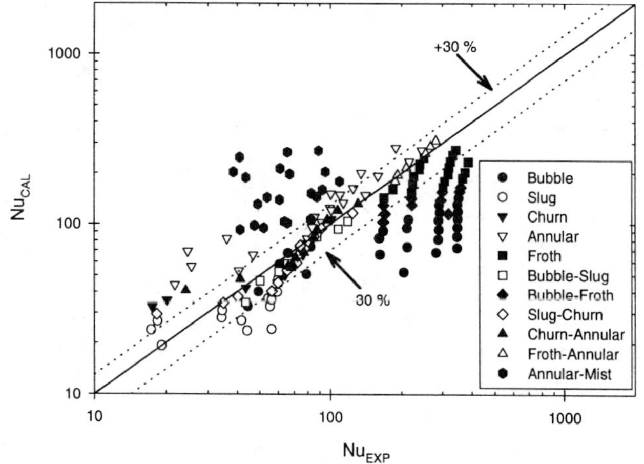

Figure 6. Comparison of Ravipudi & Godbold [15] Correlation with Rezkallah's [41] Silicone-Air Experimental Data

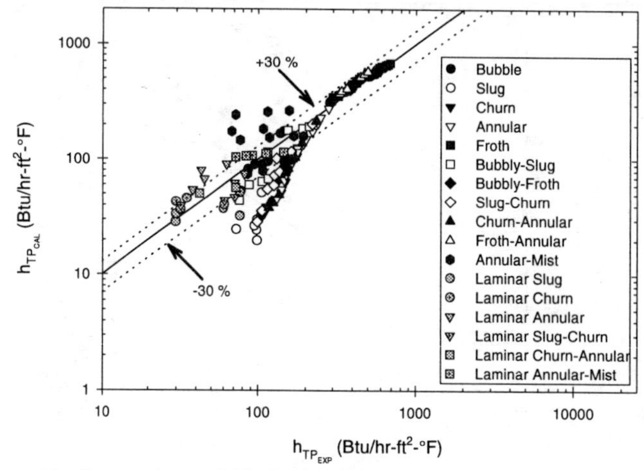

Figure 7. Comparison of Shah [18] Correlation with Rezkallah's [41] Silicone-Air Experimental Data

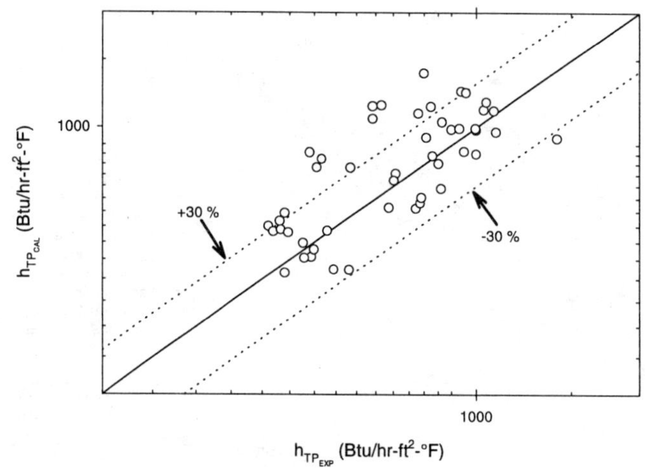

Figure 8. Comparison of Shah [18] Correlation with Pletcher's [42] Water-Air Experimental Data

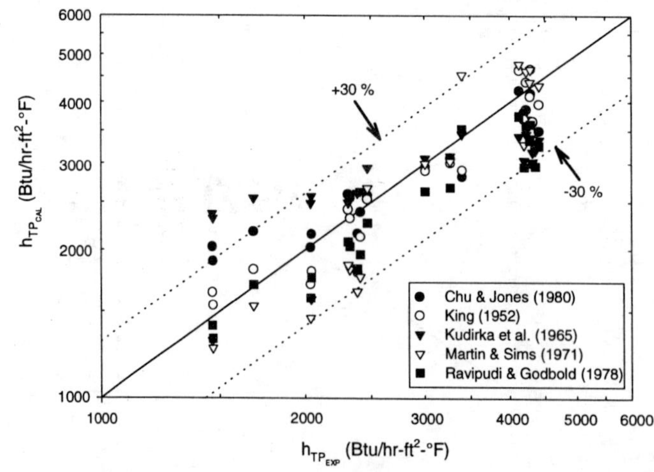

Figure 9. Comparison of Correlations of [2, 10, 12, 13, 15] with King's [10] Water-Air Experimental Data

SUMMARY AND CONCLUSIONS

We have studied the ability of 20 two-phase heat transfer correlations to predict five sets of experimental data that are available in the open literature. Three of these experimental data sets are for the flow of air-water [40], air-glycerin [40], and air-silicone [41] in various flow patterns within vertical pipes. The other two data sets are for the flow of air-water in slug [10] and annular [42] flow patterns within horizontal pipes.

With the data at hand, we make the following recommendations. For air-water flow within vertical pipes, we recommend use of the Knott et al. [11] correlation for bubbly, froth, bubbly-froth, froth-annular, and annular-mist flow patterns; use of the Ravipudi and Godbold [15] correlation for annular, slug-annular, and froth-annular flow patterns; and use of the Aggour [1] correlation for bubbly and slug flow patterns. For air-glycerin flow within vertical pipes, we recommend use of the Aggour [1] correlation for bubbly, slug, froth, annular, bubbly-slug, and slug-annular flow patterns. For air-silicone flow within vertical pipes, we recommend use of the Rezkallah and Sims [16] correlation for bubbly, slug, churn, bubbly-slug, bubbly-froth, slug-churn, and churn-annular flow patterns; use of the Ravipudi and Godbold [15] correlation for churn, annular, bubbly-slug, slug-churn, churn-annular, and froth-annular; and use of the Shah [18] correlation for bubbly, froth, bubbly-froth, froth-annular, and annular-mist flow patterns. With regard to air-water flow in horizontal pipes, we recommend use of the Shah [18] correlation for annular and use of the Kudrika et al. [12] correlation for slug flow patterns.

The above recommended correlations all have the following important parameters in common: Re_{SL}, Pr_L, μ_B/μ_W and either void fraction (α) or superficial velocity ratio (V_{SG}/V_{SL}). It appears that void fraction and superficial velocity ratio, although not directly related, may serve the same function in two-phase heat transfer correlations. However, since there is no single correlation capable of predicting the flow for all fluid combinations in vertical pipes, there appears to be at least one parameter [ratio], which is related to fluid combinations, that is missing from these correlations. In addition, since, for the horizontal data available, the recommended correlations differ from those of vertical pipes, there must also be at least one additional parameter [ratio], related to pipe orientation, that is missing from the correlations.

In our future work, we plan to continue this study by investigating the development of a correlation which is robust enough to span all or most of the fluid combinations, pipe orientations, and flow patterns. This may require experimental data parameters which are not in the currently available data sets. In order to aid in this heat transfer correlation development, we plan to obtain additional horizontal flow pattern data, and to obtain experimental data, for other fluid combinations which are applicable to the oil/gas industry.

ACKNOWLEDGMENTS

The authors gratefully acknowledge the financial assistance provided by the members of the University of Tulsa Joint Industry Project, which are: AGIP, ARC, Amoco, Arco, BDM, BHP, BP, Chevron, Conoco, Elf Aquitaine, Exxon, GRI, Japan National, Kerr-McGee, Marathon, Micro Motion, Mineral Management Service, Mobil, NKK, Norsk Hydro, Oil & Natural Gas Corp., Petro Canada, Petrolite, PETRONAS, Phillips, Shell, Texaco, TOTAL, and UNOCAL. The authors are also grateful to Mr. K.K. Ryali for assistance in verifying numerical results.

REFERENCES

1. Aggour, M.A. (1978), "Hydrodynamics and Heat Transfer in Two-Phase Two-Component Flow," Ph.D. Thesis, University of Manitoba, Canada.
2. Chu, Y-C. and B.G. Jones (1980), "Convective Heat Transfer Coefficient Studies in Upward and Downward, Vertical, Two-Phase, Non-Boiling Flows," AIChE Symp. Series, Vol. 76, pp. 79-90.
3. Davis, E.J. and M.M. David (1964), "Two-Phase Gas-Liquid Convection Heat Transfer," I&EC Fundamentals, Vol. 3, No. 2, pp. 111-118.
4. Dorresteijn, W.R. (1970), "Experimental Study of Heat Transfer in Upward and Downward Two-Phase Flow of Air and Oil Through 70 mm Tubes," Proc. 4th Int. Heat Transfer Conf., Vol. 5, B 5.9, pp. 1-10.
5. Dusseau, W. T. (1968), "Overall Heat Transfer Coefficient for Air-Water Froth in a Vertical Pipe," M. S. Thesis, Chemical Engineering, Vanderbilt University.
6. Elamvaluthi, G. and N.S. Srinivas (1984), "Two-Phase Heat Transfer in Two Component Vertical Flows," Int. J. Multiphase Flow, Vol. 10, No. 2, pp. 237-242.
7. Groothuis, H. and W.P. Hendal (1959), "Heat Transfer in Two-Phase Flow," Chemical Engineering Science, Vol. 11, pp. 212-220.
8. Hughmark, G.A. (1965), "Holdup and Heat Transfer in Horizontal Slug Gas Liquid Flow," Chem. Engng Sci., Vol. 20, pp. 1007-1010.
9. Khoze, A.N., S. V. Dunayev and V. A. Sparin (1976), "Heat and Mass Transfer in Rising Two-Phase Flows in Rectangular Channels," Heat Transfer Soviet Research, Vol. 8, No. 3, pp. 87 - 90.
10. King, C.D.G. (1952), "Heat Transfer and Pressure Drop for an Air-Water Mixture Flowing in a 0.737 Inch I.D. Horizontal Tube," M.S. Thesis, University of California.
11. Knott, R.F., R.N. Anderson, A. Acrivos and E.E. Petersen (1959), "An Experimental Study of Heat Transfer to Nitrogen-Oil Mixtures," Ind. and Engineering Chemistry, Vol. 51, No. 11, pp. 1369-1372.
12. Kudirka, A.A., R. J. Grosh and P.W. McFadden (1965), "Heat Transfer in Two-Phase Flow of Gas-Liquid Mixtures," I & EC Fundamentals, Vol. 4, No. 3, pp. 339-344.
13. Martin, B. W. and G. E. Sims (1971), "Forced Convection Heat Transfer to Water With Air Injection in a Rectangular Duct," Int. J. Heat Mass Transfer, Vol. 14, pp. 1115-1134.
14. Oliver, D. R. and S. J. Wright (1964), "Pressure Drop and Heat Transfer in Gas-Liquid Slug Flow in Horizontal Tubes," British Chem. Engrg., Vol. 9, No. 9, pp. 590 - 596.
15. Ravipudi, S.R. and T.M. Godbold (1978), "The Effect of Mass Transfer on Heat Transfer Rates for Two-Phase Flow in a Vertical Pipe," Proc. 6th Int. Heat Transfer Conf., Toronto, Vol. 1, pp. 505-510.

16. Rezkallah, K.S. and G.E. Sims (1987), "An Examination of Correlations of Mean Heat Transfer Coefficients in Two-Phase and Two-Component Flow in Vertical Tubes," AIChE Symp. Series, Vol. 83, pp. 109-114.
17. Serizawa, A., I. Kataoka and I. Michiyoshi (1975), "Turbulence Structure of Air-Water Bubbly Flow-III. Transport Properties," Int. J. Multiphase Flow, Vol. 2, pp. 247-259.
18. Shah, M.M. (1981), "Generalized Prediction of Heat Transfer During Two Component Gas-Liquid Flow in Tubes and Other Channels," AIChE Symp. Series, Vol. 77, No. 208, pp. 140-151.
19. Ueda, T. and M. Hanaoka (1967), "On Upward Flow of Gas-Liquid Mixtures in Vertical Tubes: 3rd. Report, Heat Transfer Results and Analysis," Bull. JSME, Vol. 10, pp. 1008-1015.
20. Vijay, M.M., M.A. Aggour and G.E. Sims (1982), "A Correlation of Mean Heat Transfer Coefficients for Two-Phase Two-Component Flow in a Vertical Tube," Proceedings of 7th Int. Heat Transfer Conference, Vol. 5, pp. 367-372.
21. Sieder, E.N. and G.E. Tate (1936), "Heat Transfer and Pressure Drop of Liquids in Tubes," Ind. Eng. Chem., Vol. 28, No. 12, p. 1429.
22. Akimenko, A.D., G.A. Zemskov and A.A. Skvortsov (1970), "Study of Heat Transfer to Water-Air Flow," Heat Transfer-Soviet Research, Vol. 2, No. 2, pp. 47-49.
23. Barnea, D. and N. Yacoub (1983), "Heat Transfer in Vertical Upwards Gas-Liquid Slug Flow," Int. J. Heat Mass Transfer, Vol. 26, No. 9, pp. 1365-1376.
24. Davis, E.J., S.C. Hung and S. Arciero (1975), "An Analogy for Heat Transfer with Wavy/Stratified Gas-Liquid Flow," AIChE Journal, Vol. 21, No. 5, pp. 872-878.
25. Domanskii, I. V., V.B. Tishin and V.N. Sokolov (1969), "Heat Transfer During Motion of Gas-Liquid Mixtures in Vertical Pipes," J. of Applied Chemistry of the USSR, Vol. 42, No. 4, pp. 809-813.
26. Fedotkin, I.M. and L.P. Zarudnev (1970), "Correlation of Experimental Data on Local Heat Transfer in Heating of Air-Liquid Mixtures in Pipes," Heat Transfer-Soviet Research, Vol. 2, No. 1, pp. 175-181.
27. Fried, L. (1954), "Pressure Drop and Heat Transfer for Two-Phase, Two-Component Flow," Chem. Eng. Prog. Symp. Series, Vol. 50, No. 9, pp. 47 - 51.
28. Ivanov, M. Y. and E.S. Arustamyan (1971), "Study of Heat Transfer in an Ascending Gas-Liquid Flow," Heat Transfer Soviet Reasearch, Vol. 3, No. 2, pp. 149-153.
29. Johnson, H. A. (1955), "Heat Transfer and Pressure Drop for Viscous-Turbulent Flow of Oil-Air Mixtures in a Horizontal Pipe," ASME Trans., Vol. 77, pp. 1257-1264.
30. Johnson, H. A. and A. H. Abou-Sabe (1952), "Heat Transfer and Pressure Drop for Turbulent Flow of Air-Water Mixture in a Horizontal Pipe," ASME Trans., Vol. 74, pp. 977-987.
31. Kapinos, V.M., A.F. Slitenko, N.B. Chirkin and L.V. Povolotskiy (1975), "Heat Transfer in the Entrance Section of a Pipe with a Two-Phase Flow," Heat Transfer-Soviet Research, Vol. 7, No. 2, pp. 126-128.
32. Lunde, K. E. (1961), "Heat Transfer and Pressure Drop in Two-Phase Flow," Chemical Engineering Progress Symposium Series, Vol. 57, No. 32, pp. 104 - 110.
33. Michiyoshi, I. (1978), "Heat Transfer in Air-Water Two-Phase Flow in a Concentric Annulus," Proc. 6th Int. Heat Transfer Conf., Toronto, Vol. 6, pp. 499-504.
34. Novosad, Z. (1955), "Heat Transfer in Two-Phase Liquid - Gas Systems," Collection Czecholav. Chem. Commun., Vol. 20, pp. 477- 499.
35. Oliver, D.R. and A. Young Hoon (1968), "Two-Phase Non-Newtonian Flow. Part II: Heat Transfer," Trans. INSTN Chem. Engrs., Vol. 46, pp. T116-T122.
36. Ovchinnikov, Y.V. and A.N. Khoze (1970), "Heat and Mass Transfer in Two-Component, Two-Phase Flows Inside of Cylinders," Heat Transfer Soviet Research, Vol. 2, No. 6, pp. 130-135.
37. Ozbelge, T.A. and T.G. Somer (1994), "A Heat Transfer Correlation for Liquid-Solid Flows in Horizontal Pipes," The Chemical Engineering Journal, Vol. 55, pp. 39-44.
38. Shaharabanny, O., Y. Taitel and A.E. Dukler (1978), "Heat Transfer During Intermittent/Slug Flow in Horizontal Tubes: Experiments," Proceedings of the 2nd CSNI Specialists Meeting, June 12-14, Paris, pp. 627-649.
39. Shoham, O., A.E. Dukler and Y. Taitel (1982), "Heat Transfer During Intermittent/Slug Flow in Horizontal Tubes," Ind. Eng. Chem. Fundam., Vol. 21, pp. 312-319.
40. Vijay, M.M. (1978), "A Study of Heat Transfer in Two-Phase Two-Component Flow in a Vertical Tube," Ph.D. Thesis, University of Manitoba, Canada.
41. Rezkallah, K.S. (1987), "Heat Transfer and Hydrodynamics in Two-Phase Two-Component Flow in a Vertical Tube," Ph.D. Thesis, University of Manitoba, Canada.
42. Pletcher, R.H. (1966), "An Experimental and Analytical Study of Heat Transfer and Pressure Drop in Horizontal Annular Two-Phase, Two-Component Flow," Ph.D. Thesis, Cornell University.
43. Lockhart, R. and R.C. Martinelli (1949), "Proposed Correlation of Data for Isothermal Two-Phase, Two-Component Flow in Pipes, Chem. Eng. Prog., Vol. 45, No. 1, pp. 39-48.
44. Govier, G.W. and K. Aziz (1973), The Flow of Complex Mixtures in Pipes, Van Nostrand Reinhold Company, New York.
45. Griffith, P. and G.B. Wallis (1961), "Two-Phase Slug Flow," J. Heat Transfer, Trans. ASME, Ser. C 83, pp. 307-320.
46. Hewitt, G.F. and N.S. Hall-Taylor (1970), Annular Two-Phase Flow, Pergamon Press Ltd., Oxford.
47. Taitel, Y., D. Barnea and A.E. Dukler (1980), "Modeling Flow Pattern Transitions for Steady Upward Gas-Liquid Flow in Vertical Tubes," AIChE Journal, Vol. 26, No. 3, pp. 345-354.
48. Taitel, Y. and A.E. Dukler (1976), "A Model for Predicting Flow Regime Transitions in Horizontal and Near Horizontal Gas-Liquid Flow," AIChE Journal, Vol. 22, No. 1, pp. 47-54.

HTD-Vol. 342, National Heat Transfer Conference
Volume 4
ASME 1997

TRANSIENT TEMPERATURE MEASUREMENTS IN A SUSPENDED DROPLET

Cill D. Richards and Robert F. Richards
School of Mechanical and Materials Engineering
Washington State University
PO Box 642920
Pullman, Washington 99164-2920
(509) 335-7753 Phone
(509) 335-4662 FAX
cill@mme.wsu.edu

ABSTRACT

The transient cooling of an evaporating water droplet, suspended in a jet of dry air, has been studied experimentally, using thermochromic liquid crystal thermography. Microencapsulated beads of thermochromic liquid crystals, suspended in the water droplets, enabled the visualization of the transient temperature fields within the droplets. Digital movies of the convectively cooled droplets reveal spatial and temporal temperature gradients resolved down to length scales of ~100 μm and time scales of ~0.03 seconds. The transient temperature measurements were analyzed to yield total droplet convective heat transfer rates. Droplet heat transfer rates determined from a heat balance on the droplets compare favorably to previously published measurements.

INTRODUCTION

The rate of heat transfer to droplets in sprays is a critical issue in the design of many practical spray systems. The control of such diverse processes as spray combustion, spray drying of agricultural products, and spray casting of metals depends on a knowledge of the heat transfer rates to the atomized droplets of evaporating fuel or milk, or solidifying liquid metal. Yet despite much attention, the measurement of heat transfer to droplets in sprays continues to be challenging problems.

The major stumbling block in determining droplet heat transfer is the lack of a suitable means to measure transient droplet temperatures during the heat transfer process. As a result, most heat transfer correlations available in the literature have been based on the observation of vaporizing droplets rather than direct measurements of temperature. The first of these experiments were reported by Ranz and Marshall (1952), in which water droplets were suspended in a jet of dry air, on fine-wire thermocouples. The mass flux from the evaporating droplets was determined from sequential size measurements of backlighted photographs of the shrinking droplets. Performing a heat balance on the droplets yielded the heat transfer rate to the droplets and a Nusselt - Reynolds correlation. Variations on these experiments have been performed on either suspended droplets (Charlesworth and Marshall, 1959; Trommelen and Crosby, 1970, Yearling and Gould, 1995) or on a linear array of droplets (Nishiwaki, 1955) Charlesworth and Marshall (1959) used micro-thermocouples imbedded in the drop to measure the droplet temperature, while the vaporization rate of suspended droplets was deduced from weight changes in the droplets as measured by a sensitive balance.

Yao & Schrock (1976) determined heat transfer rates for freely falling droplets by measuring the temperatures of the droplets before and after they fell. Droplet temperatures were found with one thermocouple placed in the droplet generator and a second thermocouple placed in a small dewar into which the droplet fell. An energy balance on the droplets yielded the droplet heat transfer. Moresco and Marschall (1979) conducted a similar experiment, but instead of collecting the droplets in a small dewar, placed a fine-wire thermocouple in the path of the droplets. Moresco and Marschall were able to extract both mean and surface temperatures for the droplets as they impacted on the thermocouple from the transient thermocouple signal.

Recently efforts to overcome the difficulties involved in determining droplet temperatures have resulted in a variety of new techniques in droplet thermography. Optical techniques have held out the most promise for nonintrusive measurements of droplet temperature. Two-color pyrometry has been used to measure the temperature of falling drops of molten metal in an evacuated drop tower (Hofmeister et al., 1989). However, the technique is not practicable at temperatures below about 600 C. Melton and coworkers have developed a temperature measurement technique based on laser induced exiplex fluorescence (Melton et al., 1986; Wells and Melton, 1990), and used it to measure the temperature of drops in free fall. More recently the use of liquid crystal thermography as a means to measure droplet temperatures has been reported by several groups, Hu et al. (1994), Nozaki et al. (1995), Peterson et al. (1995), Treuner et al. (1995), and Richards and Richards (1997).

Thermochromic liquid crystals (TLC) have already been employed in a wide variety of heat transfer experiments, and shown to be versatile and accurate means of temperature measurement (Akino 1989, D'abiri 1992, Ozawa 1992). Farina et al. (1993) demonstrated that with careful control of the disposition of the TLC material, illumination, and photographic or video recording of the TLC color play, temperature measurements with uncertainties of +/-0.25 K (95% confidence intervals) are possible. Thermochromic liquid crystals can be used in two forms: in their pure form (neat liquid crystals) or in the micro-encapsulated form, where the TLC is encapsulated in hollow plastic beads ranging in diameter from 5 to 3000 microns. The neutrally buoyant micro encapsulated beads can be suspended in a fluid where they act as tracer particles and follow the fluid flow.

The use of neat liquid crystals in droplet heat transfer studies has been reported by Hu et al. (1994) and Peterson et al (1995). Peterson et al. suspended droplets of neat liquid crystal from a micro-thermocouple and recorded their color play with an rgb (red-green-blue) video camera. Hu et al. atomized neat TLC and photographed the droplets in flight using a 35 mm camera.

The use of the microencapsulated form of TLC in droplet thermography has been reported by Nozaki et al. (1995) and Treuner et al. (1995), and Richards and Richards (1997). Nozaki et al. used microencapsulated beads to resolve transient temperature fields within water droplets as their buoyancy carried them up through hot, silicone oil. A color video camera recorded the transient temperature field in the water droplets as they convected through the immiscible oil. The droplets studied ranged in size from 2.9 to 8.2 mm in diameter. Treuner et al. used microencapsulated TLC beads to resolve transient temperatures within liquid droplets held between two opposed tubes. A temperature gradient was applied between the opposed tubes and the thermocapillary flow in the droplets imaged under microgravity conditions. The droplets studied, glycerol/water solutions, had diameters in the range of 13 mm to 15 mm. Richards and Richards imaged a 960 µm water droplet suspended from a slender glass capillary. An RGB video camera recorded the color play of 15 µm microencapsulated beads of TLC suspended in the water droplet as the droplet cooled in dry air. The sequence of images of the droplet reveals the transient temperature field in a slice through the center of the droplet.

In the present work liquid crystal thermography is extended to address heat transfer issues relevant to practical spray devices. The use of the microencapsulated form of TLC is applied to the investigation of the transient cooling of very small droplets (diameters less than 1000 µm), in order to map temperature fields and determine transfer rates for the cooling droplets. Video images of convectively cooled water droplets suspended in a jet of dry air are acquired to document the transient temperature fields in the droplets. The images reveal both temporal and spatial temperature gradients within the cooling droplets with resolution down to ~100 µm and ~0.03 seconds. The transient temperature data are then analyzed to yield convective heat transfer rates for the droplets.

Experimental Facility

The experiments were conducted in a facility designed to allow video imaging of suspended droplets exposed to precisely controlled convective conditions. Droplets consisting of dilute suspensions of microencapsulated TLC beads in distilled water were hung from a capillary tube in the potential core flow of a vertically oriented jet. A step change in temperature was introduced to the air flow to heat the TLC beads within the droplets above their clearing point. The heating source was then removed and video images were acquired as the droplets cooled. A schematic of the experimental facility is shown in Figure 1.

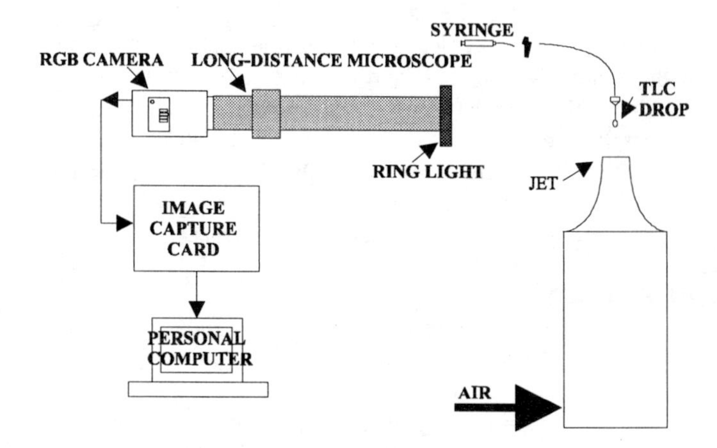

Figure 1. Schematic of experimental facility

The microencapsulated liquid crystal used in all experiments was Hallcrest R30C5W with a nominal color-play range from 30 to 35 C. The beads ranged in diameter from 5 to 15 µm, and were neutrally buoyant in water. The droplets under study were formed of a dilute suspension of the beads of microencapsulated TLC in distilled water. The dilute suspensions were made by mixing the beads in distilled water with concentrations of beads in water of 0.1 % on a volume basis.

Droplets with nominal diameters of 1 mm were hung from a Pyrex capillary tube with an OD of 80 µm and an ID of 40 µm. The capillary was bonded to a 24 gauge stainless steel needle with epoxy. The needle was coupled to a Teflon delivery tube using a Luer-Lok fitting. A syringe filled with the dilute suspension of distilled water and microencapsulated TLC was connected to the other end of the delivery tube. Droplets were formed by forcing a small amount of the dilute suspension from the syringe, throughout the delivery tube and out the end of the capillary tube. The syringe provided control over the delivery and size of the droplets. The needle and capillary assembly were fastened to a three-axis micro-positioning device for accurate positioning of the droplet in the potential core of the jet.

An axisymmetric air jet was produced by a nozzle with an ID of 20 mm and a contraction ratio of 9. The inner nozzle contour is a fitted cubic. The contraction is preceded by a flow straightening and conditioning section consisting of honeycomb and screens. Hot-wire anemometry was used to quantify the jet flow produced. Measurements showed that 'top hat' velocity profiles were produced at the jet exit with low turbulence intensity levels (< 0.5 %). A regulated compressed air source was used to supply the jet flow. The flowrate was monitored using a rotometer calibrated against hot-wire measurements. The air was filtered to remove moisture and particulate matter. Dry and wet bulb thermometers were used to monitor the jet air temperature and relative humidity.

The jet exit velocity was varied from 0 to 2 m/s in the experiments which resulted in Reynolds numbers of 0 to 2666 based on the jet diameter. Droplet diameters were nominally 1 mm which resulted in droplet Reynolds numbers of 0 to 133. The upper range of Reynolds numbers was restricted because excessive vibration of the droplets (due to vortex shedding) made imaging difficult.

Transient heat transfer experiments were run by suspending a droplet in the potential core of the jet flow and then introducing a step change in temperature to the air flow. This was done by inserting a heated coil of resistance wire (24 gauge Nickel-Chromium wire) briefly into the jet flow upstream of the droplet. The droplets were heated well above the TLC upper event temperature with this method. The elapsed time between removal of the heated wire and the onset of data acquisition was on the order of 2 seconds which was sufficient to ensure that any flow disturbances introduced by the presence of the wire had been convected downstream of the droplet.

A computer based digital image acquisition system was used to obtain images of the cooling droplets. The system consisted of a long-distance microscope, a CCD camera, an image capturing card, and a personal computer (PC). A schematic of the imaging system is provided in Figure 1.

The color Hitachi VK-C370 CCD RGB camera acquired RGB signals at 30 frames per second. An Infinity K2 long-distance microscope provided sufficient magnification to resolve individual beads of micro-encapsulated TLC in the water droplets. The spatial resolution was 2 μm at a stand-off distance of 120 mm. The depth of field for this condition was approximately 100 μm. On-axis lighting was supplied by a fiber optic ring light mounted to the microscope. The camera and microscope were mounted on three-axis micro-positioners.

The 30 Hz video RGB output from the camera was received by an IC-PCI/AM-CLR image capturing card from Imaging Technology, Inc. The image capturing card was capable of digitizing 24-bit RGB signals (8 bits for each channel) at a rate of 30 Hz as movies or as individual snapshot images. Digitized images were then sent to a PC with a P5-90 processor and 73 MB of RAM for image viewing, processing, analysis, and storage. The expanded RAM in the PC allowed over 40 frames of digitized RGB images (640 x 480 pixels) to be saved. Thus, transient events with durations of up to 1.5 seconds could be recorded.

Image Processing

Experimental results were stored as uncompressed Audio Video Interleave (AVI) files. In this form, frames could be viewed in sequence as a movie. In addition, each frame was saved in the Tag Image File Format (TIFF) format for individual processing. The time between successive frames is 1/30 of a second.

The first step in image processing was to manually tag each TLC bead in the droplet image. The tagged images were then digitally processed using software developed in-house to yield bead locations and hue values within the droplet.

The Red, Green, and Blue (RGB) values associated with each pixel in an image were transformed into Hue, Saturation, and Intensity (HSI) values using the transformation given by Gonzalez & Woods (1992). Hue was then converted to temperature using an experimentally derived calibration relationship.

The entire movie was processed in this manner to yield a series of files containing point measurements of temperature within a droplet as a function of time. From this information contour maps of temperature within a droplet at a specific time were made. In addition, the change of droplet temperature with time was analyzed to obtain heat transfer rates.

To do this the average droplet temperature was determined at each time (i.e., in each frame) using a volume averaging technique. The image was divided up into j rings of equal thickness. A mean temperature and volume was calculated for each ring. A volume-weighted mean droplet temperature was then calculated from

$$T_v = \sum_{j=1}^{n} T_j \frac{V_j}{V_{drop}} \tag{1}$$

where T_v is the volume-average temperature, T_j is the average temperature of ring j, V_j is the volume of ring j, and V_{drop} is the volume of the droplet.

Calibration Facility and Procedure

A relationship between the hue and temperature was developed experimentally. The calibration was conducted in a quiescent chamber in which both temperature and humidity were controlled and monitored. The chamber has optical access for the imaging system described above. The chamber is described in detail in Peterson(1996). Droplets produced from the same suspensions used in the experiments were suspended from capillaries and inserted into the chamber. The chamber was then heated above the clearing point temperature of the TLC. The air in the chamber was saturated (100% rh) by the introduction of steam. The chamber was then allowed to slowly cool to ambient temperature while the air remained saturated. The cooling rate was 0.2 °C per minute. Wet and dry bulb thermometers placed in close proximity (within 1 cm) to the suspended droplet were used to monitor the temperature and humidity. Images were acquired every 2 1/2 minutes (or 0.5 °C) as the droplet cooled from 36.0 to 29.0 °C.

The acquired images were digitized and processed to yield the hue-temperature (wet-bulb) calibration curve shown in Figure 2. The hue versus radius is shown plotted in Figure 3 for isothermal droplets at three different temperatures. The mean hue values are indicated by dashed lines on the figure. The figure shows that the hue throughout the droplets was uniform, as expected for an isothermal droplet.

Optical Considerations

The lighting and observation axes were coincident as shown in Figure 1. The ring light was mounted concentrically on the microscope to provide the on-axis lighting. The focus was adjusted so that a 100 μm slice (i.e., corresponding to the depth of field) through the middle of the droplet was imaged. Due to refraction effects the position of beads within the droplet is distorted. By assuming that 1) the droplet is spherically symmetric, 2) the incident light rays are parallel, and 3) the beads lie in the diametric plane of the droplet then a correction to bead position may be made through the application of Snell's Law. Because of refraction effects beads near the surface in the outer radius of the droplet are not imaged.

The issue of whether or not chromatic aberration due to refraction of the illuminating and reflected light had an impact on the observed color (hue) can be addressed by referring to Figure 3. The plot shows that hue is essentially independent of radius in the isothermal droplets. As noted above, due to refraction effects, beads on the outer edge of the droplet are not imaged.

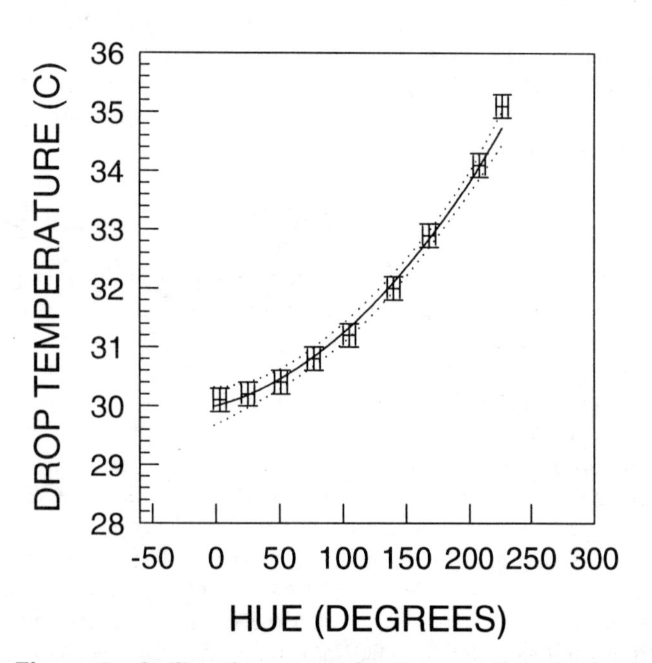

Figure 2. Calibration curve for thermochromic liquid crystal.

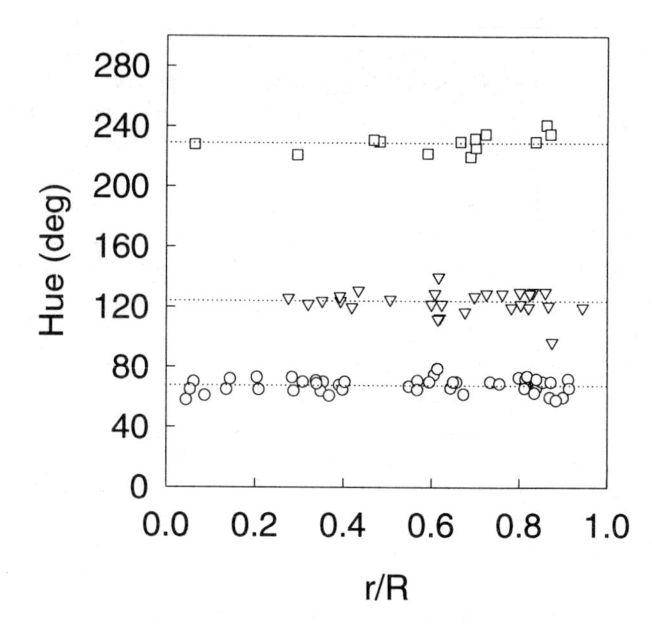

Figure 3. Hue versus radius for isothermal droplets.

RESULTS AND DISCUSSION

Spatial Variation of Droplet Temperature

A sequence of three images from a cooling-droplet movie is shown in Figure 4. Images have been taken from odd numbered frames in the sequence so that the elapsed time between images is 1/15 of a second. The diameter of the droplet is 960 μm and the Reynolds number (based on the droplet diameter) is 16. The dry and wet-bulb temperatures of the jet air were 20 and 8 °C, respectively. The TLC beads can be clearly identified in the images. A reflection of the ring light is seen on the front surface of the droplet. Red TLC bead colors correspond to the lower end of the color play region (30.0 C) and blue TLC beads correspond to the upper end of the color play region (35 C).

The temperature gradient in the droplet can be identified by the variation of TLC bead color in each of the images. The progression of the transient cooling process is evident in the sequence of raw images. In the final frame shown only a portion of the droplet temperature remains within the color play range of the TLC beads.

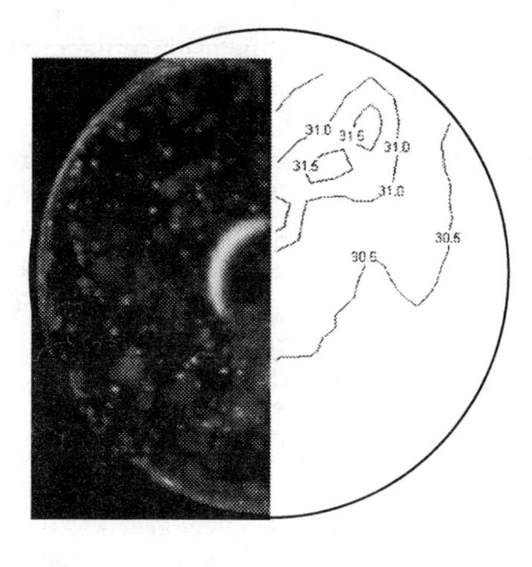

(4a)

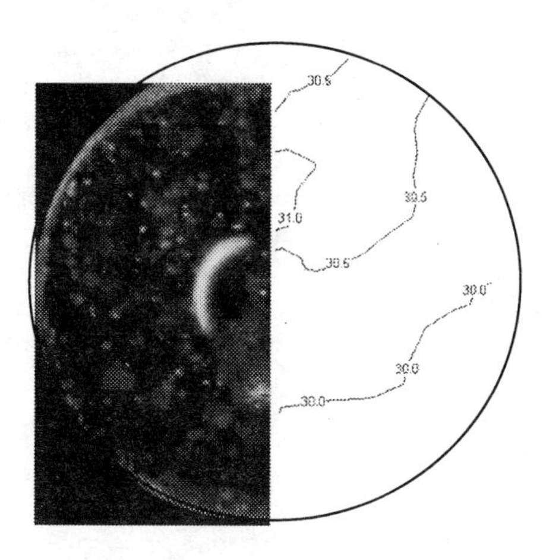

(4b)

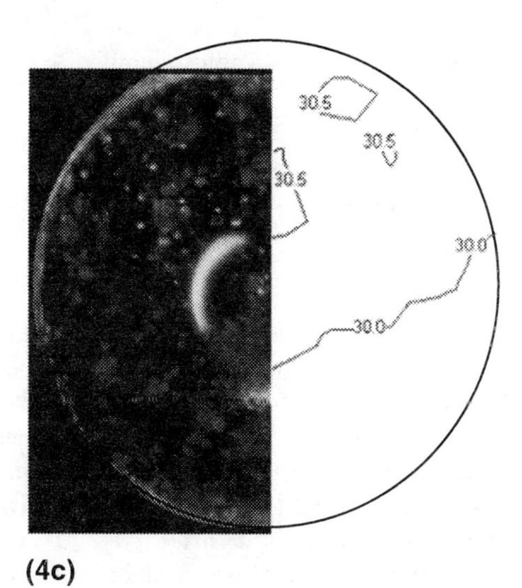

(4c)

Figure 4. Images and contours of a cooling droplet.

Shown adjacent to the droplet images are contours of temperature within the droplet acquired by processing the images. The temperature contours were mapped by dividing the droplet image into smaller subregions. For the present work square subregions 100 μm on a side were chosen The hue values of TLC beads over the entire image were sampled, and converted into bead temperatures using the calibration curve. Bead temperatures were then averaged over each subregion in the droplet image. Contours were then drawn on the basis of the average temperature for each subregion in the image.

It is important to note that bead positions were not corrected for refraction effects. As a result, there is a direct correspondence between the images shown and the contour plots.

An example of a typical distribution of sampled bead temperatures and locations is shown in Figure 5. In the figure TLC bead temperatures are plotted on a grid corresponding to pixel location in the image, for the first of the frames shown in Figure 4. Comparison to the contour in Figure 4 shows the correspondence between the two representations of temperature within the droplet. Although temperature trends within the droplet can be identified in Figure 5, there is some scatter in the data. The scatter is due mainly to the fact that bead images from throughout the 100 μm depth of field in the droplet are sampled. For example, in the image (a 2D projection), two beads which appear next to each other may in fact be 100 μm apart. The averaging process used in the mapping of the contours seen in Figure 4 smoothes out the scatter visible in the individual bead temperatures in Figure 5.

The uncertainty in the temperature contours produced in

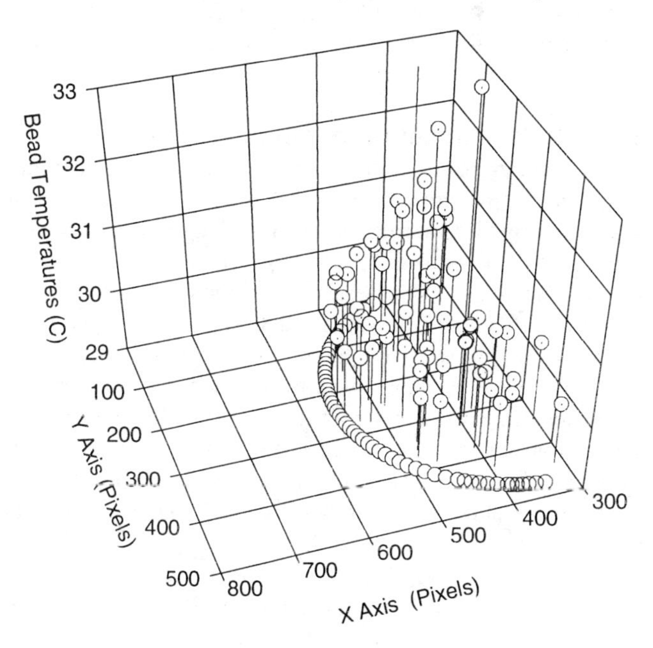

Figure 5. Scatter plot of TLC bead temperatures from Fig. 4a.

this way is estimated to be ± 0.3 C, based on 95% confidence intervals determined from the uncertainty in the individual microencapsulated TLC bead temperature measurement, and the scatter in the ensemble of averaged bead temperatures. The spatial resolution of this method of determining temperature contours is dependent on the size of the subregion chosen for averaging over and on the depth of field of the optics used. In this work, the size of the subregion chosen and the depth of field were both 100 μm. The size of the subregion chosen depends in on the density of beads in the droplet images. A higher density of beads allows the use of smaller subregions for the averaging process and finer spatial resolution. On the other hand a wider depth of field will introduce more scatter into the temperatures of the individual bead sampled in each subregion.

The contours quantify the phenomena identified by eye in the droplet images. The droplet is convectively cooled by air which flows from the upward from bottom of the images. The images show the droplet cooling from the bottom, with the isotherms moving upward and inward over time. The temperature difference within the droplet is seen to be as much as one and one half degrees. The temperature contours do not show a fore-aft symmetry. The highest temperatures in the droplet occurring off-center, towards the back of the droplet. The location of the high temperature region is stable from frame to frame. In the sequence of images shown, the maximum temperature drops from 31.5 to 30.5 °C in an elapsed time of 1/5 of a second. In the first frame all the beads are in the color-play range (i.e., above 30.0 °C), as seen in the contour plot. However as the droplet cools, the 30.0 °C contour progresses up through the droplet, causing those beads below the line to fall below the color-play range of the TLC. In the final frame, the majority of TLC beads in the droplet are below the color-play range.

Temporal Variation of Droplet Temperature

In addition to providing spatially resolved temperature measurements, the experiment provided a means to extract heat transfer information during transient cooling events. Nusselt numbers were calculated for cooling droplets for a range of Reynolds numbers. The results were then compared to the correlation obtained from Ranz and Marshall's experiments with water droplets evaporating in a laminar jet of dry air.

In Ranz and Marshall's experiments mass transfer rates from evaporating droplets were measured and used to determine the heat transfer to a suspended droplet by invoking a steady-state energy balance. The resulting droplet Nusselt number was found to correlate with droplet Reynolds number and the Prandtl number as:

$$Nu = 2.0 + 0.60\,\mathrm{Re}^{1/2}\,\mathrm{Pr}^{1/3} \tag{2}$$

Although a comparison will be made to the results of Ranz and Marshall (1952), there were some important differences between the two sets of experiments that should be pointed out. First the time scales of the experiments performed by Ranz and Marshall were relatively long. The time scales in the present set of experiments were very short. Second, the experiments by Ranz and Marshall were steady-state; the temperature of a droplet was constant at the wet-bulb temperature. The present set of experiments were transient, with the droplet temperature constantly decreasing. The droplets never reached the wet-bulb temperature of the air stream. Third, the experiments by Ranz and Marshall were performed by placing a cool droplet in a hot air stream. Under these conditions heat was transferred by conduction into the droplet while latent heat was transported by evaporation out of the droplet. Conduction and evaporation were opposed. The present set of experiments were performed by placing a hot droplet in a cool air stream. Under these conditions both sensible heat and latent heat were transported out of the droplet. Conduction and evaporation acted in the same direction. Finally, Ranz and Marshall's heat transfer measurements were made by determining the change in the mass of the evaporating droplets, and invoking a balance between latent and sensible heat transfer to the droplet. In the present work, the heat transfer measurements were made by determining the change in temperature of the evaporating droplets, and invoking a balance between the latent and sensible heat out of the droplet and the change of enthalpy of the droplet.

In the present experiments then, the heat balance applied to the droplet was

$$\underbrace{mC_p\frac{dT}{dt}}_{\substack{\textit{droplet enthalpy}\\ (\textit{rate of change})}} = \underbrace{h_c A_s (T_\infty - T_s)}_{\textit{convection}} + \underbrace{h_m A_s (m_\infty - m_s) h_{fg}}_{\textit{evaporation}} + \underbrace{\sigma\varepsilon A (T_\infty^4 - T_s^4)}_{\textit{radiation}} + \underbrace{kA\frac{dT}{dx}}_{\textit{conduction (along capillary)}} \tag{3}$$

where h_c and h_m were the total droplet heat and mass transfer coefficients, C_p and h_{fg}, were the constant pressure specific heat and enthalpy of vaporization for water, k, was the thermal conductivity of air, σ was the Stefan-Boltzmann constant. The symbols A_s and m were the surface area and mass of the droplet, and A was the cross-sectional area of the capillary. The driving potentials for heat and mass transfer were T and m the temperature and mass fraction of water vapor in the free stream and T_s and m_s, the temperature and mass fraction of water vapor at the droplet surface.

The contributions from radiation and conduction along the capillary were small effects and thus were neglected in the analysis. The terms were rearranged to solve explicitly for the Nusselt number by invoking the definitions of the Nusselt, Sherwood, and Lewis numbers:

$$Sh = \frac{h_m D}{\rho D_v}, \quad Nu = \frac{h_c D}{k}, \quad \frac{Sc}{\Pr} = Le \tag{4}$$

The Nusselt number was then obtained from

$$Nu = \frac{mC_p \dfrac{dT}{dt}}{\dfrac{A_s}{D}\left[\rho D_v Le^{1/3}(m_\infty - m_s)h_{fg} + k(T_\infty - T_s)\right]} \tag{5}$$

where C_p and h_{fg}, were evaluated at the mean droplet temperature. The binary diffusivity of water vapor in air, D_v, and the conductivity, k, and Lewis number, Le, for air were evaluated at the film temperature. The free stream temperature, T_∞, was measured by the dry-bulb thermocouple. The droplet surface temperature, T_s, was determined by plotting the temperatures of all the TLC beads in a given image versus radius, applying a linear fit to the data, and extrapolating to the droplet surface. Water vapor mass fractions in the free stream and at the droplet surface, m_∞ and m_s, were found from

$$m = \frac{P_{H_2O}}{1.61P_{tot} - 0.61P_{H_2O}} \tag{6}$$

where the water vapor partial pressure $P_{H2O} = P_{H2O,\infty}$ in the free stream and $P_{H2O} = P_{H2O,s}$ at the droplet surface, respectively. P_{tot} was the total pressure. Water vapor partial pressure in the free stream was determined from the relative humidity

$$P_{H_2O,\infty} = \psi_\infty P_{sat}(T_\infty) \tag{7}$$

where the free stream relative humidity, ψ_∞, was found from the wet and dry bulb temperatures. The water vapor at the surface of the droplet was taken to be saturated so that the partial pressure there was

$$P_{H_2O,s} = P_{sat}(T_s) \tag{8}$$

The rate of change of the droplet temperature with time, dT/dt, was determined by plotting the volume average temperature of the droplet versus time and then fitting a line to the data points. The slope of this line was taken to be dT/dt. In Figure 6 a plot of volume average temperature versus time is shown for a cooling droplet with a Reynolds number, Re = 0. The change in temperature of the droplet is seen to be linear, with the temperature of the droplet falling from 32.1 to 30.2 C during an elapsed time of 0.38 seconds.

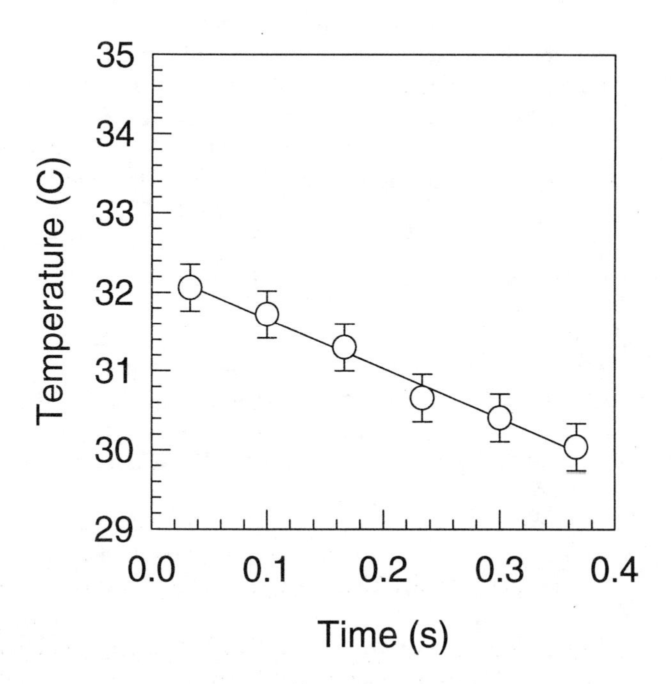

Figure 6. The change in average droplet temperature over time (Re = 0).

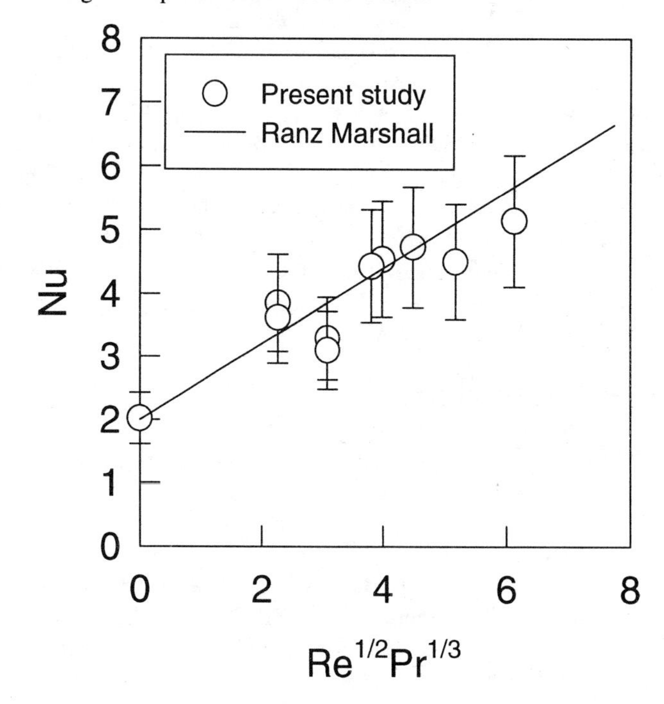

Figure 7. Heat transfer correlation.

Droplet diameter was measured directly from the images. The mass and surface area of the droplet were determined the measured droplet diameter, assuming a spherical droplet. The droplet seen in Figure 4 was typical of the droplets used in the experiments, which were all very close to spherical. Droplet size did not measurably change during the runs, which were never longer than 1.5 seconds.

The results of the heat transfer measurements are shown in Figure 7. In the figure, Nusselt number is plotted against $Re^{1/2}Pr^{1/3}$. Bars on each datum indicate estimated error in Nu and in $Re^{1/2}Pr^{1/3}$. The largest contribution to the uncertainty in the Nusselt number arises from the estimate of dT/dt. Other contributions to uncertainty in the Nusselt number arise from uncertainty in the estimate of droplet surface temperature and measurements of droplet diameter, free stream dry and wet bulb temperatures. The uncertainty in droplet surface temperature is hard to assess since no independent means of measurement are available. However, the Nusselt number is relatively insensitive to changes in the surface temperature. The uncertainty in droplet diameter is estimated at 5%. The uncertainty in the wet and dry bulb temperatures is ± 0.2 C.

The heat transfer correlation reported by Ranz and Marshall (1952) is shown with a solid line in the figure. The comparison between the present set of measurements and the measurements made by Ranz and Marshall is quite good.

CONCLUSIONS

Thermochromic liquid crystal thermography has been used to study transient temperature events in a convectively-cooled droplet. The work has demonstrated that imaging of individual microencapsulated beads suspended within a droplet can be used successfully to determine heat transfer and temperature information. Both spatially and temporally resolved temperature measurements were acquired in cooling droplets from Reynolds numbers ranging from 0 to 100. Transient temperature gradients have been imaged in the cooling droplet with resolutions down to ~100 μm and ~0.03 seconds. Heat transfer rates determined from transient temperature data extracted from the images compare favorably to previous measurements by Ranz and Marshall.

ACKNOWLEDGMENTS

This work was supported in part by Washington State University and by the National Science Foundation NYI program grant CTS-9457108.

REFERENCES

Akino, N. Kunugi, T., Ichimiya K., Mitsushiro, K., and Ueda, M., 1989, "Improved Liquid-Crystal Thermometry Excluding Human Color Sensation," *ASME J. Heat Transfer*, vol. 111, pp.558-565.

Charlesworth, D. H., and Marshall, W. R., 1959, "Evaporation of Drops Containing Dissolved Solids," *AIChE Journal,* Vol. 6, pp. 9.

D'abiri, D., and Gharib, M., 1992, "Digital Particle Image Thermometry: The Method and Implementation," *Experiments in Fluids*, vol. 11, pp.765-775.

Farina, D. J., Hacker, J. M., Moffat, R. J., and Eaton, J. K., 1993, "Illuminant Invariant Calibration of Thermochromic Liquid Crystals," Visualization of Heat Transfer Processes, HTD-Vol. 252, *ASME 29th National Heat Transfer Conference*, Atlanta, Georgia, pp. 1-11.

Gonzalez, R. C., and Woods, R. E., 1992, Digital Image Processing, Addison-Wesley, pp 229 - 235.

Hofmeister, W. H., Bayuzick, R. J., and Robinson, M. B., 1989, "Noncontact Temperature Measurement of a Falling Drop," *International Journal of Thermophysics*, Vol. 10, pp. 279 - 292.

Hu, S. H., Richards, R. F., and Richards, C. D., 1994, 'Thermography of Atomized Droplets in Flight Using Thermochromic Liquid Crystals,' *ILASS Conference*, Seattle, WA (1994).

Melton, L. A., Murray, A. M., and Verdieck, J. F., 1986, "Laser Fluorescence Measurements in Fuel Sprays," *Society of Photo Optical Instrumentation Engineering*, Vol. 664, pp. 40.

Moresco L L, and Marschall E., 1979, "Temperature measurements in a liquid-liquid direct-contact heat exchanger," *AIChE Symp* Ser 75: 266-272.

Nishiwaki, N. , 1955, "Kinetics of Liquid Combustion Processes: Evaporation and Ignition Lag of Fuel Droplets," *5th Symposium (Intl.) on Combustion*, pp. 148.

Nozaki, T., Mochizuki, T., Kaji, N., and Mori, Y. H., 1995, "Application of liquid-crystal thermometry to drop temperature measurements," *Experiments in Fluids*, Vol. 18, pp. 137 - 144.

Ozawa, M., Muller, U., Kimura, I., and Takamori, T., 1992, "Flow and Temperature Measurement of Natural Convection in a Hele-Shaw Cell Using a Thermo-sensitive Liquid-Crystal Tracer," *Experiments in Fluids*, Vol. 12, pp. 213-222.

Peterson, D., Hu, S. H., Richards, C. D., and Richards, R. F., 1995, "The measurement of droplet temperature using thermochromic liquid crystals," *ASME HTD*, Vol. 308, pp. 39-46.

Ranz, W. E. and Marshall, W. R., 1952, "Evaporation from Drops: Part II," *Chem. Eng. Prog.*, Vol. 48, No. 4, pp. 173-180.

Richards, C. D., and Richards, R. F., 1997, "Convective cooling of a suspended water droplet," *ASME J. Heat Trans.*, Vol. 119, no. 2.

Treuner, M., Rath, H. J., Duda, U., Siekmann, J., 1995, "Thermocapillary flow in drops under low gravity analysed by the use of liquid crystals," *Experiments in Fluids*, Vol. 19, pp. 264-273.

Trommelen, A. M. and Crosby, E. J., 1970, "Evaporation and Drying of Drops in Superheated Vapors," *AIChE Journal*, Vol. 16, pp. 857.

Wells, M. R. and Melton, L. A., 1990, "Temperature Measurements of Falling Droplets," *ASME Journal of Heat Transfer*, Vol. 112, pp. 1008 - 1013.

Yao, S-C., Schrock, V. E., 1976, "Heat and Mass Transfer From Freely Falling Drops," *ASME J. Heat Trans.*, Vol. 98, no.1, pp. 120-126.

Yearling, P. R. and Gould, R. D., 1995, "Convective Heat and Mass Transfer from Single Evaporating Water, Methanol and Ethanol Droplets," ASME Fluids Engineering Conference, FED-223, pp. 33-38.

HTD-Vol. 342, National Heat Transfer Conference
Volume 4
ASME 1997

PHASE CHANGE HEAT TRANSFER DURING CYCLIC HEATING AND COOLING WITH INTERNAL RADIATION AND TEMPERATURE DEPENDENT PROPERTIES

Bedru Yimer
University of Kansas
Mechanical Engineering Department
Lawrence, Kansas 66045

ABSTRACT

This paper presents the results of a numerical investigation of the effects of internal radiation on the heat transfer process during cyclic heating and cooling modes of a phase change energy storage system. The effects of internal radiation on the solidification/melting , temperature distribution and the energy stored and extracted were investigated. An absorbing, emitting, and isotropically scattering, finite, and semi-trasparent gray phase change medium bounded between two concentric cylinders was studied. The phase change medium employed in this study was a fluoride salt with a weight composition of LiF - 41.27% , MgF_2 - 48.76% and KF - 8.95%. The results show that radiative heat transfer significantly affects the dynamics of solidification/melting as conduction-radiation interaction parameter decreases. The results of this study will lead to a better design and optimization of phase change thermal energy storage systems particularly for space applications.

NOMENCLATURE

- a = absorption coefficient, 1/m
- A = area, m^2
- C = specific heat, J/kg-K
- F_r = dimensionless radial radiative heat flux
- F_z = dimensionless axial radiative heat flux
- H = dimensionless enthalpy
- h = specific enthalpy, J/kg
- h_{sl} = latent heat of fusion, J/kg
- K = thermal conductivity, W/m-K
- K_1 , K_2 = thermal conductivities
- K_3 , K_4 of element surfaces, W/m-K
- N = conduction/radiation parameter
- R = dimensionless radius
- r = radial distance, m
- S = dimensionless parameter
- T = temperature, K
- t = time, s
- v = volume, m^3
- Z = dimensionless axial distance
- z = axial distance, m

Greek Letters

- α = thermal diffusivity, m^2/s
- η = axial optical coordinate variable
- θ = dimensionless temperature
- ξ = dimensionless time
- ρ = density, kg/m^3
- σ = Stefan-Boltzmann constant, W/m^2-K^4
- τ = radial optical coordinate variable
- ϕ = dimensionless Planck function
- ψ_0 = dimensionless zeroth moment of intensity

ω = scattering coefficient, 1/m

Subscripts

f = fusion
I = inner
l = liquid
m = time level
o = outer
r = radial
s = solid at fusion temperature
w = wall
z = axial

INTRODUCTION

Eventhough solidification and melting of materials by heat transfer has been of importance in many technical fields and a subject of interest for over a century, the analysis of the problem including internal thermal radiation has been done only over the past twenty five years. Studies have shown that internal radiation can significantly affect the dynamics of melting and solidification of many optical materials and neglecting it can cause serious error.

The contribution of thermal radiation acting simultaneously with conduction during the phase change process imposes mathematical difficulties. The resulting integrodifferential equations of radiative transfer do not lend themselves to exact analytical methods. As a result a number of approximate methods including higher-order differential approximations have been developed to approximate the intensity integral of the radiation field. Simultaneous heat transfer by conduction and radiation in semi-transparent participating media has been studied by several investigators. Yuen and Wang (1980) studied a one-dimensional system bounded by two parallel gray, diffuse and isothermal walls. Assuming a physical model of linear anisotropic scattering, the resulting integral-difeerential equation was solved by a successive approximation technique similar to the method of undetermined parameters. Other investigators include Viskanta and Hirleman (1978), Ratzel and Howell (1982), and Bayazitoglu and Hegenyi (1980) . For phase change in cylindrical geometry , Kim and Yimer (1888) and Yimer (1996) analysed , using the enthalpy model and P_1 differential approximation, the effects of thermal radiation on the solidification of a finite concentric cylindrical medium.

ANALYSIS

In this study, the describing equations are developed for a two-dimensional radial and axial system which contains a semi-transparent, gray, absorbing, emitting, and isotropically scattering phase change medium bounded by two finite concentric cylinders. In the analysis the enthalpy method which is used to determine the enthalpy and the temperature distributions without tracking the solidification front was employed.

Application of the law of conservation of energy to a control volume including conduction and radiation results in:

$$d/dt\iiint \rho h dv = \iint K \operatorname{grad} T \bullet n dA - \iint q_r \bullet n dA \tag{1}$$

For a two-dimensional element that contains a semi-transparent phase-change medium,the describing equation in dimensionless form is written as:

$$K_s\ \delta\tau(dH/d\xi) = [1+(\delta\tau/2\tau)]K_1\ (\partial\theta/\partial\tau)|_{\tau+\delta\tau/2} - [1-(\delta\tau/2\tau)]K_2\ (\partial\theta/\partial\tau)|_{\tau-\delta\tau/2} + K_3\ (\delta\tau/\delta\eta)(\partial\theta/\partial\eta)|_{\eta+\delta\eta/2} - K_4\ (\delta\tau/\delta\eta)(\partial\theta/\partial\eta)|_{\eta-\delta\eta/2} + \{-\ [1+(\delta\tau/2\tau)]F_r|_{\tau+\delta\tau/2} + [1-(\delta\tau/2\tau)]F_r|_{\tau-\delta\tau/2} - (\delta\tau/\delta\eta)F_z|_{\eta+\delta\eta/2} + (\delta\tau/\delta\eta)F_z|_{\eta-\delta\eta/2}\}K_s \times (S/N)\ . \tag{2}$$

The dimensionless temperature and enthalpy, for various values of enthalpy, are related by:

$$\theta = (\alpha K_s / K\alpha_s)H \quad \text{for } H < 0 \text{ (solid region)}, \tag{3}$$

$$\theta = 0 \quad \text{for } 0 < H < 1 \text{ (during phase change)}, \tag{4}$$

$$\theta = (\alpha K_s / K\alpha_s)(H-1) \quad \text{for } H > 1 \text{ (liquid region)}. \tag{5}$$

The dimensionless variables are defined as follows:

$$H = (1/\rho_s\ \delta v)\iiint[\rho(h - h_s)/h_{sl}]dv,\ \theta = C_s\ (T - T_f)/h_{sl} = (K_s/\alpha_s\ \rho_s)(T - T_f)/h_{sl},\ \tau = r(a + \omega) = r\beta,\ \xi = \alpha_s\ \beta^2\ t,\ F_r = q_r/\sigma T^4,\ F_z = q_z/\sigma T^4,\ N = K_s\ \beta/4\sigma T_r^3,\ S = K_s\ T_r/4\alpha_s\ \rho_s\ h_{sl}, \tag{6}$$

where H = dimensionless enthalpy, θ = dimensionless temperature, τ = dimensionless radial optical distance, ξ = dimensionless time, F_r = dimensionless radial radiative heat flux, F_z = dimensionless axial radiative heat flux , N = Stark number (conduction-radiation parameter), S = dimensionless parameter.

The radiative heat flux term in the energy equation was obtained from the differential form of the equation of transfer. The P-1differential approximation to the equation of radiative transfer was utilized to derive the necessary moment equations. Marshak type boundary conditions were used to solve the differential equation describing the radiative transfer. The following moment equations were then conveniently coupled with the other terms in the energy equation.

$$\partial^2 \psi_0 /\partial\tau^2 + 1/\tau(\partial\psi_0 /\partial\tau) + \partial^2 \psi_0 /\partial\eta^2 = 3(1-\omega)(\psi_0 - 4\pi\phi) \quad (7)$$

$$F_r = \psi_r /\pi = -(1/3\pi)\partial\psi_0 /\partial\tau \quad (8)$$

$$F_z = \psi_z /\pi = -(1/3\pi)\partial\psi_0 /\partial\eta \quad (9)$$

The describing equation, Equation 2, is modified to include the boundary conditions. When heat at the inner surface is extracted at a constant rate, the equation takes the form:

$$K_s\, \delta\tau(dH/d\xi) = [1+(\delta\tau/2\tau)]K_1\, (\partial\theta/\partial\tau)|_{\tau+\delta\tau/2} - (Ste)K_s + K_3\, (\delta\tau/\delta\eta)(\partial\theta/\partial\eta)\,|_{\eta+\delta\eta/2} - K_4\, (\delta\tau/\delta\eta)(\partial\theta/\partial\eta)\,|_{\eta-\delta\eta/2}$$
$$[1+(\delta\tau/2\tau)]F_r\,|_{\tau+\delta\tau/2} + [1-(\delta\tau/2\tau)]F_r\,|_{\tau-\delta\tau/2} - (\delta\tau/\delta\eta)F_z\,|_{\eta+\delta\eta/2} + (\delta\tau/\delta\eta)F_z\,|_{\eta-\delta\eta/2} \}K_s \times(S/N) \quad (10)$$

where

$Ste = q/(\alpha_s\, \rho_s\, h_{sl}, \beta)$ = Stefan Number

When heat at the inner surface is extracted by convection, Equation 1 is written as:

$$K_s\, \delta\tau(dH/d\xi) = [1+(\delta\tau/2\tau)]K_1\, (\partial\theta/\partial\tau)|_{\tau+\delta\tau/2} - B_I\, \theta_W K_s$$
$$B_I\, \theta_0\, K_s + K_3\, (\delta\tau/\delta\eta)(\partial\theta/\partial\eta)\,|_{\eta+\delta\eta/2} -$$
$$K_4\, (\delta\tau/\delta\eta)(\partial\theta/\partial\eta)\,|_{\eta-\delta\eta/2} + \{-\ [1+(\delta\tau/2\tau)]F_r\,|_{\tau+\delta\tau/2}$$
$$+ [1-(\delta\tau/2\tau)]F_r\,|_{\tau-\delta\tau/2} - (\delta\tau/\delta\eta)F_z\,|_{\eta+\delta\eta/2} + (\delta\tau/\delta\eta)F_z\,|_{\eta-\delta\eta/2}$$
$$\}K_s \times(S/N)\,. \quad (11)$$

where

$B_I = h/(\beta K_s)$ = Biot Number

$\theta_W = C(T_W - T_F)/h_{sl}$, = dimensioless inner wall temperature

$\theta_0 = C(T_F - T_0)/h_{sl}$, = dimensionless secondary fluid temperature

NUMERICAL RESULTS AND DISCUSSION

The resulting describing equations which include the radiative flux term, along with the appropriate boundary and semi-transparent moving interfacial conditions were written in finite difference form. The Gauss-Seidel iterative method with successive over-relaxation was used to solve the non-linear simultaeous equations. The phase change medium employed in this study was a fluoride salt with a weight composition of LiF - 41.27% , MgF_2 - 48.76% and KF - 8.95%. This phase change material was selected primarily because of the availability of thermophysical property data including experimentally determined temperature dependent thermal conductivity values.The thermophysical property data of the medium is presented in Table 1. In this two-dimensional transient system, heat at the inner surface was added at a constant rate as well as by convection from a secondary heating fluid. The phase change material was heated to well above the fusion temperature and the contributions of both latent and sensible heats were included. For the discharge mode, heat at the inner surface was also extracted at a constant rate and by a low temperature secondary fluid. The medium was cooled sufficiently below the fusion temperature. Heat losses from outer and end surfaces were considered negligible.

The transient temperature distribution, solid/liquid interface location, radiative heat flux and energy storage/recovery capacity of the semi-transparent phase change medium were then determined using the numerical scheme. Using the transient multi-dimensional model, the effects of parameters such as optical and thermophysical properties of the medium, conduction-radiation interaction parameters and geometric dimensions on the rate of solidification/melting and energy storage/recovery capacity were studied. For heat extraction at the inner surface at a constant rate, Figure 1 shows the radial dimensionless temperature distribution at two different time levels during the heat extraction mode of the cyclic process. The figure shows the significance of the effects of internal radiation on the temperature distribution. For the same condition and for the heat extraction mode, Figures 2 and 3 respectively show the radial and axial solidification front positions. It should be noted that these figures show that internal radiation accelerates the rate of solidification. It should also be pointed out that due to the symmetry of the system, the solidification front does not show any significant axial variation.

For the case where heat at the inner surface is added/extracted by convection via secondary fluids, Figure 4 shows the radial dimensionless temperature distribution at two time levels. These results are for the heating mode of the process. For the same boundary condition and heating mode, Figure 5 shows the dimensionless axial temperature distribution. Again, eventhough internal radiation significanly affects both the radial and axial temperature profiles, axial temperature variation is not significant. Figure 6 shows the heat stored and recovered during the cyclic heating and cooling process where heat at the inner surface is transferred by convection.

In general, the results show that radiative heat transfer significantly affects the dynamics of solidification/melting , the temperature profile and the energy storage/recovery capacity of the system. This is especially terue when conduction-radiation interaction parameter , N, is small. In this study this parameter had a value of 0.2215. The results of this study will lead to a better design and optimization of phase change thermal energy storage systems particularly for space applications.

TABLE 1

Thermophysical and Optical Property Data for Flouride Salt

Property	$LiF-MgF_2$ -KF
Solid density at fusion temperature, kg/m^3	2900
Latent heat of fusion, kJ/kg	814
Fusion temperature, K	983
Temperature dependent thermal conductivity, kJ/hr-m-K	$K(T) = 2.4407 - 0.0053T + 0.000013T^2$
Emissivity	0.5
Scattering coefficient	0.2

REFERENCES

Bayazitoglu, Y. and Higenyi, J., 1980, "Differential Approximation of Radiative Heat Transfer in Gray Medium-Axially Symmetric Radiation Field," ASME *Journal of Heat Transfer*, Vol. 102, pp. 719-723.

Kim, K. S. and Yime, B., 1988, "Thermal Radiation Heat Transfer Effects on Solidification of Finite Concentric Cylindrical Medium-Enthalpy Model and P-1 Approximation," *Numerical Heat Transfer Journal*, Vol. 14, pp. 483-498.

Ratzel, A. C. and Howell, J. R., 1982, "Heat Transfer by Conduction and Radiation in One-Dimensional Planar Media Using the Differential Approximation", ASME *Journal of Heat Transfer, Vol. 104, pp. 388-391.*

iskanta, R. and Hileman, E. D., 1978, "Combined Conduction-Radiation Heat Transfer Through Irradiated Semi-Transparen", ASME *Journal of Heat Transfer*, Vol. 100, pp. 169-172.

Yimer, B., 1996, "Multi-Dimensional Solidification with Internal Radiation and Temperature Dependent Properties," Paper No. AIAA-96-3978.

Yuen, W. W. and Wong, L. W., 1980, "Radiative Transfer in Rectangular Enclosures with Gray Medium ," ASME Paper No. 80-HT-101.

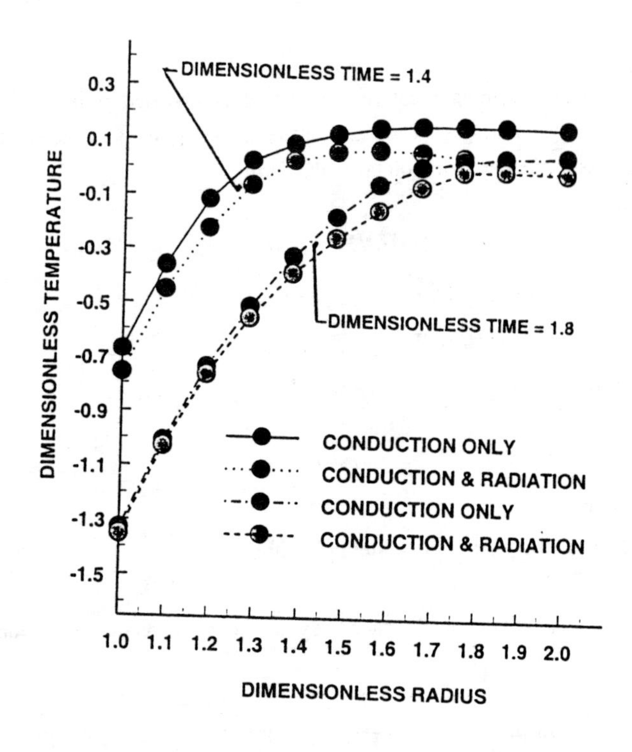

Fig. 1. Transient radial temperature distribution during heat extraction at a constant rate, Stefan number = 2.0.

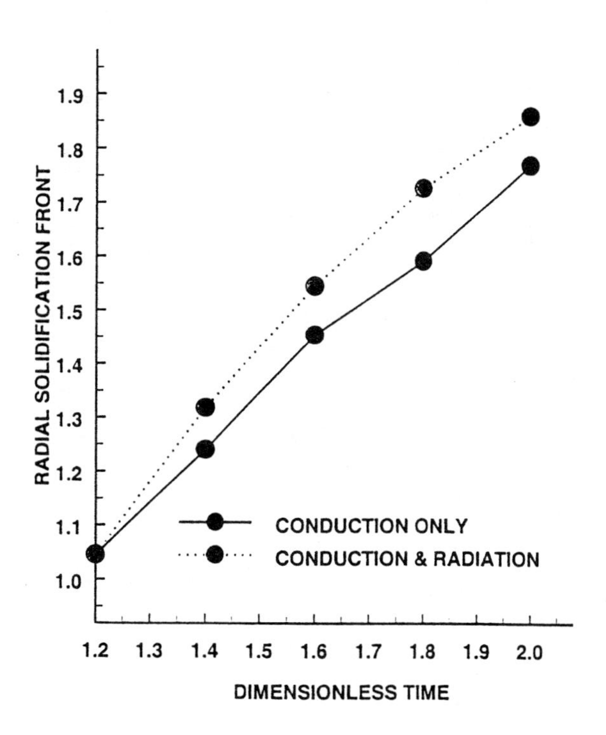

Fig. 2. Radial solidification front position vs dimensionless time during heat extraction at a constant rate, Stefan number = 2.0.

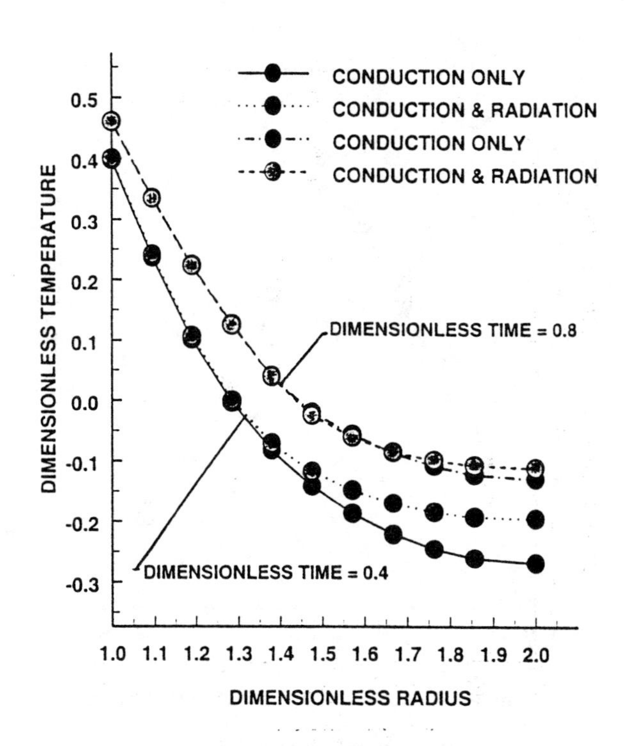

Fig. 4. Transient radial temperature distribution during heat addition by convection: Biot number = 3.0, dimensionless secondary fluid temperature for heating = 2.0.

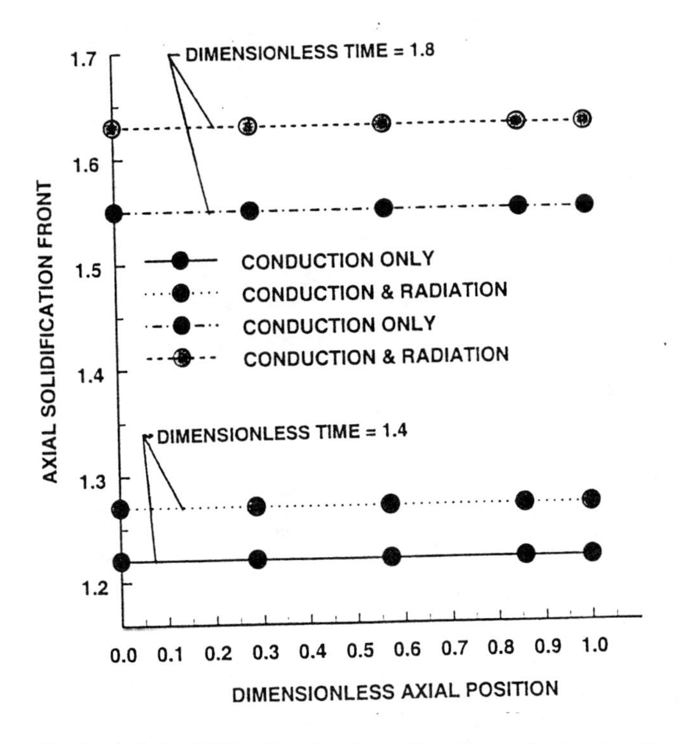

Fig. 3. Axial solidification front position during heat extraction at a constant rate, Stefan number = 2.0.

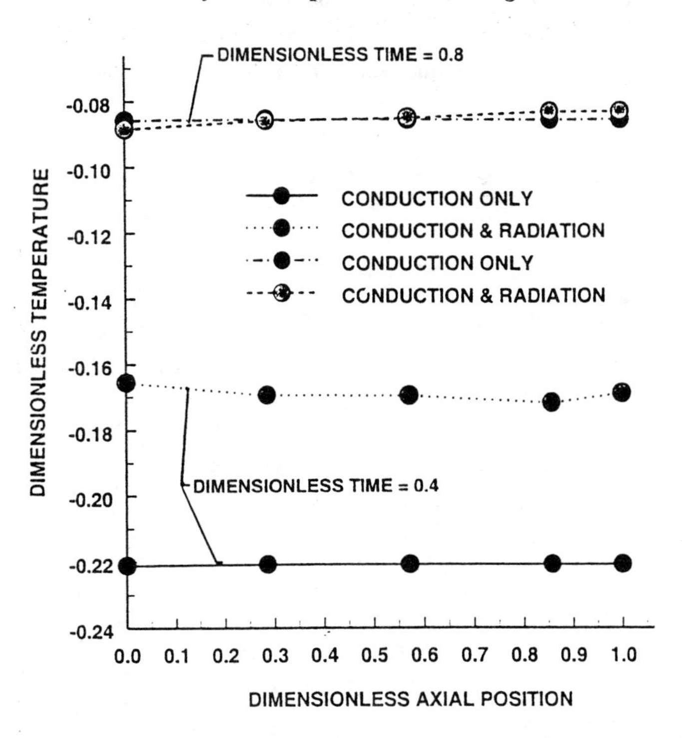

Fig. 5. Dimensionless axial temperature distribution during heat addition by convection: Biot number = 3.0, dimensionless secondary fluid temperrature for heating = 2.0.

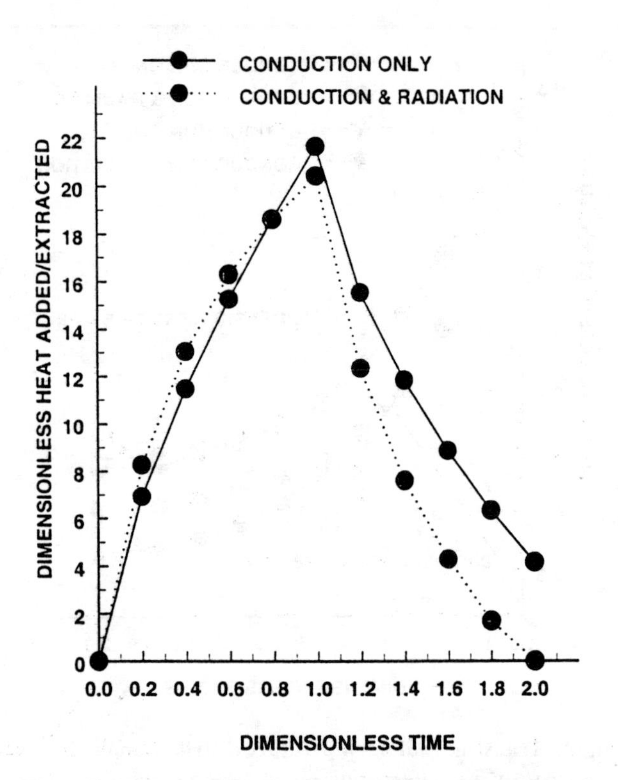

Fig. 6. Dimensionless heat added/extracted during cyclic heating and cooling by convection: Biot number = 3.0, dimensionless secondary fluid temperture for heating = 2.0, dimensionless secondary fluid temperature for cooling = -2.0.

HTD-Vol. 342, National Heat Transfer Conference
Volume 4
ASME 1997

AN EXPERIMENTAL STUDY OF PULSE BOILING IN A TWO-PHASE THERMOSYPHON

Donald P. Shatto, Kimberly I. Renzi, G. P. Peterson
Department of Mechanical Engineering
Texas A&M University
College Station, Texas

ABSTRACT

Pulse boiling occurs in two-phase thermosyphons operating at low saturation pressures and at low heat transport rates. This phenomenon, a direct result of the long waiting periods associated with high nucleation superheats, results in temperature and pressure oscillations which vary in amplitude and frequency. An experimental investigation has been conducted to measure the amplitude and period of pulse boiling temperature oscillations in a stainless steel thermosyphon using water and ethanol as the working fluids. Two methods of predicting the pulse period have been applied to the data of the current study. Recommendations have been made regarding the applicability of pulse period and amplitude predictions to the design of thermosyphon evaporators.

NOMENCLATURE

b superheated liquid layer thickness [m]
c_P specific heat at constant pressure [J/kg·K]
h_{fg} enthalpy of vaporization [J/kg]
k thermal conductivity [W/m·K]
P pressure [Pa]
q heat flux [W/m^2·K]
R gas constant [J/kg·K]
R' evaporator inner radius [m]
r_b distance from tube centerline to nucleate bubble [m]
R_c nucleation site radius [m]
$\bar{r}$ r_b / R'
T temperature [K]
V^+ fill ratio

Greek
α thermal diffusivity [m^2/s]
ΔP^* bubble pressure differential at critical superheat [Pa]
ΔT pulse amplitude [K]
λ pulse period [s]
υ pulse frequency [1/s]
θ^* critical superheat [K]
ρ density [kg/m^3]
σ surface tension [N/m]
τ_w waiting time [s]

subscripts
L liquid
s saturation
V vapor
w evaporator wall

INTRODUCTION

A thermosyphon is a form of wickless heat pipe in which the working fluid flows from the condenser back to the evaporator due to gravity. Thermosyphons can work in two different ways depending on the fill ratio, defined as the volume of liquid in the evaporator divided by the total volume of the evaporator: At low fill ratios, evaporation takes place from the falling film, and the heat transport is limited by dryout of the heated wall. At high fill ratios heat is transferred to the working fluid through nucleate boiling, and the heat transport is usually limited by flooding of the condenser due to shear stress at the liquid-vapor interface, and subsequent dryout of the evaporator. High fill ratios typically result in a higher maximum heat transport rate than lower fill ratios.

Thermosyphons have been employed in a broad range of applications, including thermal stabilization of the permafrost below the Alaskan oil pipeline, air-to-air heat exchangers for industrial regeneration applications, and electronics thermal management.

During normal operation a thermosyphon will continuously transport heat from the evaporator to the condenser, with the working fluid flowing steadily and maintaining constant saturation conditions. However, at high fill ratios, low saturation pressures, and low heat

Table 1 - Previous Experimental Studies

Study	Working Fluid(s)	Evaporator	D (mm)	L (m)	V^+ (%)
Casarosa et al. (1983)	water	brass	42	0.87	112
"	"	glass			112
Negishi et al. (1987)	water	copper	22	1.3	30
Liu et al. (1988, 1989)	water, methanol, ethanol, acetone	glass	15, 20, 30	1.25	
Liu and Dong (1990)	water	steel	20	1.18	13 - 50
Niro and Beretta (1990)	water, acetone	glass	12, 30	1.0	80
Zhou (1992)	water	glass	10	0.6	100
Liu et al. (1992)	water, ethanol (mixtures)	glass	27	0.98	
Lin et al. (1995)	water, ethanol	stainless stl.	(annulus)	0.55	30 - 140

transport rates, several researchers have observed a peculiar phenomenon. Under certain thermal conditions a large amount of working fluid is periodically propelled from the evaporator to the condenser with significant velocity, resulting in temperature and pressure fluctuations. This phenomenon has previously been referred to as the geyser, pulse boiling or intermittent boiling effect. The pulse boiling phenomenon is a direct result of the intermittent suppression of nucleation in the evaporator associated with low saturation pressures and low heat fluxes. Under such conditions high liquid superheats are required for nucleation, and the resulting bubbles can grow to a relatively large size. These effects combine to bring about the pulse boiling regime, in which long periods of liquid quiescence in the evaporator (up to several hundreds of seconds) alternate with sudden and almost explosive boiling.

Thermosyphons are used extensively in applications in which the evaporator must be maintained at a constant temperature. The occurrence of pulse boiling in such applications could result in damage to electronics or the malfunction of sensitive instrumentation.

The current experimental investigation seeks to measure the amplitude and frequency of evaporator temperature fluctuations during pulse boiling in a stainless steel thermosyphon with different working fluids over a range of condenser temperatures and heat transport rates. The resulting data are used to evaluate the accuracy of two previous methods to predict the pulse boiling frequency.

Pulse Boiling Mechanisms

The mechanisms which constitute the pulse boiling cycle have been described in detail by several researchers, including Casarosa et al. (1983), Negishi et al. (1987), Liu et al. (1988), Niro and Beretta (1990), and Lin et al. (1995). The resulting observations are summarized here as a sequence of three phases of transient operation.

(1) In the initial phase of the pulse boiling cycle, the evaporator is filled with a quiescent liquid pool. Heat is transferred from the evaporator wall to the liquid through single-phase natural convection, with some evaporation occurring at the free surface of the pool. The liquid begins this phase at (or near) its saturation temperature and is gradually heated to a temperature which may be several degrees above saturation. At low pressure a fluid with a relatively high surface tension in contact with a smooth surface will not boil until a relatively high superheat has been attained.

(2) Eventually the temperature in the evaporator becomes high enough for a single nucleation site to become activated. Thermal energy stored in the superheated liquid surrounding the resulting bubble is transferred to the vapor, causing the bubble to grow rapidly. In many cases this bubble will quickly fill the cross-section of the evaporator and continue its rapid growth. This sudden evaporation causes the liquid above the nucleation site to be thrust upward into the condenser region of the thermosyphon. In experimental investigations this explosive vaporization can be recognized by sharp decreases in evaporator temperature and sudden rises in condenser temperature. The increase in pressure associated with the growth of the vapor bubble causes a corresponding rise in the saturation temperature in the thermosyphon. Combined with the sudden loss of thermal energy in the evaporator, this results in the immediate cessation of nucleate boiling.

(3) Finally the liquid and vapor, which have been thrust upward, contact the cold walls of the condenser. The vapor condenses as the liquid is cooled to a temperature below its initial saturation temperature and flows (as a film on the thermosyphon wall) back to the evaporator, cooling the liquid pool back to its initial temperature. When most of the liquid has returned to the evaporator, the cycle begins again.

Though the pulse boiling effect described above represents the most common instability mechanism observed in thermosyphon operation, two other types of cyclical boiling mechanisms have been encountered in experimental investigations. Fukano et al. (1983) observed large-amplitude oscillations which began at the flooding limit. Another type of oscillation mechanism was observed by Drolen and Fleischman (1995) in which boiling in the liquid pool produced a bubbly flow with a sufficiently low liquid volume fraction to cause the pool height to rise to the level of the condenser, causing an increase in the rate of condensation and a subsequent surge of liquid flow back to the evaporator. These two types of instability mechanisms are not considered in the current study.

PREVIOUS EXPERIMENTAL STUDIES

Several previous researchers have reported measurements of the frequency and magnitude of temperature oscillations in experimental investigations of pulse boiling in thermosyphons. More often than not, these studies incorporated visualization of the physical mechanisms of heat pipe operation. Often water was chosen as the working fluid in these experiments. The high surface tension of water and the smooth evaporator surfaces in glass visualization thermosyphons combine to produce high liquid superheats at the onset of nucleate boiling. This results in a strong propensity for glass/water thermosyphons to exhibit pulse boiling behavior.

Table 1 lists the previous experimental studies that observed pulse boiling and includes information regarding the evaporator material, internal diameter, and overall length of each test article used. The range of fill ratios, V^+, for which pulse boiling was observed are also listed. In some cases the cited studies offered no information regarding the fill ratio, and in one study the dimensions of the test article were not given. Liu et al. (1988) also conducted experiments with refrigerant-113 and detected no pulsations with this working fluid. Lin et al. (1995) used an annular thermosyphon with a transparent outer tube. The inner surface of the annulus was made of a hollow stainless steel cylinder with an outside diameter of 3.2 cm. The outer envelope was glass with an outside diameter of 5.3cm and 0.16cm thick.

Previously Observed Trends

In each of the experimental investigations described above, researchers have measured the frequency of the pulses during thermosyphon operation. In most cases the qualitative variation of the frequency has been reported as a function of one or more governing parameters, such as the working fluid, heat transport rate, and fill ratio. The observations from these previous studies are summarized below.

Virtually all of the previous investigators noted that the pulse frequency tends to rise with increasing heat transport rate. Both Zhou (1992) and Lin et al. (1995) reported that the increase in frequency is nearly linear. Casarosa et al. (1983) reported that the amplitude of the pulses remained nearly constant regardless of heat transport rate, but Niro and Beretta (1990) found that increases in heat transport resulted in larger pulses. Though few researchers made mention of it, it can be observed from previous results that pulse boiling occurs more irregularly at high heat load, in terms of both the frequency and amplitude of temperature oscillations. Lin et al. (1995) attributed this irregularity to multiple bubble nucleation caused by the activation of more than one nucleation site at higher evaporator wall temperatures.

A few of the studies considered here noted changes in the frequency and/or amplitude of pulse boiling with changes in the saturation condition as dictated by the condenser coolant temperature and measured as the pressure in the thermosyphon. Most researchers found that the thermosyphons had to be operated at relatively low saturation pressures for oscillatory boiling to occur at all. Casarosa et al. (1983) and Niro and Beretta (1990) determined that the amplitude of temperature pulsations decreases when pressure is increased. Lin et al. (1995) found that, in some cases, the pulse period is insensitive to the saturation condition, however, in general, the pulse frequency is higher when the condenser is at a lower temperature.

Those studies which used more than one working fluid in their experimental investigations generally found that, at a given saturation temperature, the amplitude of pulsations is higher for fluids with lower vapor pressures and higher values of surface tension. Typically these studies compared water with other more volatile fluids and found that water yielded larger pulses. For some fluids (i.e: refrigerants) pulse boiling did not occur at all.

Some studies have concluded that pulse boiling can only occur if the evaporator is flooded (fill ratio of 100% or more). This may be due to the fact that evaporation will occur preferentially at the liquid-vapor interface at lower fill ratios, eliminating the possibility of nucleation at low heat fluxes. In their experiments with water, Lin et al. (1995) found that at low fill ratios, the frequency of the geyser boiling is higher, and the amplitude of the oscillation is smaller. In experiments with ethanol as the working fluid, however, no pulse boiling was observed when the liquid fill V^+ was at 50% or 30%.

It might be expected that the pulse boiling phenomenon would be affected by the size and relative dimensions of the evaporator. Casarosa et al. (1983) conjectured that the violent expulsion of liquid from the evaporator should decrease in amplitude when the thermosyphon diameter increases, because the quantity of vapor that is released from the liquid-vapor interface during the quiet phase is related to the size of the free surface, and this can determine the rate of overheating of the liquid. Lin et al. (1995) found that the frequency of pulse boiling was higher for shorter evaporator lengths. This is probably due to the decreased amount of energy required to heat up a smaller evaporator.

PREDICTIONS OF PULSE FREQUENCY AND AMPLITUDE

Though several experimental investigations have been conducted to determine the behavior of the pulse boiling phenomenon, few researchers have proposed methods of predicting the magnitude and/or frequency of pulsations. Of the previous investigations considered in the current study, some have presented empirical correlations to describe the data trends, but the ranges of applicability of such correlations are often clearly limited to the data from the study in which they are presented. Of the correlation methods proposed in these previous studies, only two may be expected to be applicable to a broader range of conditions.

Liu and Wang (1992) presented a theoretical prediction of pulse boiling frequency applicable to any thermosyphon with a uniform heat flux applied at the evaporator. This model is based on the superheat at which the first nucleate bubble would form, as predicted by Hsu (1962). The model makes three assumptions: (1) At the beginning of the pulse period, the liquid pool is heated from the saturation temperature. (2) Before the first bubble creation, the liquid pool is motionless and the heating process can be treated as transient conduction. (3) The height of the liquid pool is much larger than its diameter, so evaporation from the liquid surface can be neglected. The result was presented as a correlation to predict the frequency, ν, of the pulse boiling cycle.

$$\upsilon = \frac{60\,\alpha}{(R')^2}\cdot \left\{\frac{T_S\, k_L}{1.57\,\bar{r}^{5.71}\, R'\, q}\left[\left\{1-\frac{R\, T_S}{h_{fg}}\ln\left(1+\frac{2\cdot\sigma}{P_S\, R'\,(1-\bar{r})}\right)\right\}^{-1}-1\right]\right\}^{-1.756} \tag{1}$$

$\bar{r}$ is defined as r_b / R', where r_b is approximately equal to the tube radius minus the radius of the nucleation site, so $\bar{r}$ is very close to 1.0; r_b is determined empirically from the data for a given surface/fluid combination. Liu and Wang (1992) found that Eqn. (1) tends to over-predict the pulse frequency in most cases. The pulse periods measured in the current experimental study will be compared with those predicted by Eqn. (1) with the value of $\bar{r}$ determined empirically from the current data. This will be referred to as the Liu and Wang (1992) prediction.

Niro and Beretta (1990) proposed a model in which the critical superheat for nucleation, $\theta^* \equiv T_w - T_s$, depends on the heated surface characteristics and the thermodynamic properties of the fluid. The critical superheat is predicted by:

$$\theta^* \approx \frac{\Delta P^*}{dP_s / dT} \tag{2}$$

where ΔP^* is the difference between vapor and liquid pressures when the nucleus reaches the critical dimensions. The critical superheat is reached by a liquid layer of thickness b near the heated wall, where b is related to the radius of the nucleation site:

$$b = 1.5\, R_C \tag{3}$$

For intermittent boiling, the waiting time required to reach the critical superheat is much longer than the bubble growth time. Assuming a purely conductive heat transfer mechanism in the liquid close to the heated wall, the waiting time can be approximated by:

$$\tau_w = \frac{\pi}{4}\left(1 + \frac{b\,q}{k_L\,\theta^*}\right)^2 k_L\,\rho_L\,c_{pL}\,\frac{\theta^{*2}}{q^2} \tag{4}$$

obtained by linearizing the pure conduction solution for a semiinfinite geometry and fixed thermal flux boundary conditions.

To predict the critical superheat, the denominator of Eqn. (2) can be estimated from the Clausius-Clapeyron equation:

$$\frac{dP_s}{dT} = \frac{\rho_V\, h_{fg}}{T} \tag{5}$$

Niro and Beretta (1990) suggested that the numerator of Eqn. (2) can be estimated from the Young-Laplace equation, which relates the surface tension at a bubble interface and the radius of curvature of the bubble, to the pressure differential across the phase interface. Equation (2) then becomes:

$$\theta^* = \frac{2\sigma}{R_C} \cdot \frac{T}{\rho_V\, h_{fg}} \tag{6}$$

This prediction of the critical superheat can then be inserted into Eqn. (4) to yield a prediction of the waiting time, which is assumed to be approximately equal to the pulse boiling period.

In the current study it is assumed that the liquid in the evaporator begins the first phase of the pulse boiling cycle as a saturated liquid. The measured pulse amplitude, ΔT, is then approximately equal to the critical superheat, θ^*. In practice, it would be difficult to predict the pulse magnitude using Eqn. (6), as the critical superheat is inversely proportional to the initial nucleation site radius, R_C, which depends on the maximum surface roughness of the evaporator material and the geometry of the evaporator. In the current study, experimentally measured values of the amplitude of evaporator temperature pulses are inserted in Eqn. (6) to determine an average value of R_C, and the resulting average value of R_C, and the corresponding predicted value of the pulse amplitude, are used in Eqn. (4) to predict the pulse period.

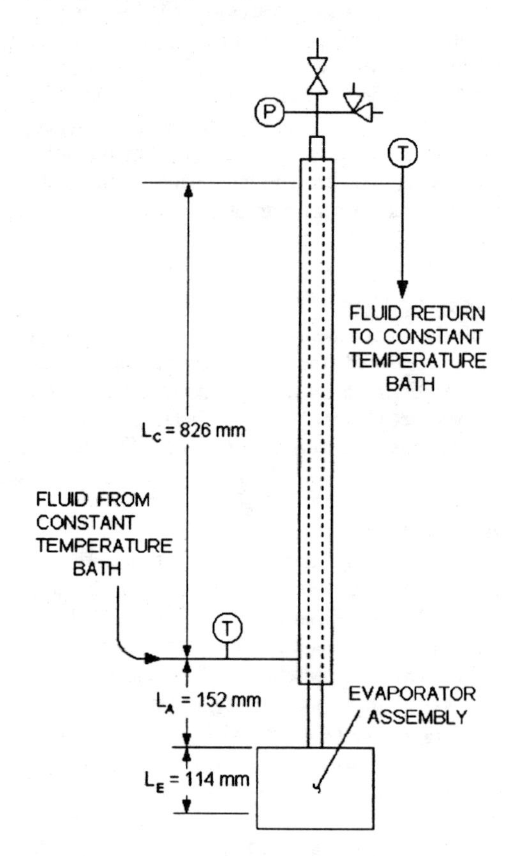

Figure 1 - Schematic of experimental apparatus

EXPERIMENTAL INVESTIGATION

An experimental investigation of thermosyphon pulse boiling has been conducted in the Two-Phase Heat Transfer Lab in the Department of Mechanical Engineering at Texas A&M University. The purpose of these experiments is to measure the amplitude and frequency of temperature pulsations using a variety of working fluids in a stainless steel thermosyphon.

Experimental Apparatus

A schematic representation of the test apparatus used in the current study is shown in Fig. 1. The thermosyphon consists of a stainless steel tube with an outside diameter of 12.7 mm and a wall thickness of 0.89 mm. The overall length of the thermosyphon tube is 1.092 m. The relative lengths of the evaporator, condenser and adiabatic sections are shown to scale in Fig. 1. A pressure transducer, a pressure relief valve, and a fill valve are connected to the upper end of the thermosyphon tube. The condenser consists of a tube with an inside diameter of 19 mm, which forms an annulus through which a water and ethylene glycol mixture is pumped from a constant temperature bath.

Figure 2 is a cross-sectional detail of the evaporator assembly. The evaporator is sealed by a stainless steel plug welded into the lower end of the tube. Three band heaters, each with a maximum heat output of 150 Watts, are clamped onto an aluminum heat spreader with an outside diameter of 25.4 mm. A T-type thermocouple is inserted into a slit in the heat spreader to measure the temperature of the outer surface of the stainless steel thermosyphon tube. The evaporator and adiabatic sections of the thermosyphon are covered with 7 cm of urethane foam insulation. The pressure transducer and evaporator thermocouple are connected to a computerized data acquisition system which was typically set to log data at a rate of 2 Hz.

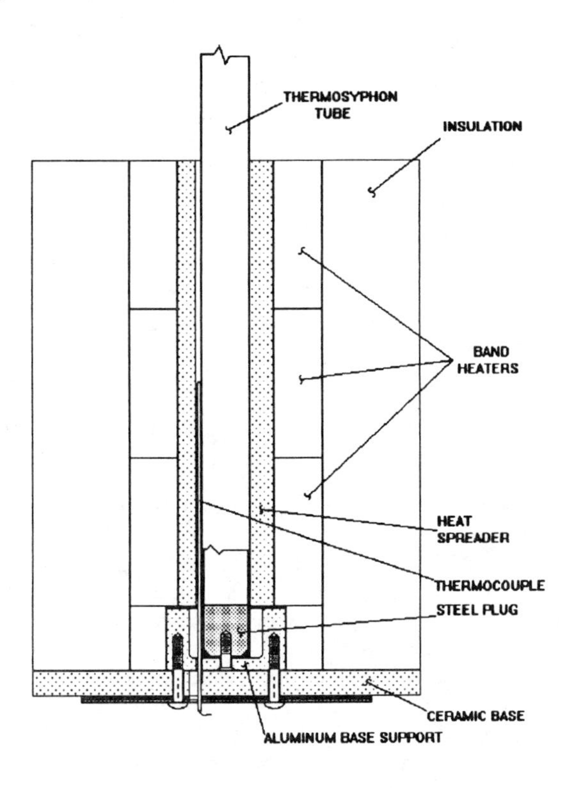

Figure 2 - Detail of thermosyphon evaporator

The thermosyphon tube was rinsed thoroughly with refrigerant-11 to remove any contaminants. The seamless stainless steel tube was used "as received". The roughness of the inside surface was not altered from its original condition.

Experimental Procedures

Prior to charging the thermosyphon between experimental test runs, it was important to remove any fluid left over from previous experiments, and it was also necessary to prevent any non-condensable gases from being introduced during the charging procedure.

To remove fluid from previous experiments, the thermosyphon was turned upside down, and compressed air was repeatedly blown through the fill valve and expelled. To remove any moisture that was introduced during this procedure, the thermosyphon fill line was connected to a vacuum pump and evacuated for approximately two hours, during which time the heaters were turned on to a very low power setting to maintain the thermosyphon at a temperature above the ambient temperature. At the end of this two hour period, the measured pressure and temperature of the thermosyphon were compared against the saturation curve of water to confirm that all moisture had been removed.

The required amount of working fluid was then introduced through the fill valve. The volume of working fluid was carefully measured with a burette. The condenser coolant was then allowed to flow through the condenser and the evaporator heaters were set to an intermediate power setting, causing the fluid in the thermosyphon to boil in the evaporator and condense in the condenser. The thermosyphon was operated for approximately thirty minutes, during which time non-condensable gases collected at the top of the condenser. With the fill valve connected to the vacuum pump, the valve was opened very briefly (approximately half a second) to draw the non-condensable gases out of the thermosyphon condenser without removing a significant amount of working fluid vapor. The power setting was then reduced to the initial setting for the pulse boiling experiments. After several minutes of operation, the measured pressure was compared with the evaporator temperature to confirm that the non-condensable gases had been removed.

At the initial power setting, the thermosyphon usually took between two and three hours to reach steady-state conditions, at which time pulsations would be observed. For subsequent power settings, approximately one hour was required to reach steady-state. The data acquisition system was set to log data after the pulses began. For each power setting, thirty minutes of data would be logged at a rate of 2 Hz. Data acquisition took two days for each combination of working fluid and condenser temperature setting.

RESULTS

Experiments have been performed with refrigerant-11, water, and ethanol. Tests with refrigerant-11 did not exhibit pulsations. All of the results shown here are for a fill ratio, V^+, of 100%. Experiments were conducted with water with a fill ratio of 50%, but no pulses were detected in these tests.

Figure 3 shows typical evaporator temperature pulse measurements from experiments with water at a condenser temperature of 20°C and a heat transport rate of 29.5 W. These pulses, with rapid temperature changes at the beginning and end of each pulse cycle and more gradual heating as the fluid approaches the onset of nucleation, are similar to those reported by Liu and Wang (1992).

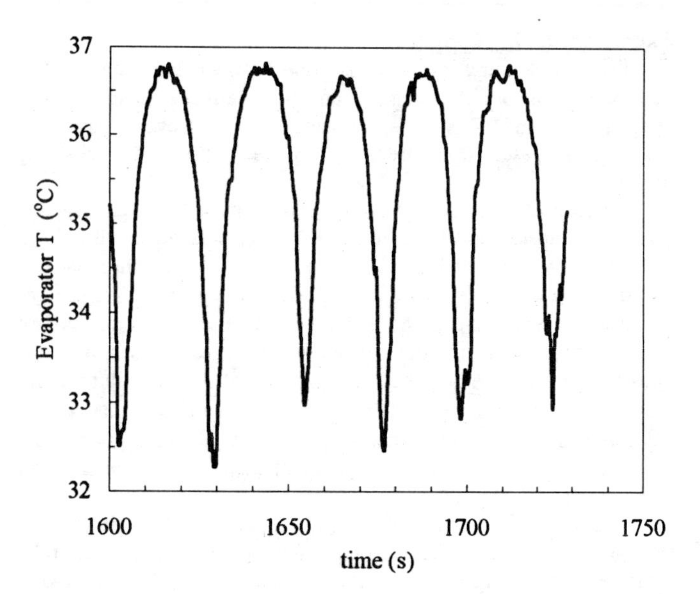

Figure 3 - Evaporator temperature pulses, water @ 20°C, $V^+ = 1.0$, Q = 29.5 W

Experimental Uncertainty

In the current experimental investigation the heat transport rate was determined from the voltage applied to the band heaters surrounding the evaporator. Taking into account the uncertainty in the voltage and electrical resistance measurements and the variation of electrical resistance with temperature, the uncertainty in the measured heat transport rate is ± 1.4 W. The evaporator temperature was measured by a T-type thermocouple with special limits of error, for which the uncertainty in an individual temperature measurement is ± 0.5°C. When measuring small-amplitude changes in temperature, however, the accuracy is limited by the resolution of the computerized data acquisition system, which corresponds to an uncertainty of ± 0.06°C. The uncertainty of the pulse wavelength measurements is ± 0.7 seconds, determined by the rate at which measurements were logged. It is evident from Fig 3 that when several pulses are recorded, the variations in both the amplitude and period of the pulses are greater than the limits of uncertainty specified above. Therefore, the error bars in Fig's. 4 - 9 represent the standard deviation of the measured values of pulse period or amplitude, which, in all cases, was greater than the limits of uncertainty.

Under steady-state operating conditions, the measured condenser heat transport rate agreed with the heat input at the evaporator to within the limits of uncertainty of the condenser heat transport measurements. This indicates that the evaporator was sufficiently insulated to justify the assumption of negligible heat loss.

Trends in Experimental Data

Figure 4 shows the measured pulse amplitude for the 19 test conditions considered in the current study. The tests with water at low condenser temperatures yielded the highest pulse amplitudes. Ethanol exhibited much smaller evaporator temperature pulses. The amplitude of pulsations was slightly lower for water at a condenser temperature of 40°C than for water at 20°C. These observations are consistent with the trends predicted by Eqn. (6). As the saturation temperature increases, the surface tension of a fluid will decrease, while the density of the saturated vapor increases. This will cause the critical superheat prediction of Eqn. (6) to decrease rapidly with increasing temperature for any fluid. Also, Eqn. (6) predicts that water, because of its high values of surface tension, will exhibit large pulses compared to ethanol. The decrease in pulse magnitude with increasing heat transport rate noted by some previous researchers was not evident in the current study.

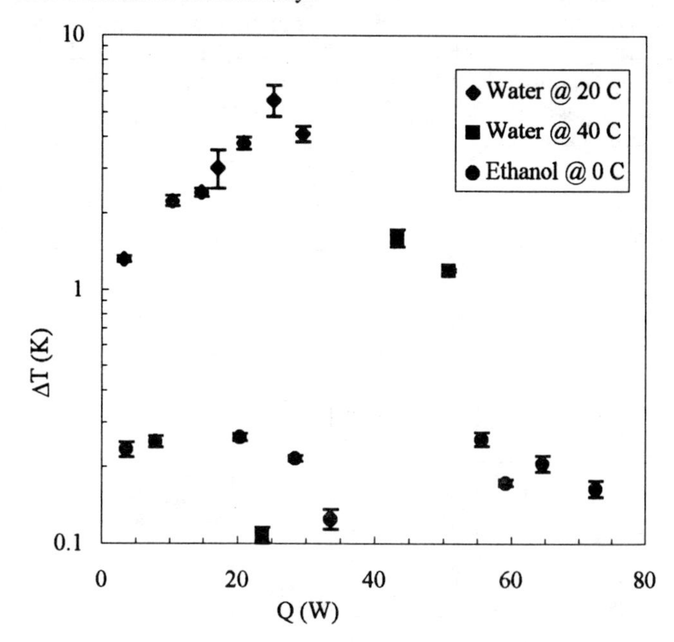

Figure 4 - Measured pulse amplitudes

In the current experiments the evaporator temperature was measured at only one location on the outside of the tube, approximately half-way up the length of the evaporator, as shown in Fig. 2. It is reasonable to assume that in most cases the temperature measurement was not in close proximity to the nucleation site. The seemingly random variation in the pulse magnitude, especially evident in the ethanol data in Fig. 4, may be due to movement of the nucleation site. The resulting changes in the proximity of the nucleation site and the evaporator thermocouple would then appear as variations in the measured pulse magnitude due to transient conduction in the evaporator material.

Figure 5 shows the measured pulse periods as a function of heat transport rate. All of the test conditions resulted in approximately the same range of pulse periods, with some of the test conditions exhibiting large scatter due to irregular pulses at moderate heat transport rates. The increase in frequency with heat transport rate observed by some previous researchers was not evident in the current study.

Comparison with Predictions

Figures 6 and 7 show the pulse amplitudes predicted by Eqn. (6) for the experiments with water and ethanol, respectively. In Fig. 6 the line to the lower right corresponds to a condenser temperature of

40°C, and the other line is the prediction for a condenser temperature of 20°C. The optimum value of R_C was determined, through least-squares linear regression, to be approximately 1.1 mm. This nucleation site radius is too large to be attributable to the roughness of the stainless steel evaporator surface, but it could correspond to the a nucleation site at the lower end of the evaporator, where the stainless steel plug is welded to the thermosyphon tubing. Equation (6) correlated the ethanol data well, and, as mentioned above, correctly predicts the general trends with respect to changes in working fluid and condenser temperature. The amplitude of temperature pulsations for water, however, especially at low condenser temperatures, was greatly under-predicted by this correlation.

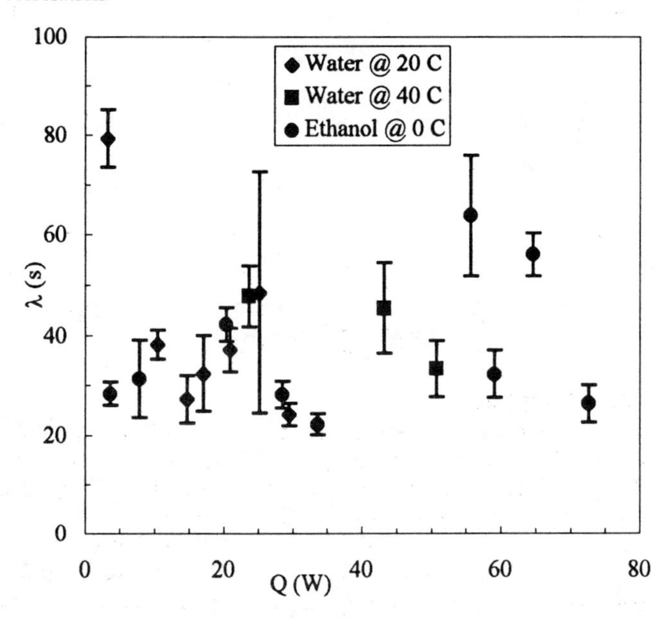

Figure 5 - Measured pulse periods

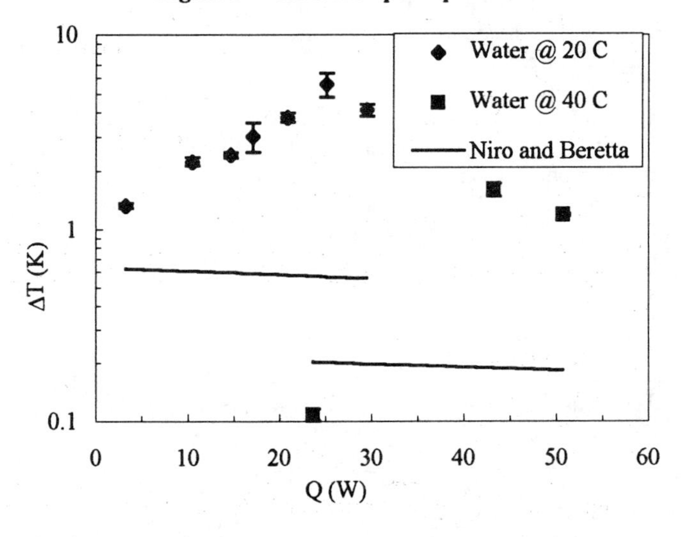

Figure 6 - Predicted pulse amplitude - water

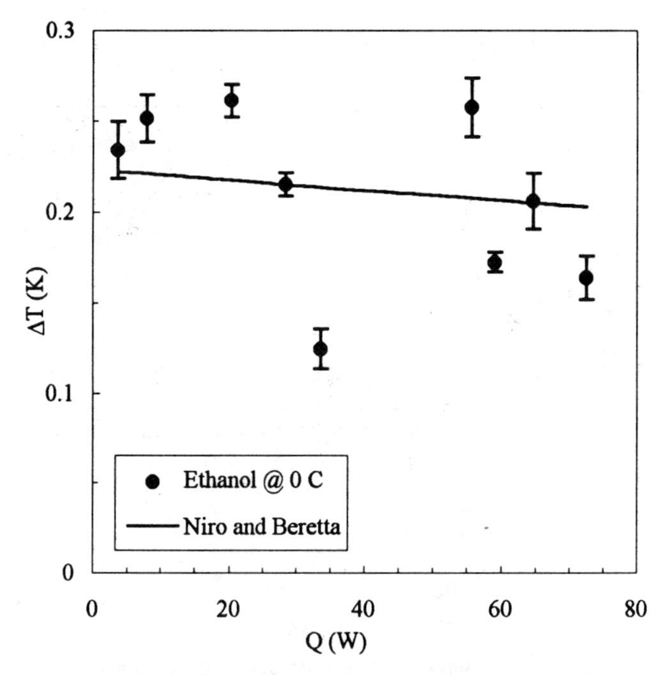

Figure 7 - Predicted pulse amplitude - ethanol

It is interesting to note that Eqn. (6), based on the analysis presented by Niro and Beretta (1990), predicts a slight decrease in the pulse amplitude with increasing heat transport rate, while the data of the current study, especially those for water in Fig. 6, exhibit an increase. The decrease in amplitude is predicted because of the change in fluid properties associated with the increase in saturation pressure with increasing heat transport rate due to the condenser thermal resistance. As noted above, Niro and Beretta observed the same amplitude increase with increasing heat transport in their experimental results. This trend is not predicted by any theoretical model, and the mechanisms associated with this behavior are unknown.

Figures 8 and 9 show the pulse wavelengths predicted by Eqn. (1) (Liu et al.) and by Eqn. (4) (Niro and Beretta, 1990) for water and ethanol, respectively. Again, the two lines in Fig. 8 correspond to different condenser temperatures. The empirical constant, b, in the Liu et al. prediction was found (through iteration to minimize the percent standard error) to be equal to 0.9998. Liu et al. predict a drastic decrease in pulse period with increasing heat transport rate, which, as mentioned above, was not observed in the current experimental investigation. This sensitivity to heat transport rate caused Eqn. (1) to exhibit a standard error of 255 percent when compared to the data of the current study. It is evident from Fig's. 8 and 9 that the pulse period prediction of Niro and Beretta (1990) is much more accurate. The only empirical constant in this prediction, R_C, was adjusted to fit the pulse amplitude data rather than the pulse period. Nevertheless, Eqn. (6) predicts the pulse period with a standard error of only 48 percent. The accuracy of this predictive method is further depicted in Fig. 10, from which it is evident that 68 percent of the pulse period data are predicted to within ± 50 percent.

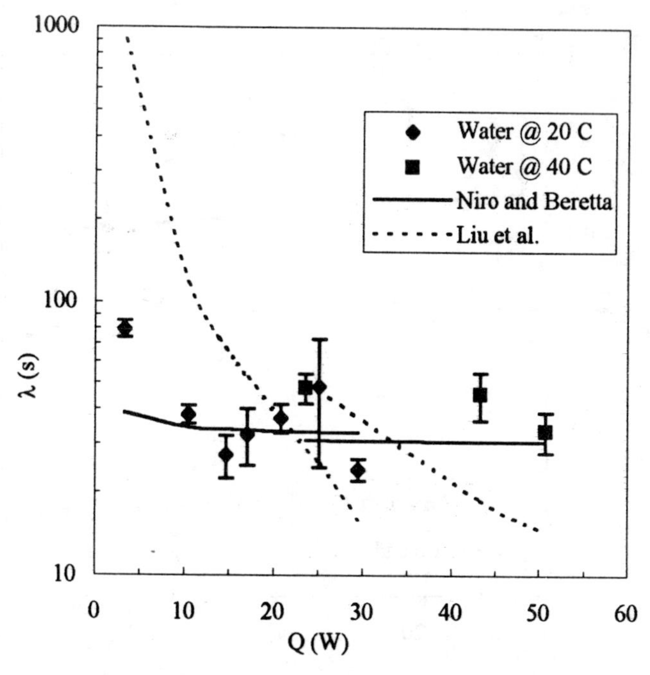

Figure 8 - Comparison of pulse period predictions - water

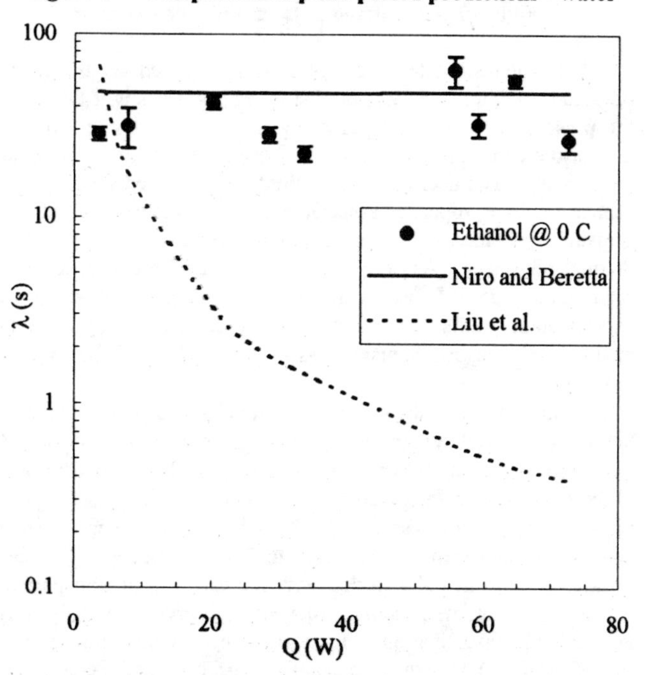

Figure 9 - Comparison of pulse period predictions - ethanol

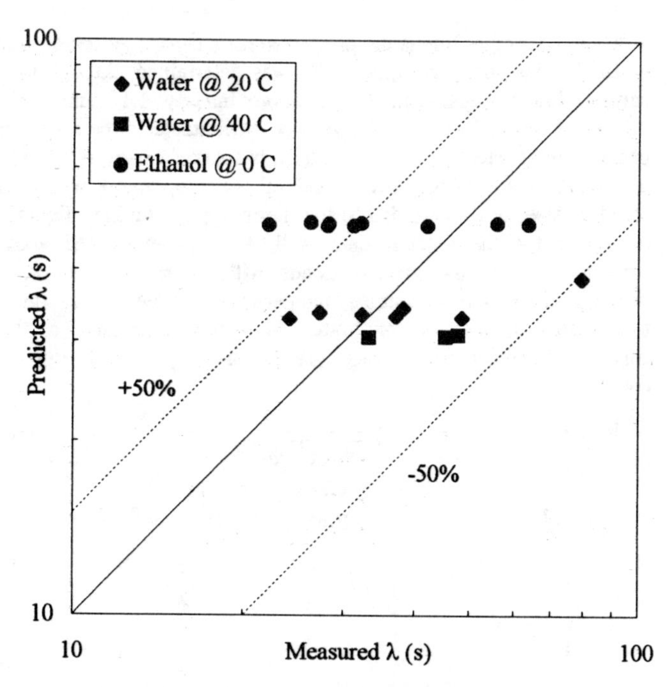

Figure 10 - Accuracy of Niro and Beretta (1990) pulse period prediction

CONCLUSIONS

An experimental investigation has been conducted to measure the amplitude and period of pulse boiling in a stainless steel thermosyphon. Water and ethanol were used as the working fluids, and pulse boiling was observed in 19 experimental test runs, covering a range of condenser temperatures and heat transport rates.

Two methods of correlating and/or predicting the period of pulse boiling oscillations have been applied to the data of the current study. The predictive method of Niro and Beretta (1990) yielded much more accurate predictions than the correlation presented by Liu et al. which predicted a sharp decrease in pulse period with increasing heat transport rate.

Equation (6), though it tends to under-estimate the pulse amplitude in some cases, can be used as a qualitative means of identifying situations in which pulse boiling may occur in a proposed thermosyphon design. This equation also predicts that the amplitude of temperature pulsations is inversely proportional to the size of the nucleation site. It may, therefore, be possible to eliminate pulse boiling in some applications by artificially roughening the evaporator surface.

REFERENCES

Casarosa, C., and Latrofa, E., 1983, "The Geyser Effect in a Two-Phase Thermosyphon," *International Journal of Heat and Mass Transfer*, Vol. 26, No. 6, pp. 933-941.

Drolen, B. L., and Fleischman, G. L., 1995, "Reflux Performance of Cryogenic Oxygen and Methane Space Heat Pipes," *9th International Heat Pipe Conference, Albuquerque, New Mexico, May 1-5*, Paper No. 10-2.

Fukano, T., Chen, S. J., and Tien, C. L., 1983, "Operating Limits of the Closed Two-Phase Thermosyphon," *ASME-JSME Thermal; Engineering Joint Conference Proceedings, Honolulu, Hawaii, March 20-24*, Ed: Y. Mori and W. Y. Yang, Vol. 1, pp. 95-101.

Hsu, Y. Y., 1962, "On the Size Range of Active Nucleation Cavities on a Heating Surface," *ASME Journal of Heat Transfer*, Vol. 84, pp. 207-216.

Lin, T. F., Lin, W. T., Tsay, Y. L., Wu, J. C., and Shyu, R. J., 1995, "Experimental Investigation of Geyser Boiling in an Annular Two-Phase Closed Thermosyphon," *International Journal of Heat and Mass Transfer*, Vol. 38, No. 2, pp. 295-307.

Liu, F., Liu, J., Wang, K., Wang, Z., and Guan, L., 1992, "The Visualization Studies on the Boiling Phenomena of Mixture Fluid in Thermosyphons," *8th International Heat Pipe Conference*, Paper No. A-P4.

Liu, J. F., Ang, X., and Yamamoto, T., 1988, "The Theoretical and Experimental Study on the Frequency of Geyser Boiling in Thermosyphon," *3rd International Heat Pipe Symposium, Tsukuba*, pp. 67-72.

Liu, J. F., and Wang, J. C. Y., 1992, "On the Pulse Boiling Frequency in Thermosyphons," *Journal of Heat Transfer*, Vol. 114, February, pp. 290-292.

Liu, J. F., Wang, J. C. Y., Dong, J. C., and Ang, X. Y., 1989, "Studies on the Pulse Boiling Phenomenon in Thermosyphons," *National Heat Transfer Conference*, HTD-Vol. 108, pp. 57-63.

Negishi, K., Kaneko, K., and Kusumoto, F., 1987, "Analysis of Pulsation in Two-Phase Thermosyphons," *6th International Heat Pipe Conference*, pp. 436-440.

Niro, A., and Beretta, G. P., 1990, "Boiling Regimes in a Closed Two-Phase Thermosyphon," *International Journal of Heat and Mass Transfer*, Vol. 33, No. 10, pp. 2099-2110.

Zhou, J., 1992, "Study on the Pulse Boiling Mechanism in the Gravity Heat Pipe," *8th International Heat Pipe Conference*, Paper No. A-17.

HTD-Vol. 342, National Heat Transfer Conference
Volume 4
ASME 1997

VAPOR-CONDENSATE INTERACTIONS DURING COUNTERFLOW IN INCLINED REFLUX CONDENSERS

A. Zapke

D. G. Kröger

Department of Mechanical Engineering, University of Stellenbosch,
Stellenbosch 7600, Republic of South Africa
Tel: +27 (0)21 808 4259, Fax: +27 (0)21 808 4958, E-Mail: dgk@ing.sun.ac.za

ABSTRACT

Direct air-cooled steam condensers often make use of a secondary condenser or dephlegmator, which acts as a de-aerator. The ability to reject heat in such a reflux condenser may be limited by flooding. In the present investigation adiabatic countercurrent flow and flooding experiments are conducted. The results are applied to predict the steamside pressure drop and flooding vapor flow rate of an existing dephlegmator. A thorough performance evaluation is carried out on a full scale dephlegmator. Cold zones are observed on the finned tube bundles and the heat transfer to the air, calculated from the measurements, is only 50 to 60% of the design heat rejection rate. The corresponding vapor flow rates are, however, well below the predicted flooding vapor rates. It is postulated that droplet entrainment and subsequent liquid hold-up may contribute to the ineffective operation of the dephlegmator.

NOMENCLATURE

A	area, m^2
c_p	specific heat at constant pressure, J/kgK
d	diameter, m
e	effectiveness, dimensionless
f	friction factor, dimensionless
g	gravitational acceleration, m/s^2
h	heat transfer coefficient, W/m^{2}K
H	duct height, m
i_{lg}	latent heat, J/kg
K	loss coefficient or coefficient defined by Eq. (12)
L	length, m
m	mass flow rate, kg/s
n	exponent defined by Eq. (12)
Q	heat transfer rate, W
W	duct width, m
p	pressure, N/m^2
T	temperature, °C or K
v	velocity, m/s
z	co-ordinate

Greek symbols

β	factor defined by Eq. (10), dimensionless
μ	viscosity, kg/ms
θ	angle, °
ρ	density, kg/m^3
σ	surface tension, N/m or area ratio, dimensionless

Subscripts

a	air
c	condensate, condensation, contraction or cross-sectional
e	equivalent or expansion
f	frictional
fr	frontal
g	gas phase or gravitational
in	inside
i	inlet
(i)	row (i), i = 1 or 2
l	liquid phase
m	mean or momentum
o	outlet
s	superficial
sp	single-phase
t	tube or duct
tp	two-phase
v	vapor

Dimensionless groups
$Fr_{sg} = \rho_g v_{sg}^2 / [g H (\rho_l - \rho_g)]$, superficial gas Froude number
$Fr_{sl} ,= \rho_l v_{sl}^2 / [g H (\rho_l - \rho_g)]$, superficial liquid Froude number
$Re_{sg} = \rho_g v_{sg} d_e / \mu_g$, superficial gas Reynolds number
$Zk = (\rho_l H \sigma)^{0.5} / \mu_l$, flooding parameter (Zapke and Kröger, 1996a)

INTRODUCTION

In a power generating cycle high pressure steam expands as it drives a turbine. The exhaust steam is condensed at sub-atmospheric pressures before it is returned to the boiler. Forced draft direct air-cooled condensers are often employed in arid areas as an alternative to wet-cooling systems to achieve the desired cooling. Such a direct air-cooled condenser or dry-cooling system consists of inclined bundles of finned tubes in which the steam and condensate flow cocurrently downwards, followed by a secondary condenser in which the flow is countercurrent. The secondary condenser which acts as a de-aerator is also referred to as a dephlegmator. An ejector is connected to the top header of the dephlegmator to remove non-condensable gases, such as ambient air which leaks into to the cooling system. A condenser-dephlegmator layout is illustrated in Fig. 1. The purpose of the dephlegmator is to facilitate a net outflow of steam at the bottom of the primary or main condenser. In the absence of such a net outflow, secondary steam flow patterns or backflow may exist due to the tube-row-effect present in multi-row condensers (Schulenberg, 1977; Larinoff et al., 1978; Berg and Berg, 1980). The backflow of steam may trap non-condensable gases at the bottom of the main condenser.

Under ideal operating conditions the temperature distribution of the dephlegmator finned tube bundles will be uniform due to the essentially isothermal condensation process. Cold zones were, however, observed on an existing dephlegmator. Infrared measurements showed that at times approximately 40% of the dephlegmator frontal area was close to the ambient temperature. The dephlegmator frontal area forms 17% of the entire cooling system under consideration and thus the net loss in heat rejection capability due to the presence of cold or dead zones amounts to approximately 7%. This presents a significant reduction in performance and it is the purpose of the present investigation to find the possible causes of the dephlegmator ineffectiveness. Although substantial progress has been made, the observed phenomena cannot be fully explained at this point in time. The investigation has, however, revealed important findings so far, which are reported in this paper.

BACKGROUND ON REFLUX CONDENSATION

Reflux condensation of steam inside inclined and vertical ducts and tubes has been studied by, amongst others, Russell (1980), Banerjee et al. (1983), Fürst (1989), Girard and Chang (1992) and Obinelo et al. (1994). For the purpose of the present study the findings by Banerjee et al. (1983) and Obinelo et al. (1994) will be discussed briefly.

At low bottom plenum pressures or steam flow rates the condensate drained downwards in an annular wavy film. As the bottom plenum pressure was increased, the steam flow rate increased and condensate droplets were entrained into the gas core. At sufficiently high vapor flow rates a condensate plug, or single-phase region, was formed above the two-phase condensing region. An important observation at this stage was that once the plug had been formed, the steam flow rate remained fairly constant while the pressure drop was increased up to $\approx$30 000 N/m^2. The temperature was measured inside the condenser tube along the axial direction. In the two-phase region it was found to be close to the saturation temperature but a sharp drop was observed as the single-phase region was entered. The condensate was significantly sub-cooled. Banerjee et al. (1983) concluded that flooding at the steam inlet controls the maximum steam rate during reflux condensation.

DEPHLEGMATOR DETAIL AND DIMENSIONS

The finned tubes under investigation are arranged in an A-frame configuration as shown in Fig. 1. The dephlegmator A-frame contains eight finned tube bundles, four on each side. A bundle consists of two rows of finned tubes, each row having a different fin pitch. Air at ambient temperature enters row (1). It is heated as it passes through row (1) and the warm air enters row(2). The fin pitch of row (2) is smaller than that of row (1) so that approximately equal heat transfer rates are obtained in each row. A cross-sectional view of the flattened tubes or rectangular ducts employed is shown in Fig. 2. The relevant heat exchanger details and dimensions are as follows:

Number of bundles: $n_b = 8$
Number of tube rows: $n_{tr} = 2$
Number of tubes per bundle, row (1): $n_{tb(1)} = 57$
Number of tubes per bundle, row (2): $n_{tb(2)} = 58$
Dephlegmator frontal area, row (2): $A_{fr} = 185.6$ m^2
Bundle inclination angle: $\theta_b = 60°$ to the horizontal
Duct inside height: $H_t = 100 \times 10^{-3}$ m
Duct inside width: $W_t = 12.941 \times 10^{-3}$ m
Duct inside surface area per unit length: $A_{tin} = 215 \times 10^{-3}$ m^2/m
Duct cross-sectional area: $A_{tc} = 1330 \times 10^{-6}$ m^2
Duct length: $L_t = 8$ m
Equivalent or hydraulic diameter: $d_e = 24.77 \times 10^{-3}$ m

STEAMSIDE FLOW MODELLING

Consider the schematic of a double row bundle illustrated in Fig. 3. The heat transfer characteristics of the rows are such that row (1) condenses slightly more steam than row (2). It can therefore be assumed that steam backflow from the top of row (2) into the top of row (1) will occur. The length of the backflow region is denoted as L_b. This assumption is based on the fact that the ejector suction rate is negligibly small and therefore does not influence the steamside flow distribution.

Conservation of mass

Consider each row separately. The outflow of condensate at the bottom must be equal to the net inflow of vapor, i.e.

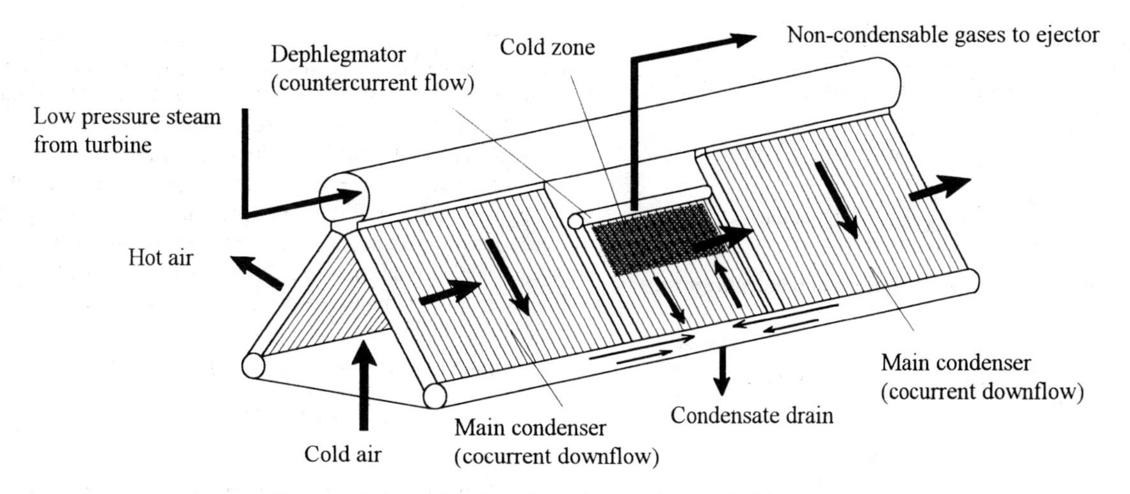

Figure 1. Schematic of an air-cooled condenser-dephlegmator layout.

$$m_{v2(1)} + m_{v3(1)} = m_{c(1)} \tag{1}$$

$$m_{v2(2)} - m_{v3(2)} = m_{c(2)} \tag{2}$$

The excess vapor at the top of row (2) enters row (1) via the top header and therefore we have

$$m_{v3(1)} = m_{v3(2)} \tag{3}$$

The heat transfer can be considered as a constant heat flux process for the purpose of this analysis and thus

$$m_{v3(1)}/m_{c(1)} = L_b/L_t \tag{4}$$

Momentum equations

Consider the vapor flow of row (1). Take a decrease in pressure as positive in the z-direction. The pressure drop across a duct is

$$\Delta p_{14(1)} = \Delta p_{12(1)} + \Delta p_{25(1)} + \Delta p_{53(1)} + \Delta p_{34(1)} \tag{5}$$

$\Delta p_{12(1)}$ and $\Delta p_{34(1)}$ are the changes in the static pressure across the contraction in flow area at the bottom and top end of the duct respectively. The pressure drop across a sudden contraction in flow area during countercurrent gas-liquid flow was determined by experiment. The data is correlated in terms of a two-phase loss coefficient K_{tp} defined by

$$\Delta p_{12} = \frac{1}{2}\rho_v v_{sv2}^2 \left[K_{tp} - \sigma_{21}^2\right] \tag{6}$$

where σ_{21} is the contraction area ratio and

$$K_{tp} = 1.5636 \, exp\left(2.9526 \, Fr_{sg}\right) \tag{7}$$

K_{tp} is expressed in terms of Fr_{sg} which, according to Zapke and Kröger (1996b), becomes the major governing dimensionless variable in the range of vapor flow rates applicable.

The resultant pressure drop $\Delta p_{25(1)}$ inside the duct across the length (L_t - L_b), i.e. is from the bottom of the duct to the point where the backflow region begins, is the sum of the frictional, gravitational and momentum components:

$$\Delta p_{25(1)} = \Delta p_f + \Delta p_g + \Delta p_m = \int_0^{L_t - L_b} \beta \left(dp/dz\right)_{ftp} dz + \int_0^{L_t - L_b} g\rho_{v(1)} sin(\theta_b)\, dz + \int_0^{L_t - L_b} -\frac{d}{dz}\left(\rho_{v(1)} v_{sv(1)}^2\right) dz \tag{8}$$

$(dp/dz)_{ftp}$ is the frictional pressure gradient experienced by a gas during adiabatic countercurrent gas-liquid flow while β is a factor which accounts for the increase in frictional pressure drop due to condensation and the subsequent deformation of the vapor boundary layer. $(dp/dz)_{ftp}$ was measured for air-water flow inside a rectangular duct inclined at 60° to the horizontal. The experiments were carried out at low liquid flow rates, i.e. high void fraction flow, typically encountered in air-cooled reflux condensers. See Zapke and Kröger (1996b) for details regarding the apparatus. In the flow range applicable to the dephlegmator the two-phase vapor frictional pressure drop does not deviate significantly from the corresponding single-phase values and may be correlated in terms of the superficial gas Reynolds number in the form of a Blasius-type equation

$$\left(\frac{dp}{dz}\right)_{ftp} = K\, Re_{sg}^n \left(1/d_e\right) 1/2\, \rho_v v_{sv}^2 \tag{9}$$

where K = 0.2259 and n = -0.2088.

Groenewald (1994) presented a correlation for β for condensation inside flattened tubes or rectangular ducts of high

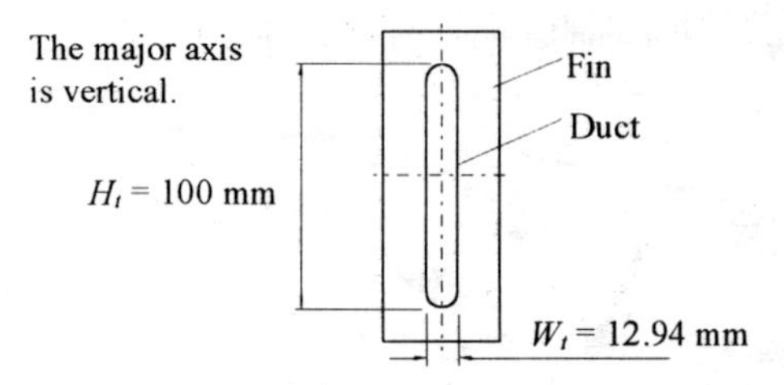

Figure 2. Cross-sectional view of the ducts used in the dephlegmator for condensation.

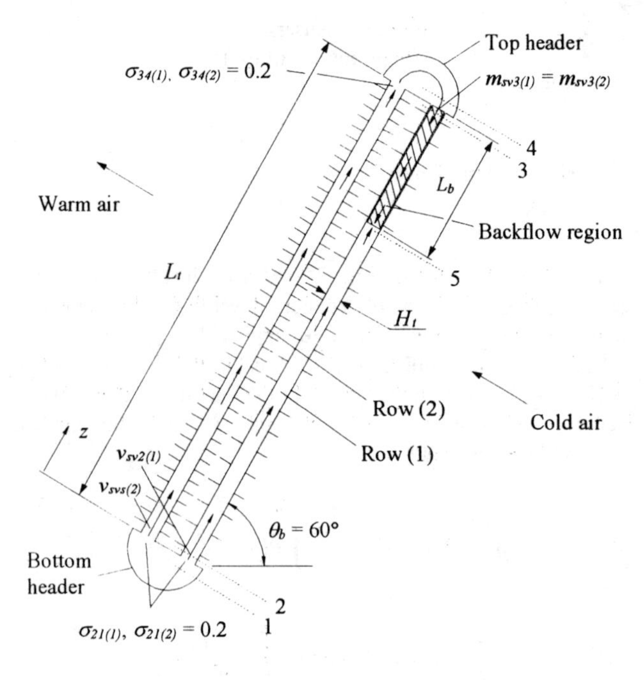

Figure 3. Steamside flow distribution inside the double row dephlegmator in the absence of liquid hold-up and non-condensable gases.

aspect ratio during turbulent flow, i.e.

$$\beta = a_1 + a_2 / Re_{sv} \tag{10}$$

where a_1 and a_2 are related to the condensation rate.

The variation in the steam velocity along the axial direction of the duct may be approximated by a linear relation

$$v_{sv(1)} = v_{sv2(1)} \left(1 - z/\left(L_t - L_b\right)\right) \tag{11}$$

Substitute Equations (9) to (11) into Eq. (8) and integrate to obtain

$$\Delta p_{25(1)} = \frac{1}{2}\rho_{v(1)} v_{sv2(1)}^2 \left[\frac{\left(L_t - L_b\right)}{d_e} K Re_{sv2(1)}^n \left(\frac{a_1}{n+3} + \frac{a_2}{(n+2)Re_{sv2(1)}}\right) - 2\right] + \rho_{v(1)} g \left(L_t - L_b\right) \sin(\theta_b) - \rho_{v(1)} v_{sv(1)}^2 \tag{12}$$

Note that a transition from turbulent to laminar flow occurs close to the backflow region. The pressure drop in the laminar region is very small compared to the total pressure drop across the headers and is for the purpose of this analysis approximated by the relations for turbulent flow without any loss in accuracy.

The same governing equations are applied for the backflow region except that the entrance pressure drop at the top of the finned tube may be calculated with a single-phase loss coefficient. Equation (5) for the total pressure drop across the headers can now be obtained by adding the respective components, resulting in

$$\Delta p_{14(1)} = \Delta p_{12(1)} + \Delta p_{25(1)f} + \Delta p_{53(1)f} + \Delta p_{34(1)} + \rho_{v(1)} g L_t \sin(\theta_b) - \left(\rho_{v(1)} v_{sv2(1)}^2 - \rho_{v(1)} v_{sv3(1)}^2\right) \tag{13}$$

A similar expression can be derived for row (2), keeping in mind that there is an outflow of steam at the top. The common bottom and top header configuration requires that

$$\Delta p_{14(1)} = \Delta p_{14(2)} \tag{14}$$

Heat transfer relations

The heat transfer from the condensing steam to the air stream is

$$Q_a = \sum_{i=1}^{n_{tr}} m_{c(i)} i_{lg(i)} = \sum_{i=1}^{n_{tr}} m_a c_{pam(i)} \left(T_{v(i)} - T_{ai(i)}\right) e_{(i)} \tag{15}$$

The effectiveness for each tube row is (Holman, 1986)

$$e_{(i)} = 1 - exp\left[-(UA)_{(i)} / \left(m_a c_{pam(i)}\right)\right] \tag{16}$$

where

$$(UA)_{(i)} = \left[1/h_{ae(i)} A_{a(i)} + 1/h_{c(i)} A_{c(i)}\right]^{-1} \tag{17}$$

The effective airside thermal conductance may be expressed as

$$h_{ae(i)} A_{a(i)} = \left[1/h_{a(i)} e_{f(i)} A_{a(i)} + \sum_n \frac{R_n}{A_n}\right]^{-1} \tag{18}$$

The summation term represents the thermal resistances other than the airside and steamside values.

A correlation developed by Groenewald (1994) for flattened tubes is employed to determine the condensation heat transfer coefficient. $h_{ae(i)} A_{a(i)}$ was obtained in a wind-tunnel for each row of finned tubes.

The rate of condensation in each row can be calculated by making use of Equations (15) to (18), provided that the air flow rate and the air inlet temperature are known. In the present investigation these two quantities were measured at the dephlegmator. Equations (1) to (4) together with the equal pressure drop condition Eq. (14)

form a set of five equations which can be solved for the five variables $v_{sv2(1)}$, $v_{sv3(1)}$, $v_{sv2(2)}$, $v_{sv3(2)}$ and L_b , once the condensation rates have been determined.

Flooding

Flooding experiments were conducted to obtain a correlation for the flooding vapor velocity of the finned tubes under consideration. The tests were carried out in the apparatus used for the entrance pressure drop and pressure gradient experiments. Due to the fact that the duct geometry and the gas and liquid inlet configuration strongly affect the flooding process, care was taken that the test section geometry and the flow conditions resembled that of the finned tubes as closely as possible. Air, water, methanol and propanol were used as working fluids. The data is correlated in terms of the phase Froude numbers and a dimensionless group Zk proposed by Zapke and Kröger (1996a, 1996c), i.e.

$$Fr_{sg}^{0.25} + 0.94\,Fr_{sl}^{0.25} = 0.73\,Zk^{0.03} \tag{19}$$

The correlation is valid for rectangular ducts inclined at 60° with a sharp-edged gas inlet at the bottom. During reflux condensation the maximum vapor and condensate flow rates are encountered at the bottom of the finned tubes. It can be expected that at sufficiently high vapor flow rates flooding will be initiated at the bottom of the finned tubes. The superficial liquid velocity at the vapor inlet required in Eq. (19) can be obtained from the condensate flow rate at the bottom of the duct.

This completes the steamside flow analysis. In the next section the full scale dephlegmator tests are described and the results are compared to the predicted heat transfer rate and pressure drop across the headers.

FULL SCALE DEPHLEGMATOR PERFORMANCE TESTS

A comprehensive performance evaluation of the dephlegmator was conducted by measuring the air inlet temperature, the air outlet temperature and velocity distribution, the steamside temperature inside the bottom and the top header and inside the ejector duct, the steamside bottom and top header pressures and the steamside pressure drop across the headers. The temperatures were measured with copper-constantan thermocouples while differential pressure transducers were used for the pressure measurements. The air outlet velocity distribution was measured with propeller-type anemometers. The instrumentation was connected to a data logger controlled by a personal computer. The analog signals were digitised by the logger and communicated to the personal computer where the data was stored on disc for later retrieval and post-processing.

The air inlet temperature was measured inside the plenum. The air outlet temperature and velocity distribution was determined by dividing the frontal area into 360 rectangular areas of 0.52 m^2 (0.89 m x 0.58 m) and taking measurements at the centre of each of the incremental areas. The air flow rate through such an area increment was calculated by the relation $\Delta m_a = \rho_{ao} v_{ao} \Delta A_{fr(2)}$. The total heat transfer to the air was then obtained from

$$Q_a = Q_{a(1)} + Q_{a(2)} = \sum_{j=1}^{360} \Delta m_{aj} c_{pamj}\left(T_{ao(2)j} - T_{ai(1)}\right) \tag{20}$$

Ten thermocouples and anemometers supported by a 6 m long aluminium beam were used to monitor the air outlet conditions by moving the beam across the frontal area above row (2) and taking readings at the predetermined grid points.

The results of two case studies are presented in this paper. Case 1 illustrates the ineffective performance and undesirable steamside flow conditions while in case 2 the dephlegmator achieved the cooling rate according to design. The steamside and heat transfer data are presented in Table 1 together with the predicted quantities. The fluid properties required for the steamside flow analysis were evaluated at the steam temperature measured at the bottom header of the dephlegmator. The measured air outlet temperature distributions are given in Fig. 4. v_{sv2} in Table 1 is the actual or measured superficial velocity entering row (1) and row (2) and is approximated by

$$v_{sv2} = \frac{Q_a(measured)}{i_{lg}\,\rho_v\left(n_{tb(1)} + n_{tb(2)}\right)n_b\,A_{tc}} \tag{21}$$

DISCUSSION

i) In case 1 the measured heat transfer rate is well below the predicted rate while in case 2 they correspond fairly closely.

ii) The steamside pressure drop is approximately six times the predicted value in case 1 while in case 2 good agreement is obtained between theory and experiment.

iii) In both cases the steam flow rate entering the dephlegmator at the bottom of the finned tubes is below the predicted flooding vapor velocity.

iv) In case 1 cold zones appear on the dephlegmator and the air outlet temperature varies from ≈50 °C at the bottom of the bundles to ≈20 °C at the top. In case 2 the air outlet temperature is uniform and all the bundles are at a temperature just below the steam saturation temperature inside the finned tubes.

The characteristic high pressure drop associated with flooding leads one to believe that flooding is the cause of the cold zones and the subsequent ineffective dephlegmator performance. Also, the significant sub-cooling occurring in the single-phase liquid region, which is formed as a result of flooding (Banerjee et al., 1983) may suggest that the cold zones are due to the presence of liquid plugs. This can, however, not be the case because firstly the length of the cold zones, measured in the direction of the finned tubes, was at times 2 to 3 metres long. The corresponding pressure drop across such a single-phase column is ≈20 000 N/m^2, but such high pressure differentials were never recorded. Secondly the prediction according to Eq. (19) undoubtedly proves that the dephlegmator is unlikely to flood.

The fact that the dephlegmator can operate satisfactorily, as found in case 2, was discovered by coincidence when measurements were taken while the turbine load was increased from partial load (≈70%) to full load. As a rule cold zones were always present when the turbine was running steady at full load. Case 2 seems to be possible but unlikely to occur during steady-state full load conditions. The heat transfer rate measured from time to time during the presence of cold zones varied between 9 and 11 MW whereas approximately 16 to 17.5 MW should ideally have been rejected.

Table 1. Comparison of the measured and predicted dephlegmator performance.

	Case 1	Case 2
m_a, kg/s (measured)	550.0	512.0
T_v, °C: Bottom header	56.6	66.5
T_v, °C: Top header	29.1	66.6
T_v, °C: Ejector duct	34.5	65.9
$T_{ai(1)}$, °C (measured)	19.6	27.1
Q_a, MW (measured)	10.9	15.7
Q_a, MW (predicted)	15.93	16.22
Δp_{14}, N/m^2 (measured)	4174	588
Δp_{14}, N/m^2 (predicted)	712	418
v_{sv2}, m/s (measured)	33.6	31.9
$v_{sv2(1)}$, m/s (predicted)	48.9	32.8
$v_{sv2(2)}$, m/s (predicted)	49.2	33.1
$v_{sv3(1)}$, m/s (predicted)	5.3	4.3
$v_{sv3(2)}$, m/s (predicted)	5.2	4.2
L_b, m (predicted)	0.78	0.92
$v_{sv(1)flooding}$, m/s	70.2	56.8
$v_{sv(2)flooding}$, m/s	71.4	58.0

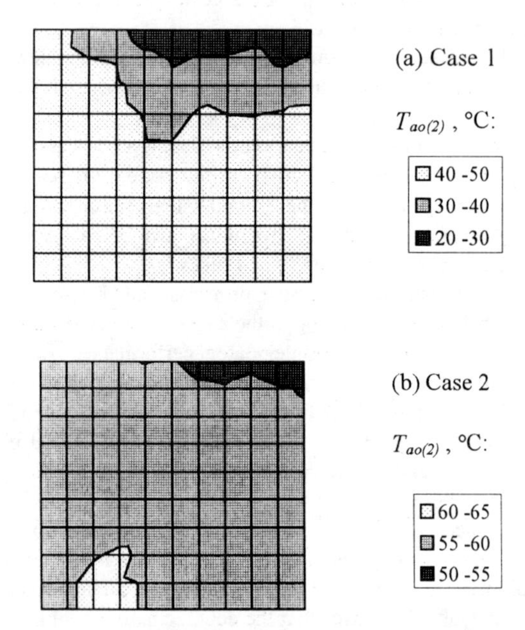

Figure 4. Air outlet temperature distribution of two adjacent bundles of case 1 and 2 respectively. The cold zone at the top of the bundles seen in Fig. 4 (a) forms part of the cold zone marked in Fig. 1 on four bundles of the dephlegmator. Note that each side of the dephlegmator A-frame arrangement contains four bundles and that cold zones can be present in all eight bundles.

CONCLUSIONS

According to Eq. 19 for adiabatic flow, flooding appears not to be the reason for the ineffective dephlegmator performance. It is postulated that droplet entrainment at the sharp-edged vapor inlet occurring at velocities below the predicted flooding rates and the subsequent accumulation of condensate inside the ducts may be the cause of the high pressure drop. Most of the ducts are then partially blocked causing a mal-distribution of the steamside flow. This renders the removal of non-condensable gases ineffective and air is trapped inside the dephlegmator. The cold zones seem to contain a two-phase mixture of mainly sub-cooled condensate and air.

REFERENCES

Banerjee, S., Chang, J. S., Girard, R., and Krishnan, V. S., 1983, "Reflux Condensation and Transition to Natural Circulation in a Vertical U-Tube," Journal of Heat Transfer, Vol. 105, pp. 719-727.

Berg, W. F., and Berg, J. L., 1980, "Flow Patterns for Isothermal Condensation in One-Pass Air-Cooled Heat Exchangers," Heat Transfer Engineering, Vol. 1, No. 4, pp. 21-31.

Fürst, J., 1989, "Kondensation in Geneigten Ovalen Rohren," VDI Fortschitt-Berichte, VDI-Verlag, Düsseldorf, Reihe 19, Nr. 36.

Girard, R., and Chang, J. S., 1992, "Reflux Condensation in Single Vertical Tubes," International Journal of Heat Mass Transfer, Vol. 35, No. 9, pp. 2203-2218.

Groenewald, W., 1994, Heat Transfer and Pressure Change in an Air-Cooled Flattened Tube during Condensation of Steam," M.Eng. Thesis, Stellenbosch Univeristy, Stellenbosch.

Holman, J. P., 1986, "Heat Transfer," McGraw Hill Book Company, New York, Chapt. 10, p. 548.

Larinoff, M. W., Moles, W. E., and Reichelm, R., 1978, "Design and Specification of Air-Cooled Steam Condensers," Chemical Engineering May 22, pp. 86-94.

Obinelo, I. F., Round, G. F., and Chang, J. S., 1994, "Condensation Enhancement by Steam Pulsation in a Reflux Condenser," International Journal of Heat and Fluid Flow, Vol. 15, No. 1, pp. 20-29.

Russell, C. M. B., 1980, "Condensation of Steam in a Long Reflux Tube," Heat Transfer and Fluid Flow Service, AERE Harwell and National Engineering Laboratory.

Schulenberg, F., 1977, "The Air Condenser for the 365 MW Generating Unit in Wyoming/USA, Special print from collected edition of the VGB-Conference "Power Station and Environment 1977".

Zapke, A., and Kröger, D. G., 1996a, "The Influence of Fluid Properties and Inlet Geometry on Flooding in Vertical and Inclined Tubes," International Journal of Multiphase Flow, Vol. 22, No. 3, pp. 461- 472.

Zapke, A., and Kröger, D. G., 1996b, "Pressure Drop during Gas-Liquid Countercurrent Flow in Inclined Rectangular Ducts," Pre-print of Proceedings, 5th International Heat Pipe Symposium, Melbourne.

Zapke, A., and Kröger, D. G., 1996c, "The Effect of Fluid Properties on Flooding in Vertical and Inclined Rectangular Ducts and Tubes," Proceedings, FED-Vol. 239, ASME Fluids Engineering Division Conference, Vol. 4, pp. 527-532.

HTD-Vol. 342, National Heat Transfer Conference
Volume 4
ASME 1997

Two-Phase Instabilities During Startup Transients in Single Heated Channel Loops

J.C. Paniagua, U.S. Rohatgi[1], and V. Prasad
Department of Mechanical Engineering
State University of New York
Stony Brook, NY 11794-2300

ABSTRACT

A new thermal hydraulics computer code was developed to simulate the geysering instability and is based on integral methods where local properties are based on local pressures and the vapor generation model is adjusted accordingly to reflect the vapor generation rate necessary to initiate the instability. This is an important modeling feature since local vapor generation rate depends on local saturation temperature. The code is designed to simulate startup transients in single heated channel natural circulation systems with subcooled water as an initial condition. The formulation of thermal hydraulics is inherently general and accounts for both single-phase liquid flow and nonhomogeneous, nonequilibrium two-phase flow. Single channel systems are investigated first because the flow dynamics of a single heated channel with the remaining sections of the flow loop are coupled and less complex as compared with multiple channel systems. (In multiple channel systems, channel dynamics can be coupled to one another and be independent of the remaining flow loop or one unstable channel may dominate the transients and make the multiple parallel channel system behave as one single channel.) Wang, et al., (1994) studied the instabilities in a low pressure natural circulation single channel loop at low powers and high inlet subcoolings. The numerical simulations illustrated channel flow reversal, which is a thermal hydraulic feature of condensation-induced geysering. This study showed that the integral method coupled with local pressure variation for the vapor generation model is suitable to predict startup or geysering transients.

INTRODUCTION

The current Light Water Reactors achieve shutdown through active safety systems. Passive safety systems have been proposed for advanced designs to enhance the reliability of safety functions. The natural circulation Simplified Boiling Water Reactor (SBWR) by General Electric has been proposed as one such concept. Because natural circulation systems require power to initiate the circulation through void generation, thereby coupling the flow to the power, thermal hydraulic instabilities could occur under low pressure conditions. If such instabilities were to transpire in the natural circulation SBWR during startup, the reactor would experience power oscillations due to strong void reactivity feedback and void fraction fluctuations in the reactor core. Therefore, it is necessary to investigate and understand properly the thermal hydraulic instabilities during startup.

Recently, some concerns have been raised about the possibility of geysering or condensation-induced instability during startup from low pressure and low flow conditions in a natural circulation plant like the SBWR (Aritomi, 1992). Aritomi (1992, 1993) and Chiang (1992) have conducted extensive research in the area of geysering under natural circulation. Aritomi, et. al. (1992) explained the driving mechanism of geysering as follows: when voids are generated in a heated channel, a large slug of bubble forms, which grows due to decrease in hydrostatic pressure head as it moves toward the exit. The vapor then mixes with the liquid in the subcooled riser or upper plenum and is condensed there. Due to the bubble collapse and subsequent decrease in pressure, the subcooled liquid reenters the channel and restores the non-boiling condition. This process repeats periodically causing flow oscillations. It is evident then that bubble formation, growth, and collapse phenomena are of importance to the geysering instability.

The RAMONA-4B code assessment, (Paniagua, et al., 1996) indicated that to correctly predict possible startup instabilities, the vapor generation should be predicted accurately. The TWOPHASE thermal hydraulics computer code was developed to address the RAMONA-4B computer code assumption of basing thermophysical properties, such as saturation temperature, on a mean system pressure (Rohatgi, et al., 1996)

[1] *Department of Advanced Technology, Brookhaven National Laboratory, Upton, NY 11973*

and its impact on the vapor generation model. In the TWOPHASE computer code, local properties are based on local pressures and the vapor generation model is adjusted accordingly to reflect the vapor generation rate necessary to initiate the geysering instability. TWOPHASE is designed to simulate startup transients in natural circulation systems with subcooled water as an initial condition. The code models single heated channel systems and the formulation of thermal hydraulics is inherently general and accounts for both single-phase liquid flow and nonhomogeneous, nonequilibrium two-phase flow.

FORMULATION AND SIMULATION SCHEME

The TWOPHASE code utilizes several of the thermal hydraulic modeling features employed in the RAMONA-4B methodology (Rohatgi, et al., 1996, and Wulff, et al., 1984). They include:

- Four fundamental balance equations - one vapor mass and liquid mass balance equation, one mixture momentum eqution, and one mixture energy equation.
- Loop momentum integral formulation.
- Drift flux model for the phasic velocities.
- Nonequilibrium vapor generation model.

All local thermophysical properties were based on system average pressure instead of local pressure. However, as an option, the TWOPHASE code calculates the pressure distribution at the initial steady-state conditions. This leads to a quasi-steady pressure distribution throughout the system. When the transients are initiated, the system pressure changes are assumed instantaneous and uniform throughout the individual computational cells. This results in an improvement over RAMONA-4B in the pressure field calculation. Whereas RAMONA-4B computes an average pressure, TWOPHASE calculates the pressure field completely for the steady-state initial conditions. The objective of the TWOPHASE code is to determine the quantitative and qualitative effect of considering the local pressure, and therefore, the local saturation temperature, instead of a single pressure. The RAMONA-4B methodology based phasic property calculations on single average pressure. This assumption was significant because it eliminated the need for predicting the spatial pressure variation or acoustic effects.

GOVERNING EQUATIONS

For the four-equation formulation for the one-dimensional two-phase flows consists of the following conservation equations (Delhaye, et al., 1981; Wulff, et al., 1984; Rohatgi, et al., 1996):

Vapor Mass Balance:

$$\frac{\partial}{\partial t}(\alpha\rho_g) + \frac{\partial}{\partial z}(\alpha\rho_g v_g) = \Gamma_g \quad (1)$$

Liquid Mass Balance:

$$\frac{\partial}{\partial t}((1-\alpha)\rho_l) + \frac{\partial}{\partial z}((1-\alpha)\rho_l v_l) = -\Gamma_g \quad (2)$$

Mixture Energy Balance:

$$\frac{\partial}{\partial t}(\rho_m u_m) + \frac{\partial}{\partial z}(\alpha\rho_g h_g v_g + (1-\alpha)\rho_l h_l v_l) = \frac{q_w^{\prime}}{A} \quad (3)$$

Mixture Momentum Balance:

$$\frac{\partial}{\partial t}G_m + \frac{\partial}{\partial z}(\alpha\rho_g v_g^2 + (1-\alpha)\rho_l v_l^2) = \frac{-\partial p}{\partial z} - \rho_m g_z - \frac{f_l \varnothing_l^2 G_m |G_m|}{2\rho_l d_h} - k_{form} k_{2\varnothing}\frac{G_m |G_m|}{2\rho_l} \quad (4)$$

By integrating Eq. (4) around the circulation loop, and utilizing the condition that $\oint \frac{\partial p}{\partial z} dz = 0$ (pressure drops at area changes and restrictions or singularities included) the result is:

$$\frac{dI}{dt} = -\oint \rho_m g_z dz - \oint \frac{f_l \varnothing_l^2 G_m |G_m|}{2\rho_l d_h} dz - L \quad (5)$$

where " I " is the total integrated momentum for the loop and the term L consists of the area change/restriction pressure drops and the momentum flow fluxes:

$$L = \sum_{singularities} \Delta p + (\alpha\rho_g v_g^2 + (1-\alpha)\rho_l v_l^2)\Big|_{inlet} - (\alpha\rho_g v_g^2 + (1-\alpha)\rho_l v_l^2)\Big|_{outlet} \quad (6)$$

In addition to the four balance equations, 10 additional constitutive relationships are required to close the formulation. The equations are listed in Wulff, et al., (1984) and Rohatgi, et al., (1996).

STEADY STATE INITIAL CONDITIONS

Steady state initial conditions are generated first to serve as the starting point for the transient simulation that follows after the boundary conditions are perturbed. The general concept of the search for steady state initial conditions for TWOPHASE is to impose a set of boundary conditions that define uniquely the steady state of the system. The entire circulatory system is assumed to be filled with subcooled liquid. The input parameters and system conditions requested by the user for TWOPHASE to calculate the steady state initial conditions are: 1) plant geometry, including flow form loss coefficients, 2) channel power 3) system pressure (dome), and 4) channel inlet subcooling.

With these parameters, TWOPHASE then searches for the steady state core channel flowrate by adjusting the input guess for core flowrate so that the pressures at the downcomer inlet and riser outlet match to within a specified tolerance.

MOMENTUM INTEGRAL METHOD

The integral momentum equation is obtained by line integrating the differential momentum equation around the loop contour of the natural circulation flow path. The contour integration begins at the dome outlet (downcomer inlet) and terminates at the dome inlet (riser outlet). The requirement that the spatial pressure gradient term vanishes as the instantaneous pressure must return to the dome pressure for a closed path is met. The major steps in the calculational procedure of the loop cell flows are the calculation of a reference mixture volumetric flow at the inlet of the downcomer and followed by the calculation of flow distributions in the individual components. The procedure for determination of the flow distribution is a variation of the derivations of RAMONA-4B (Rohatgi, et. al., 1996) and are detailed in Paniagua (1996b).

VALIDATION OF METHODOLOGY

The methodology described here has been validated with Wang, et al., (1994) test data and with a set of sensitivity calculations.

Description of Test. Wang, et al., (1994) investigated the thermal hydraulic oscillation behavior in a low pressure two-phase natural circulation loop at low powers and relatively high inlet subcoolings. The experiments were performed at atmospheric pressure with heating power ranging from 4 to 8 kW and inlet subcooling ranging from 27 to 75 degrees C. A rectangular two-phase natural circulation loop, as shown in Figure 1, included an annular heated section, riser, cooling section, downcomer and upper and bottom horizontal sections.

The annular heated section consisted of an electric heater contained in a stainless steel SUS-316 tube with an outer diameter of 10.6 mm and an outer transparent PYREX glass tube having an inner diameter of 20.2 mm. Glass tube thickness is 2.6 mm. The length of the heater tube is 1.1 m with only the lower 1.0 m being uniformly heated. Maximum heater power output was 30 kW.

The riser is also made of transparent PYREX tubing with the same thickness and with the diameter being 20.2 mm. The riser length is 3.5 m. As seen in Figure 1, the component in the cooling section is the condenser which is of a shell-tube design and is 1.8 m in length. The outlet temperature of the primary side fluid is controlled by the flow rate of the secondary cooling water. This outlet temperature is considered to be the inlet temperature of the heated section.

Assessment of Momentum Integral Method With Single Pressure. The momentum integral method utilizing a single mean system pressure for all thermophysical property calculations is the RAMONA-4B methodology. With this option, the TWOPHASE code essentially reduces to RAMONA-4B. Wang's experimental apparatus has been nodalized as shown in Figure 2. The heated channel is comprised of 24 cells because, in general, 24 nodes are used in most finite difference codes for describing the 12 ft. long core channels for stability applications. This provides a sufficiently detailed representation of the thermal hydraulic behavior in the core channels.

Effects of nodalization on code predictions have been performed by Rohatgi, et. al., (1993), where the calculations were performed with 12, 18, and 24 nodes in the heated test section for the FRIGG tests. The conclusion from this assessment was that the peak gain of the power-to-flow ratio decreased as the number of nodes increased. The 24 node model result for the peak gain converged to three significant digits, whereas the 12 node model overpredicts the gain by 5%. Therefore, 24 computational cells were chosen for the core channel for the stability analysis of Wang's experiment. Figure 3 displays the temporal evolution of mass flowrate in the experiment with a power input of 6 kW and an inlet temperature of 43.5 degrees C. From this figure it can be seen that flow reverses periodically in addition to the flowrate oscillating quasi-periodically between 500 kg/hr and -160 kg/hr. This translates into an inlet velocity oscillation between 0.434 m/sec and -0.138 m/sec. In the flow reversals, the amplitude of the forward flows was three to four times the amplitude of the reversed flows. This is because the dilatation due to vapor generation in the heated section must overcome the downcomer gravity head at the same instant. Wang, et al., (1994) notes the flow oscillation is a type of quasi-periodic one with four fundamental frequencies. Three oscillations followed another oscillation of larger magnitude. The period of the largest oscillation or first fundamental mode as seen in Figure 3 was about 50 sec. Figure 4 illustrates the effect of inlet subcooling and heater power on the first fundamental frequency. As the inlet subcooling decreases, the frequency of the oscillations increases indicating an inverse relationship. Similarly, as heater power increases, so does the frequency of the oscillations. The time period is related to the heatup of liquid in channel to the saturation point (Duffey, et al., 1996). Similar trends have been reported by Chiang, et al., (1992) for natural circulation loop geysering. It was noted that for low inlet subcoolings the period of geysering could be correlated with the time period required for boiling of subcooled liquid flowing into a heated channel.

The inlet velocity and riser void fraction predicted by the TWOPHASE code are shown in Figures 5 and 6. Flow reversal is achieved in the simulation as seen in Figure 5. The peak amplitude of the oscillations is 0.30 m/sec, and the period of oscillation is 15 sec which corresponds to a frequency of approximately 0.067 Hz. This simulation underpredicts the flow oscillation amplitude by 30% and the frequency by 25% as compared to Wang's experimental data (Figure 3). The oscillation frequency discrepancy with the test data can be attributed to the time required for liquid in the channel to boil. If local saturation temperatures are uniform throughout the system because of local properties based on a single system mean pressure, then, for a given inlet subcooling, the time for liquid to reach boiling will be greater as compared to the situation where saturation temperatures are based on local pressures. This is because the saturation temperatures are higher at the bottom of the channel since the hydrostatic pressure in a channel is a function of height. As a result, for the single pressure case, the simulation underpredicts the vapor generation rate, overpredicts the time to reach saturation and therefore, underpredicts the frequency of oscillations.

Assessment of Momentum Integral Method With Quasi-Steady Pressure Distribution. In the previous section the code with the assumption of local saturation temperature based on mean system pressure was validated. The code underpredicted the frequency and the peak velocity. In this section the option for local saturation temperature was relaxed as it was based on local pressure, estimated from the initial steady state distribution. Results from the TWOPHASE simulations are shown in Figures 7 to 10. The inlet mass flowrate is depicted in Figure 7. The first fundamental frequency occurs at approximately 40 seconds in the simulation preceded by three smaller oscillations and the inlet liquid velocity magnitude is slightly higher at 0.60 m/sec in the forward flow direction and -0.20 m/sec in the reverse flow direction. These results are approximately 33% greater than Wang's experimental data. Figure 8 exhibits the channel exit mass flowrate for the initial 60 seconds of the transient where the first fundamental frequency osciilation occurs. Similarly, Figures 9 and 10 show the core channel and riser average void fraction during the transient, respectively. The results from Figures 8 to 10 cannot be compared with any documented data from this experiment.

Void fraction measurements indicate that the flowrate increases following the increase in void fraction in the riser and heated channel. This process, however, is not strong enough to suppress boiling in the riser during the three smaller oscillations. Thus vapor voids will generate and accumulate in the riser as seen in Figure 10 since lower saturation temperatures exist there as compared to the heated channel. Once there is enough bouyancy in the channel heated section and riser, a large driving momentum force is generated and a large increase in loop flowrate will result. This large amplitude flow condenses most of the vapor bubbles in the heated section and riser and accounts for the large amplitude of the first fundamental oscillation. In addition, since there are not many vapor voids remaining in the heated section and riser after this first fundamental oscillation, such a large magnitude forward flow is not possible until three smaller amplitude flow oscillations have appeared first due to boiling in the riser. It is during these three small oscillations that vapor bubbles coalesce in the heated channel and riser regions giving rise to the bouyancy driven large loop flow. The results of the simulations are consistent with Wang's assessment (1994) of his experiments.

Inlet Subcooling Sensitivity. A sensitivity analysis was performed to determine the system behavior responses to lower inlet subcoolings. As seen in Figure 11, which represents the inlet mass flowrate, the flow amplitudes increase as well as the frequency of flow oscillations with the decrease in subcooling. The core inlet subcooling temperature is 25 degrees C at the same heater power of 6kW. The first fundamental frequency for these conditions is approximately 0.052 Hz, which corresponds to a time period of 19 sec. Figure 12 represents the channel average void fraction. Because of the lower inlet subcooling, more vapor will be generated within the channel (Figure 9) and riser as compared to the higher subcooling case where the inlet temperature was 43.5 degrees C which is equivalent to 60 degrees C subcooling. Figure 13 shows the effect of inlet subcooling on the first fundamental frequency at 6kW power. The analytical prediction is consistent with the experimental data (Wang, et al., 1994) seen in Figure 4.

Therefore, the consequence of lower inlet subcooling for fixed power input is an increase in the frequency and in the amplitude of the flow oscillations. If the inlet subcooling is low enough, saturated conditions may exist in the riser thereby suppressing the condensation-induced geysering since subcooled riser conditions are required for the instability.

System Pressure Sensitivity. An increase in system pressure at a constant heat flux and constant inlet subcooling will result in a reduction of vapor generation as evident in the channel inlet mass flowrate oscillation being damped in Figures 14 and 15 for respective system pressures of 3 and 5 atmospheres. The effect of an increase in system pressure is to stabilize the condensation-induced geysering instability. A necessary condition for geysering to occur is the existence of a large voids in the channel. Pressure acts as the damping mechanism by raising the saturation temperature thus hampering the formation of a large bubble and resulting in less oscillatory flow behavior. Figure 16 displays the effect of system pressure on peak velocity amplitude. Chiang, et al., (1992) suggested that the effect of system pressure on geysering is caused by vapor void reduction because of decreased density difference and surface tension between vapor and liquid. With an increase in the system pressure, the quality required for the same void fraction increases due to a decrease in the density difference between the liquid and vapor. At higher system pressures a large voids are hardly ever formed and churn-turbulent flow readily occurs. Consequently, the thermal equilibrium quality required for forming a large voids covering the whole flow channel becomes higher with an increase in pressure. Since the higher thermal equilibrium quality leads to a reduction in subcooling in the upper plenum, it difficult to condense the vapor, and therefore, the flow becomes stable with an increase in system pressure. In this analysis, all other parameters such as inlet subcooling and heat flux, were kept constant.

CONCLUSIONS

For the momentum integral method with single pressure, the simulation results for the 24 node channel model underpredicts by 30% the experimental results (Wang, et al., 1994) for flow oscillation amplitude and frequency, whereas, the momentum integral method with quasi-steady pressure distribution overpredicts the amplitude by 33% and is within 5% for the flow oscillation frequency. The predicted results are qualitatively consistent with the test data. The analysis can be "fined-tuned" by adjusting the vapor generation model since the model provides that flexibility. The coefficients utilized in this analysis were recommended by in RAMONA (Wulff, et al., 1984). This is the important modeling feature since condensation-induced geysering is produced by large vapor void fractions covering the entire flow area (Chiang, et al., 1992) condensing in the subcooled upper plenum or riser.

System pressure increases result in higher saturation temperatures which limit the amount of vapor generation. Reduced vapor generation dampens and supresses the geysering instability. Reducing the inlet subcooling temperature for fixed power settings leads to higher frequency oscillations since the time for heating of subcooled liquid entering the heated channel to saturation decreases. Trends indicate that if the inlet subcooling is lowered enough, saturation conditions prevailing in the upper plenum will inhibit the geysering instability.

NOMENCLATURE

General

A	Cross-sectional flow area of channel, m^2
d_h	Hydraulic diameter, m
f	Friction factor
G	Mass flux, ρv
g	Gravitational acceleration, m/s^2
h	Enthalpy, kJ/kg
I	Total loop or segment momentum, $kg/m\text{-}s$
k_{form}	Form loss coefficient

$k_{2\phi}$	Two-phase flow form loss multiplier
p	Pressure, *Pa*
q'	Linear heating rate, *W/m*
t	Time, *s*
u	Internal energy per unit volume, $kJ/kg\text{-}m^3$
v	Velocity, *m/s*
z	Axial spatial coordinate

Greek Symbols

α	Vapor void fraction
Δ	Delta (change in variable or parameter)
Γ_g	Vapor generation rate per unit volume, $kg/m^3\text{-}s$
ρ	Mass density, kg/m^3
Φ	Two-phase flow friction multiplier

Subscripts and Superscripts

f	Saturated liquid
g	Saturated vapor
l	Liquid
m	Mixture
w	Wall
z	Axial direction

REFERENCES

Aritomi, M., et. al., "Fundamental Study on Thermo-Hydraulics during Startup in Natural Circulation Boiling Water Reactors, (I)", J. of Nuclear Science and Technology, 29-7, 1992, 631-641.

Aritomi, M., Chiang, J.H., Mori, M., "Geysering in Parallel Boiling Channels", Nuc. Engrg. and Des. 141, North Holland, Amsterdam (1993), 111-121.

Chiang, J.H., Aritomi, M., Inoue R., and Mori, M., "Thermohydraulics During Startup in Natural Circulation Boiling Water Reactors", *NURETH-5*, September 1992.

Delhaye, J.M., Giot, M., Riethmuller, M.L., "Thermohydraulics of Two-Phase Systems for Industrial Design and Nuclear Engineering", McGraw-Hill, New York, 1981.

Duffy, R.B., Rohatgi, U.S., "Physical Interpretation of Geysering Phenomenon and Periodic Boiling Instability at Low Flows", *ICONE 4*, New Orleans, 1996.

Paniagua, J., et al., "Modeling of Two-Phase Flow Instabilities During Startup Transients Utilizing RAMONA-4B Methodology", *International Mechanical Engineering Congress and Exposition*, Atlanta, GA, Nov 17 - 22, 1996a.

Paniagua, J., "Thermal Hydraulic Instabilities in Parallel Channels for Natural Circulation Systems", Ph.D. thesis, State University of New York, Stony Brook, NY, December 1996b.

Rohatgi, U.S., et. al., "Assessment of RAMONA-3B Methodology with Oscillatory Flow Tests", *Nuc. Engrg. and Des.* 143, 69-82, North Holland, Amsterdam, 1993.

Rohatgi, U.S., Cheng, H.S.,and Slovik, G.C., "RAMONA-4B: A Computer Code with Three-Dimensional Neutron Kinetics for BWR System Transients - Modifications Assessment Application", NUREG/CR-6359, BNL-NUREG-52471, BNL, 1996.

Wang, S.B., et. al., "Thermal-Hydraulic Oscillations in a Low Pressure Two-Phase Natural Circulation Loop at Low Pressures and High Inlet Subcoolings", *4th International Topical Meeting on Nuclear Thermal Hydraulics, Operations and Safety*, Taipei, Taiwan, April 6-8, 1994.

Wulff, W., et. al., "A Description and Assessment of RAMONA-3B MOD.O CYCLE 4: A Computer Code with Three-Dimensional Neutron Kinetics for BWR System Transients", NUREG/CR-3664, BNL-NUREG-51746, BNL, 1984.

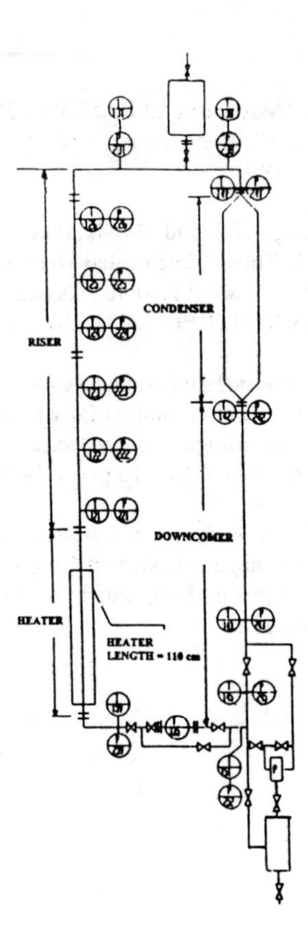

Figure 1: Natural circulation loop schematic (Wang, et al., 1994).

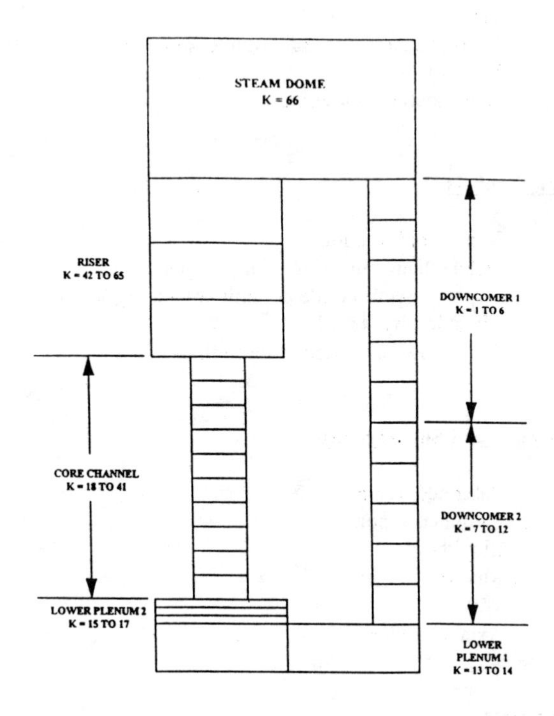

Figure 2: TWOPHASE nodalization scheme for single heated channel .

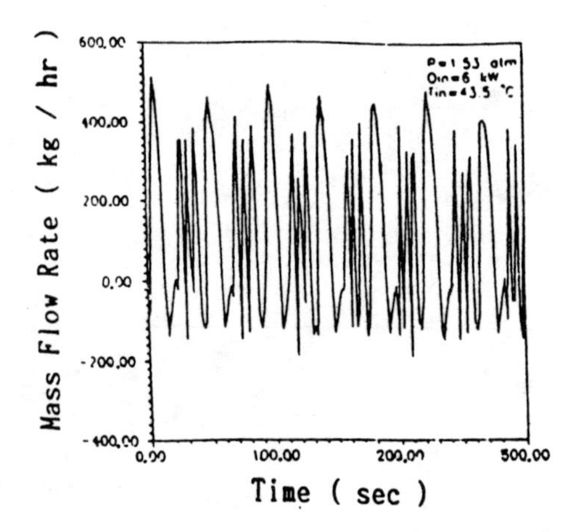

Figure 3: Transient behavior of mass flow rate (Wang, et al., 1994).

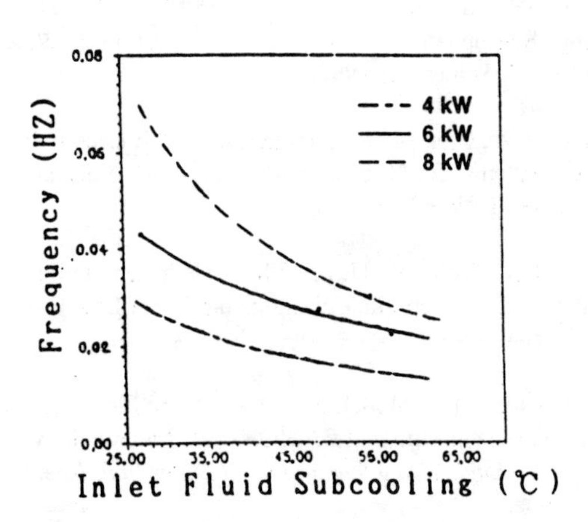

Figure 4: Effects of power and inlet subcooling on first fundamental frequency (Wang, et al., 1994).

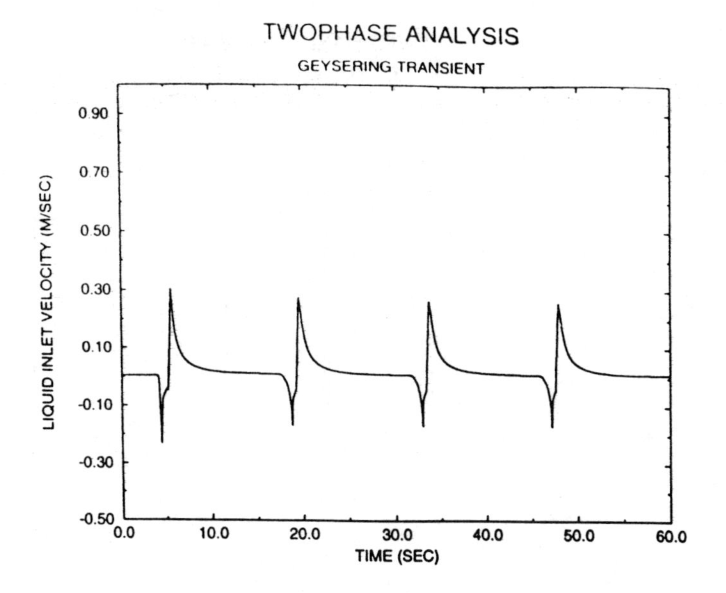

Figure 5: Channel liquid velocity at 6kW and 60 degree C inlet subcooling.

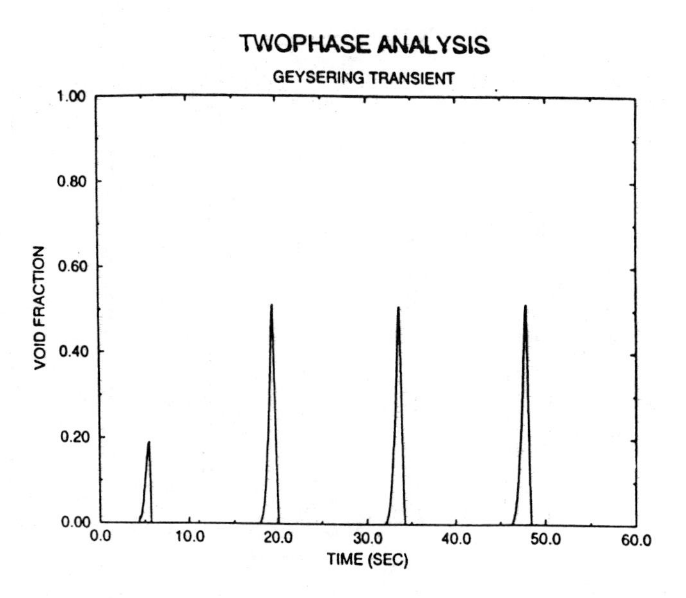

Figure 6: Riser void fraction at 6kW and 60 degree C inlet subcooling.

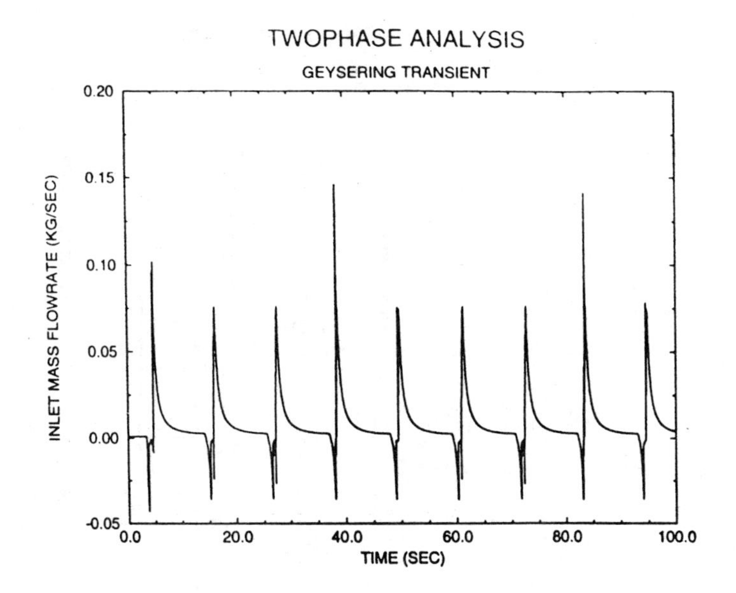

Figure 7: Channel inlet mass flowrate at 6kW and 60 degree C inlet subcooling.

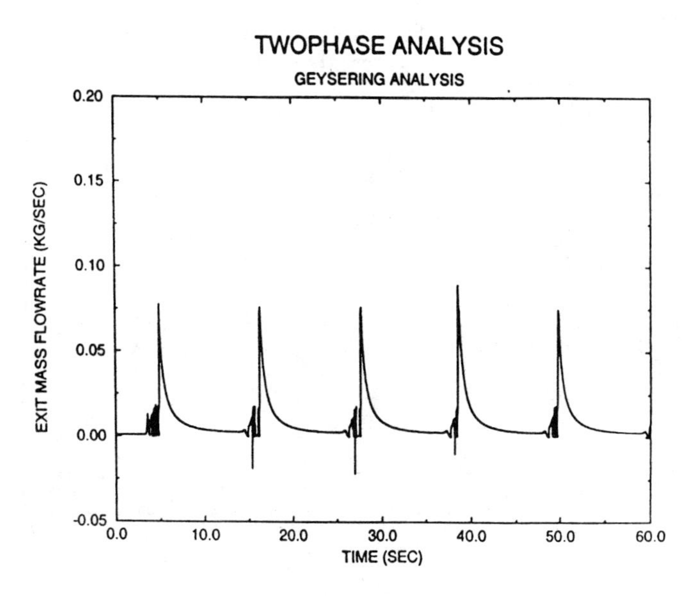

Figure 8: Channel exit mass flowrate at 6kW and 60 degree C inlet subcooling.

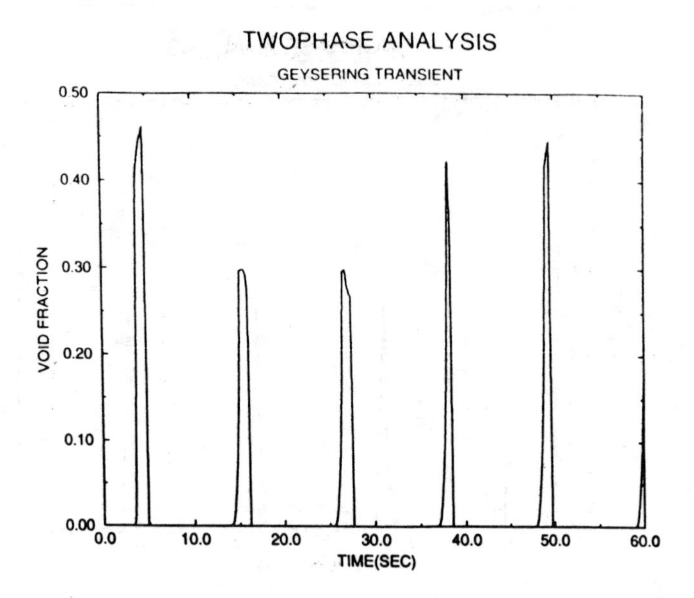

Figure 9: Channel void fraction at 6kW and 60 degree C inlet subcooling.

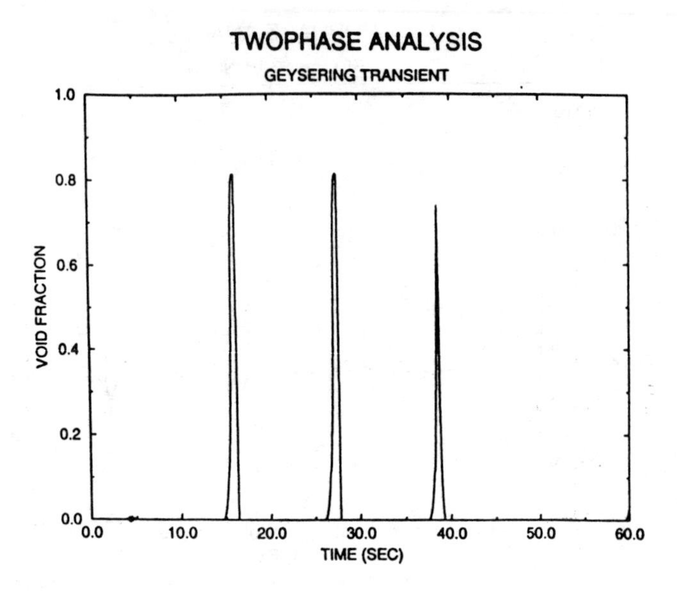

Figure 10: Riser void fraction at 6kW and 60 degree C inlet subcooling.

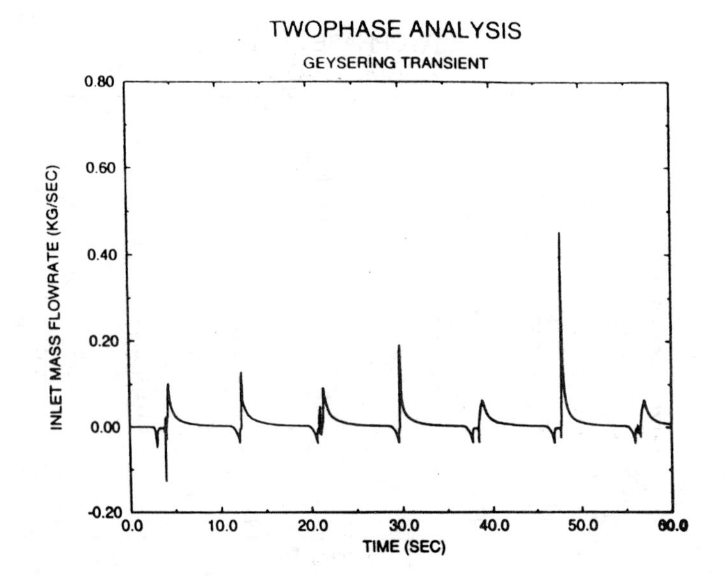

Figure 11: Channel inlet mass flowrate at 6kW and 25 degree C inlet subcooling.

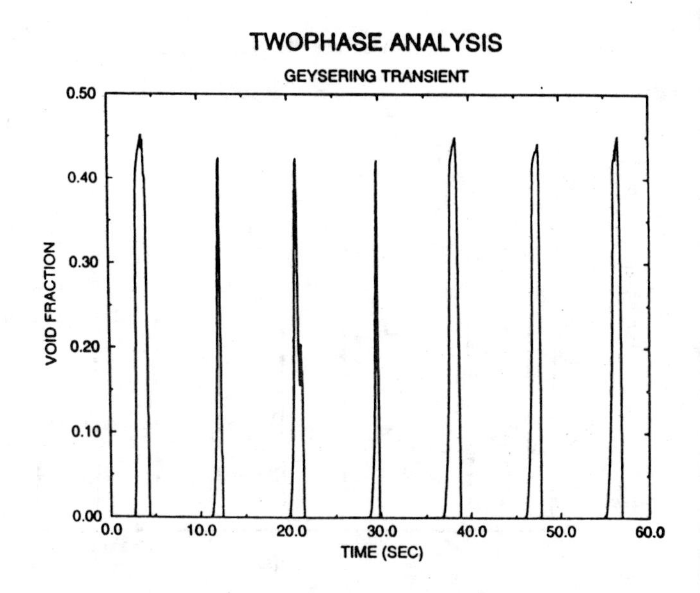

Figure 12: Channel void fraction at 6kW and 25 degree C inlet subcooling.

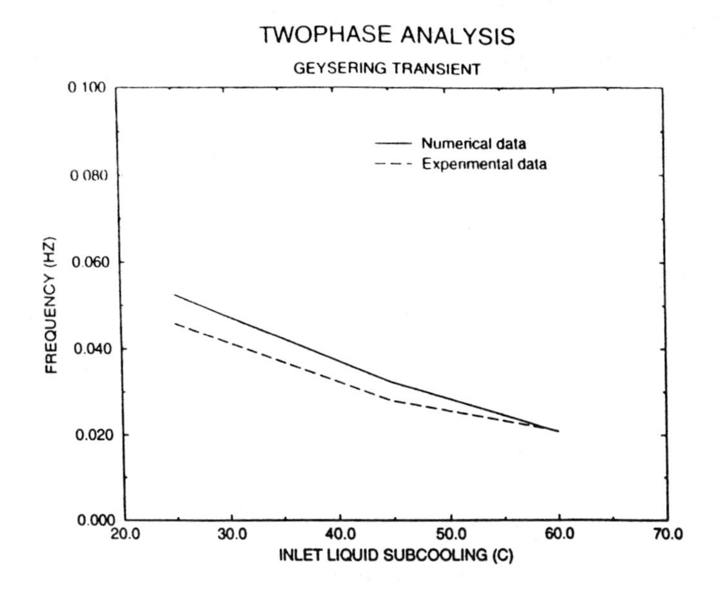

Figure 13: Effect of inlet subcooling on first fundamental frequency at 6 kW power.

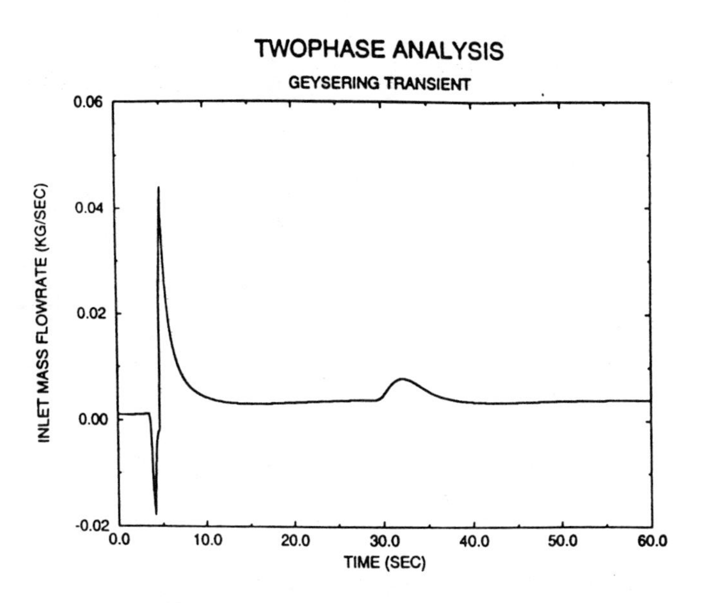

Figure 14: Channel inlet mass flowrate at 3 atm, 6kW, and 60 degree C inlet subcooling.

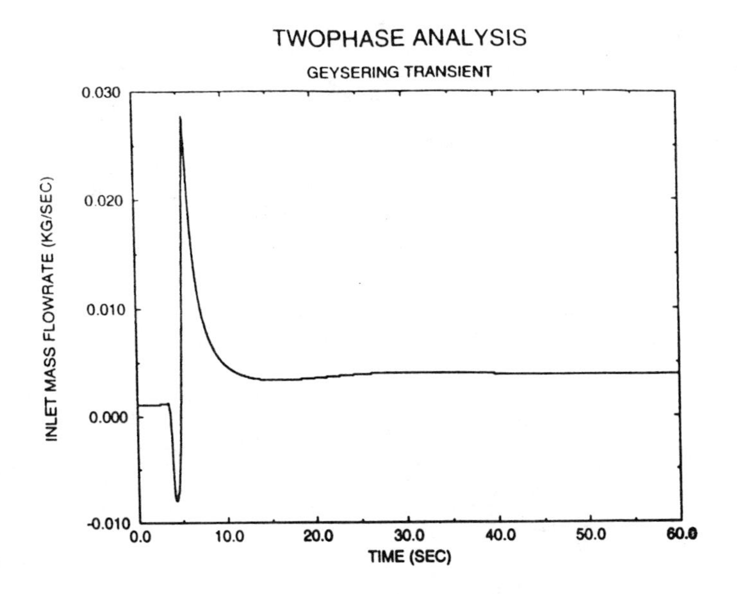

Figure 15: Channel inlet mass flowrate at 5 atm, 6kW, and 60 degree C inlet subcooling.

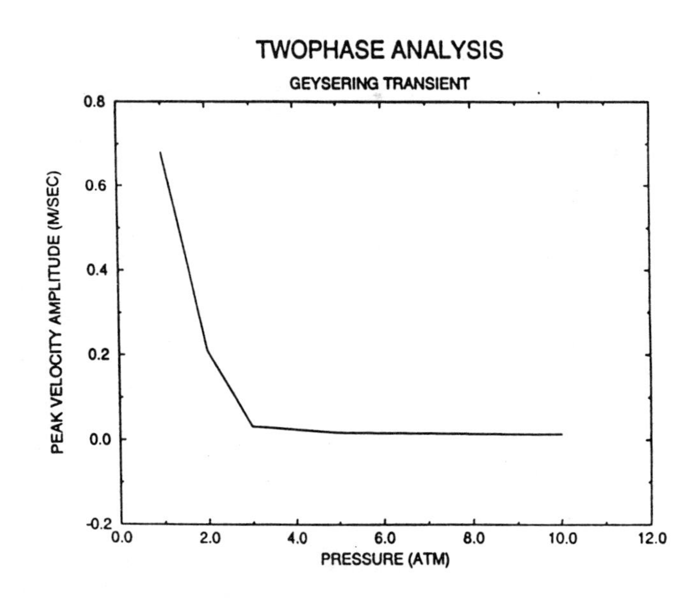

Figure 16: Effect of system pressure on peak velocity amplitude.

HTD-Vol. 342, National Heat Transfer Conference
Volume 4
ASME 1997

ADVANCES IN FILM CONDENSATION INCLUDING SURFACE TENSION EFFECT IN EXTENDED SURFACE PASSAGES

S. Q. Zhou, R. K. Shah and K. A. Tagavi
Department of Mechanical Engineering
University of Kentucky
Lexington, KY 40506-0108

ABSTRACT

Theoretical investigations of film condensation heat transfer started in 1916 by Nusselt. Initially, only the gravity and viscous force effects were taken into consideration in the condensate film, followed by the investigations of the vapor shear effect at the liquid-vapor interface. The importance of the effect of surface tension force on film condensation phenomena has been realized since 1940s (Armstrong, 1945) with the first analytical investigation by Gregorig in 1954. Surface tension drains the condensate film from the fin crest and interfin base regions into fin corner regions, thus creating thinner films and more effective condensation surfaces in these regions. On the other hand, surface tension causes the retention of the condensate in the finned tube bottom which results in a decrease of the effective heat transfer surface and in deterioration of film condensation heat transfer. The objective of this paper is to review the literature on film condensation heat transfer that emphasizes the study of surface tension and its effect on film condensation phenomena. Primary emphasis is given to condensation in extended heat transfer surfaces focusing on three fin geometries: integral-fin, longitudinal fin, and microfin tubes. The most important correlations and analytical models reported in the literature are compared and compiled in a single table. We have summarized our assessment of the influence of surface tension on condensation heat transfer in the geometries covered, and how and where it has beneficial effects. Based on this assessment, specific areas of future research for film condensation heat transfer are outlined.

NOMENCLATURE

A	total heat transfer surface area, $A = A_f + A_p$, m^2
A_{eq}	equivalent heat transfer surface area, $A = \pi D_e L$, m^2
A_f	heat transfer fin surface area, m^2
A_p	heat transfer surface area at the fin base of the tube, m^2
Bo	Bond number, $(\rho_l - \rho_v) g D_h^2/\sigma$, dimensionless
C	different constants of Eqs. (3), (15) and (21), dimensionless
C_{ft}	temperature correction factor for the fin tip region, dimensionless
C_{fv}	temperature correlation factor for the fin flank region, dimensionless
C_p	specific heat, J/gK
c	fin spacing, m
c_b	fin base length, m
c'_b	flooded fraction, dimensionless
c_t	fraction of tube circumference flooded, dimensionless, m
D_e	tube outside fin tip diameter, m
D_{eq}	equivalent diameter used in Eq. (3), m
D_h	hydraulic diameter, m
D_i	tube inside diameter (inside fin root diameter), m
D_o	tube outside diameter (outside fin root diameter), m
D_t	tube inside fin tip diameter, m
e	fin height, $e = (D_e - D_o)/2$, m
fpm	fins per meter, 1/m
g	gravitational acceleration, m/s^2
G	mass velocity, kg/m^{2}s
Ga	Galilleo number, $Ga = (S_{eq}^3 \rho_l^2 g/\mu_l^2)$, dimensionless
h	condensation heat transfer coefficient, W/m^{2}K
h_b	condensation heat transfer coefficient in the flooded region based on A_p, W/m^{2}K
h_f	condensation heat transfer coefficient on fin surface based on A_f, W/m^{2}K
h_h	condensation heat transfer coefficient for horizontal plain tube, W/m^{2}K
$\overline{h}$	average condensation heat transfer coefficient, W/m^{2}K
i_{lv}	latent heat of vaporization, J/kg
k	fluid thermal conductivity, W/mK
k_w	fin thermal conductivity, W/mK
L	tube length, m
$\dot{m}$	mass flow rate in half fin channel, or condensate rate, kg/s
N	fins/m of tube, 1/m; or number of microfin
Nu	Nusselt number, hD_h/k_l, subscript exp = experimental, st = stationary vapor
Ph	phase number, $Ph = i_{lv}\mu_l/(k_l \Delta T)$, dimensionless
Pr	Prandtl number, $Pr = \mu C_p/k$, dimensionless
p	fluid static pressure, Pa
p_t	fin pitch, m
q	heat transfer rate; q_b, q_c, q_f, q_{fl}, q_{int}, q_s, q_t, and q_v are heat transfer rates on fin base, convex profile, fin, tube flooded, unflooded interfin, channel sides, fin tip, and flank regions, respectively, W
r	radius of liquid-gas interface, m
r_t	fin tip radius, m
Re	Reynolds number, GD_h/μ, dimensionless
S_{eq}	Equivalent length of gravity flow region, m
s	coordinate along liquid-vapor interface, m
Su	surface tension number, $Su = \rho_l \sigma r_t/\mu_l^2$, dimensionless
T	temperature, K
T_o	fin root temperature, K
T_s	vapor saturation temperature, K
ΔT	temperature difference, $\Delta T = T_s - T_w$, K
t	fin tip thickness, m
t_b	fin base thickness, m
We_l	liquid Weber number, $We_l = G^2(1-\mathbf{x})^2 D_h/(\sigma\rho_l)$, dimensionless
$\mathbf{x}$	quality, $\mathbf{x}_{in}$ = inlet quality, and $\mathbf{x}_{out}$ = outlet quality, dimensionless
x, y, z	Cartesian coordinates, m
z	vertical distance from the bottom of the tube to a point on the tube surface in the interfin space, m

α	flooded angle, deg
β	helix (spiral) angle of a microfin tube, deg
δ	condensate film thickness, m
η	extended surface efficiency, dimensionless
η_f	fin efficiency, dimensionless
θ	fin tip half angle, deg
μ	fluid dynamic viscosity coefficient, Pa s
ρ	fluid density, kg/m^3
σ	coefficient of surface tension force, N/m
τ	shear stress, Pa s
Φ	unflooded angle, deg
ψ	angle in Eq. (13), $\psi = \pi/2 - \theta$, deg

Subscripts

c	cold fluid
eq	equivalent
h	hot fluid
l	liquid
v	vapor
f	fin
fl	flooded region
sh	vapor shear
st	surface tension
uf	unflooded region

INTRODUCTION

Film condensation, the process by which a vapor is converted to a liquid in the form of a film on the heat transfer surface, is one of the important heat transfer phenomena in many heat transfer equipment. As convection between wall and film in the condensate film is not very important, so the thermal resistance in film condensation is usually calculated only by heat conduction resistance across the film. Therefore, reduction of an average condensate film thickness becomes a major way to enhance film condensation heat transfer. Researchers and designers have been using different kinds of forces to drain and thin the condensate film in different kinds of condensers in order to achieve higher condensation heat transfer rates. In general, there are three major forces: viz., gravity, vapor shear and surface tension, in addition to the inertia and viscous forces that affect the heat transfer mechanism of film condensation on many different surfaces.

Film condensation controlled by the gravity force was first analyzed by Nusselt (1916), who formulated the problem in terms of simple force and energy balances within the condensate film. Many investigators, such as Minkowycz and Sparrow (1966), Dhir and Lienhard (1971), extended the Nusselt theory to take into account the effects of noncondensables, interfacial (i.e. between the film and vapor) shear, superheating, variable properties, diffusion and nonuniform gravity on film condensation. For vapor shear controlled condensation, Akers et al. (1959), Carpenter and Colburn (1951), and Traviss et al. (1973) proposed different models to predict film condensation heat transfer rate.

The pioneer works of analyzing the low-height horizontal integral-fin tubes (see Fig. 1a) used in condensation were made by Armstrong (1945) and Katz et al. (1946). In general, the effect of surface tension force on condensation may become very important for both inside and outside tubes (having circular or noncircular cross section) with extended surfaces when all of the extended surface is not inundated (flooded with the condensate) and the vapor velocity is not very high. Ever since industry has used the low-profile (low height) integral-fin tubes for condensers during the last 50 years,

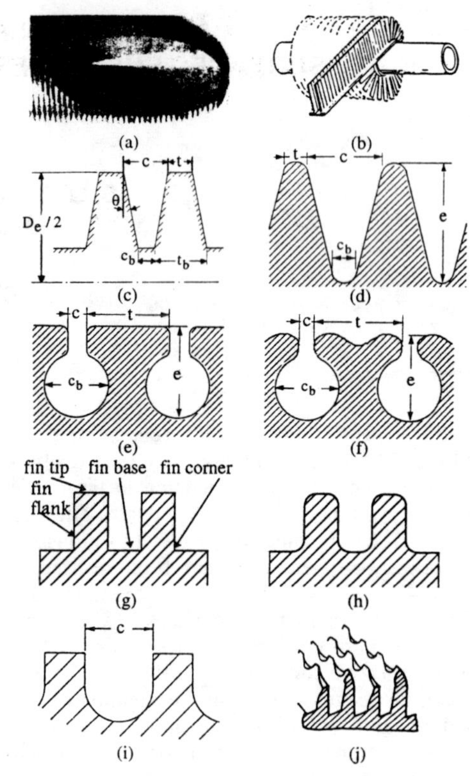

Fig. 1 Horizontal integral-fin tubes: (a) Horizontal integral-fin tube, (b) Horizontal spine-fin tube, (c) Trapezoidal-fin, (d) Trapezoidal-fin with rounded tips, (e) T-profile fin, (f) Y-profile fin, (g) Rectangular-fin, (h) Rectangular-fin with rounded corners, (i) Radiused fin, (j) 3-D fin

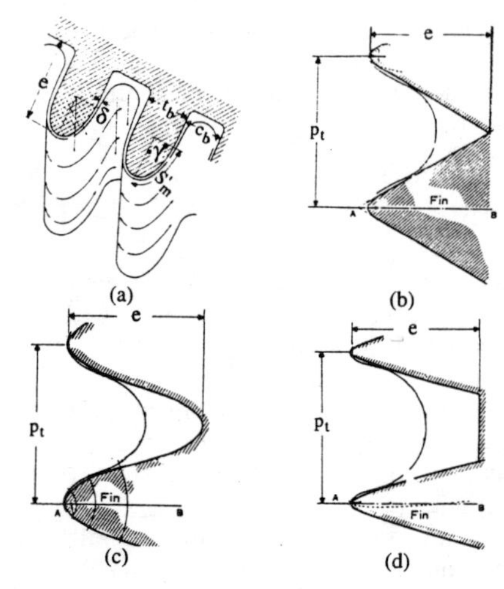

Fig. 2 (a) Longitudinal integral-fin tubes, (b) Triangular fin, (c) Wavy fin, (d) Flat bottom fin

investigators have gradually realized the importance of the effect of surface tension force on condensation in extended surface tubes. Studies of the surface tension effect have been conducted by many investigators. Although Katz et al. (1946) reported that the liquid could be retained between fins on the bottom portion of a horizontal finned tube, Gregorig (1954) was the first investigator who introduced the concept of surface tension drained condensate to actively take advantage of surface tension in designing high performance vertical longitudinal fluted tube (see Fig. 2a). After Gregorig (1954) and Lustenader et al. (1959) reported the surface tension effect on film condensation, Nabavian and Bromley (1963) and Carnavos (1965) carried out experiments to confirm and enhance the effect of surface tension on condensation on vertical fluted tubes. However, the importance of surface tension force on extended surface condensation was not recognized until the worldwide energy crisis during the 1970s. As a result of about five decades of investigation on condensation heat transfer outside integral-fin tubes by many researchers, significant advances have been made in understanding enhancement mechanisms and in the development of methods to predict the heat transfer and friction characteristics. Several reviews on condensation heat transfer have been proposed by Bergles (1978), Marto (1988), Sukhatme (1990), Webb (1994), Tanasawa (1994), and Fujii (1995). Marto (1988) provided an excellent review which summed up the significant progress made in our understanding of film condensation on horizontal integral-fin tubes (see Fig. 1a).

Compared with the application of outside integral-fin tubes in 1940s, the industrial use of inside microfin tubes (microfin tubes can be circular, Fig. 3a, or rectangular with round corners, Fig. 3b) came much later. The first microfin tube was developed by Fujie et al. (1977) of Hitachi Cable, Ltd., shown in Fig. 3a, and was described by Tatsumi et al. (1982). The microfin tubes provided enhancement of 100% or more for both condensation or evaporation heat transfer coefficients over that for plain tubes. Although different profiles of microfins were suggested and manufactured in Japan, Europe and the USA, the applied and basic research started only about ten years ago.

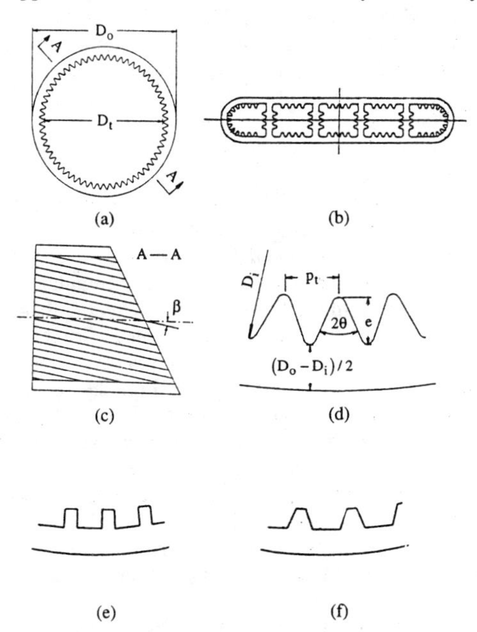

Fig. 3 Microfin tubes: (a) Circular microfin tube, (b) Flat microfin tube, (c) Section A-A of the tube, (d) Triangular fin, (e) Rectangular fin, (f) Trapezoidal fin

In the present paper, a comprehensive literature review is presented that includes most important models and correlations since 1940s which have taken into account the surface tension effect on film condensation for both inside and outside flow passages with extended surfaces (specially, integral-fin tubes, longitudinal fin tubes and microfin tubes). The importance of surface tension force on condensation heat transfer is pointed out through the assessment of various exsiting models. Finally, specific suggestions are made for future research on condensation phenomena where the effect of surface tension is important during condensation in extended surface flow passages.

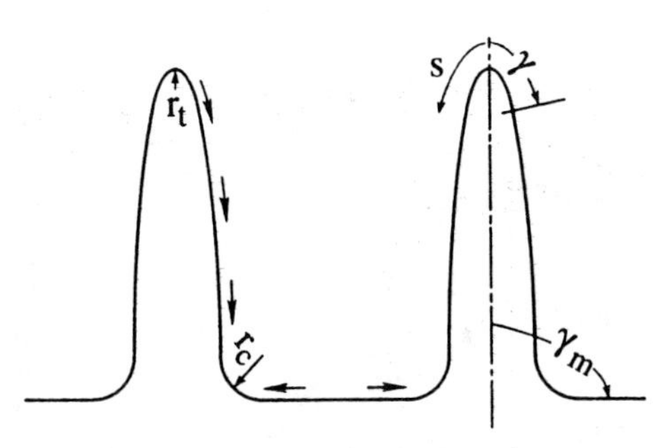

Fig. 4 A fluid flow (shown by arrows) caused by the curva-ture change of liquid-gas interface

CONDENSATION OUTSIDE INTEGRAL-FIN TUBES

Before discussing the importance of surface tension and models of condensation outside integral-fin tubes, let us briefly analyze surface tension induced fluid flows. The surface tension driven liquid flow can be due to either the curvature change of the liquid-vapor interface or the temperature gradient across the liquid film. A surface tension driven flow caused by curvature change of a liquid-gas interface in a typical fin profile is shown in Fig. 4. For the convex liquid profile, the largest curvature is at the fin tip, while the local curvature decreases down to zero with increasing distance from the fin tip. The pressure gradient in the liquid film could be obtained by following equation.

$$\frac{dp_l}{ds} = \sigma \frac{d(1/r)}{ds} \qquad (1)$$

As the curvature of liquid-gas interface decreases from the fin tip to the fin corner (see Fig. 4), so does the liquid pressure [see Eq. (1)]. Repeating a similar analysis will result in liquid flow from the fin base toward the fin corner. Thus, the liquid will be driven into the fin corner space by surface tension force caused by the curvature change of liquid-gas interface developed in the fin shape considered in Fig. 4. For most fluids, cohesion force between molecules becomes weaker with increasing temperature, and hence surface tension decreases. Thus, a temperature gradient along the liquid-vapor or liquid-solid interface will cause a change in surface tension, resulting in flow from hot to cold liquid in the condensate film, which is referred as Marangoni flow. For many condensation heat transfer problems on earth, surface tension driven flow caused by variations in the local liquid temperature is often neglected because of the dominance of other forces, such as gravity. However, when the gravity effect is negligible as in the outer space applications and orbital vehicles such as space station or in extended surface tubes, and when the vapor velocity is not very high, the effect of surface tension driven flow on condensation may become important.

Condensation in Quiescent Vapor

Fins on an integral-fin tube are mostly beneficial but sometimes detrimental for condensation heat transfer. Fins provide additional heat transfer surface area. Also, additional surface tension driven

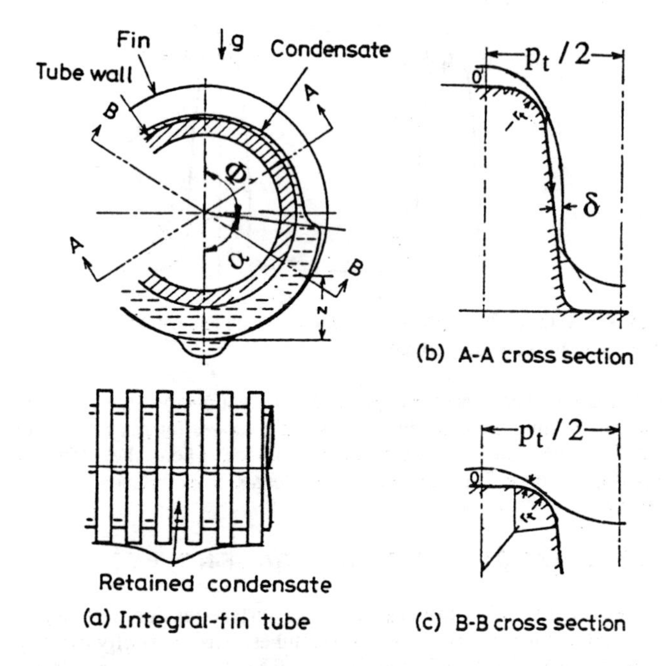

Fig. 5 A partially flooded integral-fin tube used by (Honda and Nozu, 1987)

flow in the condensate film (caused by the liquid-vapor interface curvature change) assists drainage in some parts of the surface, and thereby enhances heat transfer by reducing the film thickness. Adversely, there is little condensation taking place in bottom parts of the integral-fin tube due to capillary retention of condensate covering some or significant heat transfer surface area (see Fig. 5a). This condensate covered surface area is referred to as flooded area; the internal or external fins (when condensation is taking place) are submerged (completely filled) with the condensate. Accordingly, the study of condensate flooding and finding methods to reduce flooding is very useful for design of high performance integral-fin tubes. The measurement for the liquid film shape on the surface of integral-fin tube under no condensation condition is referred to as the "static condition measurement." This is a very convenient and effective way of measuring the film shape on the surface of integral-fin tubes. The measurement of condensate film shape during actual condensation is referred to as the "dynamic measurement." Many researchers, like Katz et al. (1946), Hirasawa et al. (1980), Rudy and Webb (1981), Honda et al. (1983), Masuda and Rose (1987), and Wen et al. (1994) observed the liquid film shapes on different surfaces of integral-fin tubes. Both under static and dynamic conditions, Rudy and Webb (1981) measured the condensate retention on integral-fin tubes and a "spine-fin" tube (see Fig. 1b) using R-11, n-pentane, and water. Honda et al. (1983) made a theoretical analysis of the static meniscus in the grooves of the low height trapezoidal fins (see Fig. 1d). They developed an approximate expression for the unflooded angle Φ (see Fig. 5a) used for $e > c/2$

$$\Phi = \cos^{-1}\left\{\left[\frac{4\sigma\cos\theta}{g\rho_l c D_e}\right] - 1\right\} \tag{2}$$

The importance of surface tension force on film condensation could be discovered not only through the qualitative analyses of physics of flow phenomena on extended surface tubes, but also through the quantitative calculation of condensation heat transfer by different models or correlations. Table 1 is a chronological listing of models and correlations for horizontal integral-fin, longitudinal fin and microfin tubes.

Beatty and Katz (1948) first applied Nusselt's equations for film condensation on horizontal tubes to the horizontal portion between the fins, and that for the vertical plates to the short vertical fins of the integral-fin tubes. By assuming that surface tension played no role in the condensation process and the condensate flow was gravity controlled, they proposed the following theoretical model to predict the condensation heat transfer coefficient on the integral-fin tubes (see Fig. 1a).

$$h = C\left\{\frac{k_l^3 \rho_l^2 g i_{lv}}{\mu_l \Delta T D_{eq}}\right\}^{1/4} \tag{3}$$

Refer to Beatty and Katz (1948) for the formula of equivalent diameter, D_{eq}. They used methyl chloride, sulfur dioxide, R-22, propane, n-butane, and n-pentane as working fluids. Although Katz et al. (1946) found experimentally that surface tension forces can cause condensate flooding between fins and thereby decreasing heat transfer rate, they did not consider this effect in their modeling. As we pointed out earlier, the surface tension effects could be both beneficial and detrimental to condensation heat transfer (on the regions above and below flooded angle of the finned tubes respectively); the two opposing effects tend to cancel each other so that their model agreed with their experimental data for low surface tension fluids and low fin density (433 to 633 fins/m) of finned tubes. However, many modern industrial condensers employ high surface tension fluids and especially high fin density of finned tubes. For these condensers, the model of Beatty and Katz (1948) may result in unacceptable large errors for the predictions of condensation heat transfer as the result of neglecting for surface tension effect on condensation.

Gregorig (1954) was the first investigator who proposed the effective utilization of surface tension forces to thin the condensation film at the tips of vertical fluted tube resulting in enhancement in condensation heat transfer, and to drain the accumulated liquid (by surface tension effects) in the fin root area. Realizing that the surface tension force on a curved surface can induce a pressure gradient many times larger than that induced by gravity, Gregorig (1954) proposed his "fluted tube," as shown in Fig. 2a, to enhance film condensation. At the fin crest, the pressure in the liquid is higher than the vapor pressure because of the convex shape of the condensate surface. At the valley, just the opposite is true because of the concave shape. Therefore, a pressure gradient exists from the crest to the valley as expressed by Eq. (1). In addition, the valley of the fin acts as a drainage channel by gravity force in order to maintain thin films near the tip and flank regions of the fin. Gregorig (1954) proposed to vary the local radii of curvature of the fin to maintain a constant pressure gradient. This yielded a constant heat transfer coefficient over the entire convex arc length, considering only the surface tension force important on the convex portion of a fluted surface. Using an approach similar to Nusselt, Gregorig (1954) replaced the gravity term by the surface tension term, and arrived at heat transfer coefficient h over the arc length S_m (see Fig. 2a) on the crest with a constant film thickness over the surface profile.

$$h = \left\{\frac{4 k_l^3 \rho_l \sigma i_{lv}}{3 \mu_l D_e S_m^{\,2} \Delta T}\right\}^{1/4} \tag{4}$$

Table 1. Chronological Summary of Models and Correlations

Year	Investigators	Fluids	Geometries and orientations	Models and Correlations	Remarks
1984	Beatty and Katz	6 fluids	Horizontal integral-fin tubes	$h = C\left\{\frac{k_l^3 \rho_l^2 i_{lv}}{\mu_l \Delta T D_{eq}}\right\}^{1/4}$	Surface tension wasn't included in the correlation which was only useful for low fin density tubes.
1954	Gregorig	—	Vertical longitudinal fluted tube	$h = \left\{\frac{4k_l^3 \rho_l \sigma i_{lv}}{3\mu_l D_e S_m^2 \Delta T}\right\}^{1/4}$	Only surface tension was included in the correlation which didn't include heat transfer in concave surface.
1971	Karkhu and Brovkov	Steam	Horizontal trapezoidal integral-fin tube	$h = \frac{\dot{m}(Z_b) i_{lv}}{A(T_s - T_o)}$	Surface tension was controlling in fin flank regions, gravity was controlling in the fin base region.
1985	Webb et al.	R-11	Horizontal integral-fin tube	$h = [(1-c_b)\left(h_h \frac{A_p}{A_{eq}} + h_f \eta_f \frac{A_f}{A_{eq}}\right) + c_b h_b] / \eta$	Surface tension was controlling on fin flank, gravity was con-trolling on fin base. Heat transfer was included in the flooded zone.
1987	Honda and Noze	11 fluids	Horizontal trapezoidal integral-fin tube	$h = \{(1-c_b) h_u \eta_u \Delta T_u + c_b h_f \eta_f \Delta T_f\} / \{(1-c_b)\Delta T_u + c_b \Delta T_f\}$	Wall temperature variation was inclu-ded in the correlation.
1989	Kaushik and Azer	Steam and R-113	Longitudinal inside rectangular fin tube	$\frac{1}{\bar{h}} = \frac{1}{x_{in} - x_{out}} \int_{x_{out}}^{x_{in}} \{\pi D_i / \left[h_b\left\{\pi D_i - Nt - 2N\left[4C(e-\delta)^2\right]^{1/4}\right\} + 2N\eta_f \bar{h}_f (e-\delta)\right]\} dz$	Only surface tension force was included in the model.
1990a	Adamek and Webb	7 fluids	Horizontal integral-fin tube	$h = 4NL(\dot{m}_{uf} + \dot{m}_{fl})/(A\Delta T)$	Flooded and unflooded area condensa-tion and fin efficiency were included in the model.
1990b	Adamek and Webb	R-11	Vertical finned plates and tubes	$q = q_c + q_s + q_b$ see original paper for each term	Flooded and unflooded area condensa-tion and fin efficiency were included in the model.
1994	Yang and Webb	R-12 R-134a	Horizontal microfin tube	$h = (h_{sh}^2 + h_{st}^2)^{1/2} \frac{A_{uf}}{A} + h_{fl} \frac{A_{fl}}{A}$	Effects of surface tension and vapor shear forces were included in the model.
1994	Briggs and Rose	R-113 steam	Horizontal integral-fin tube	$q = q_f + q_{fl} + q_{int}$ see original paper for each term	Fin efficiency effect was included in the model.
1996	Sreepathi et al.	R-123, R-11, steam, ethylene glycol	Horizontal trapezoidal fin tube	$Nu_p = Ph^{n1}\left(\frac{D_e}{p_t}\right)\{C_t + (2-c_b') C_{ft}^{0.8} Su^{n2} \psi^{1/2} C_v (1-c_b') C_{fv}^{0.8} Ga^{n3}\} + C_b(1-c_b')\left(\frac{D_o c_b}{p_t p}\right)^{n4}$	Condensation heat transfer has been considered in the tip, flank and valley regions.

Gregorig's fluted tube concept stimulated many investigators to find more accurate analytical models for predicting thermal performance of extended surface tubes and to further understand the role of surface tension forces.

After Gregorig's theory, the first analysis that recognized the importance of surface tension on horizontal finned tubes was made by Karkhu and Borovkov (1971), who studied film condensation on horizontal trapezoidal fins (see Fig. 1c). They idealized that the motion of condensation in the valley occurred due to gravity, the wall temperature remained constant across the fin height, and only the effect of surface tension surface prevailed during the motion of the film on the lateral surface of the fin toward the bottom. They found the average heat transfer coefficient as

$$h = \frac{\dot{m}(Z_b) i_{lv}}{A(T_s - T_o)} \tag{5}$$

where $\dot{m}(Z_b)$ is the mass flow rate in half fin channel. Their model predicted their experimental data of steam and R-113 within $\pm 5\%$.

Honda and Nozu (1987) described a more comprehensive and complex model for the trapezoidal fins on a horizontal integral-fin tube as shown in Fig. 1d and Fig. 5a. The condensation coefficient was predicted for two regions: the unflooded region (unflooded circumferential fraction, see Fig. 5b) and the flooded region (flooded fraction, i.e. the bottom region where the fin is completely flooded, see Fig. 5c). They divided the fin surface of unflooded region into a thin film region near the top of the fin and thick film region near the bottom. The heat transfer in the fin tip and fin flank was included in the model, but the condensation heat transfer in the thick film region was neglected. They employed Eq. (2) to obtain unflooded angle (Honda et al., 1983). On the fin surface, they idealized that the condensate flow was controlled by surface tension and gravity forces, whereas in the interfin base region the condensate flow was only controlled by gravity force. They pointed out that wall surface temperature variation should be considered as the circumferential heat conduction in the tube wall thickness. The fin efficiencies were different in the flooded region and unflooded region. Using numerical and Nusselt-type analytical approaches, they obtained the unflooded region condensation coefficient and flooded region condensation coefficient separately. Therefore the average condensation heat transfer coefficient was obtained as following.

$$h = \frac{(1-c_b)h_{uf}\eta_{uf}\Delta T_{uf} + c_b h_{fl}\eta_{fl}\Delta T_{fl}}{(1-c_b)\Delta T_{uf} + c_b \Delta T_{fl}} \tag{6}$$

The wall temperatures in unflooded region T_{wu} and in flooded region T_{wf} were obtained by solving the circumferential wall conduction equation. The model predicted condensation heat transfer rates with an error band from $\pm$ 20%.

Webb et al. (1985) developed the following model, for condensation over a horizontal integral-fin tube (see Fig. 1a).

$$h = \left\{ (1-c_b)\left(h_h \frac{A_p}{A_{eq}} + h_f \eta_f \frac{A_f}{A_{eq}} \right) + c_b h_b \right\} \Big/ \eta \tag{7}$$

The last term in the braket is for the condensate flooded region and the preceding terms in the parentheses are for the unflooded fraction of the tube circumference and the fin. The condensation coefficient h_h on the base tube in the unflooded region is calculated from the Nusselt equation for horizontal tubes, taking into account the additional condensate drainage from the fin profile to the base tube. It is given by

$$h_h = 1.514\left(\frac{\mu_l^2 Re_l}{k_l^3 \rho_l^2 g} \right)^{-1/3} \tag{8}$$

The condensation coefficient h_f on the fin is calculated using the surface tension drained model for the corresponding profile shape. The fin efficiency is calculated for a constant heat transfer coefficient h_f on the fin by using the conventional single-phase formula. The heat transfer coefficient h_b could be obtained from considering the thermal resistances of tube wall, fin portion and liquid filled interfin region. Webb et al. (1985) provided detailed information on how h_b may be modeled. As a conservative approximation, one may neglect the condensation in this region. Webb et al. (1985) used their model to predict R-11 condensation coefficient for the three finned tube geometries, within the prediction accuracy of $\pm 20\%$. Wanniarachchi et al. (1986) compared their data for steam to the model of Webb et al. (1985), and found that the model predicted their experimental data to about $\pm$ 20% percent for the not fully flooded tube, and underpredicted more than 100% for the data of a fully flooded tube.

The condensate film thickness on the trapezoidal (Fig. 1c) or rectangular (Fig. 1g) fin surface cannot be written in terms of continuous functions as used by Adamek (1981) or Kedzierski and Webb (1990). Adamek and Webb (1990a) developed a model to calculate condensation heat transfer coefficient in the concave drainage channel in the region between fins on a horizontal integral-fin tubes. Trapezoidal fins may have both surface tension and gravity drained regions. By dividing the fin surface into surface tension or gravity drained regions, they obtained following model of condensation heat transfer on a horizontal integral-fin tube.

$$h = 4NL(\dot{m}_{uf} + \dot{m}_{fl})/(A\Delta T) \tag{9}$$

The model of Eq. (9) comprised of condensation coefficients on the fin tip, fin sides, and in the drainage channel. Adamek and Webb (1990a) showed that Eq. (9) predicted their experimental data within ±15% the condensation coefficients for a variety of fin pitches, fin heights, and fluids. Since analytical models, such as those proposed by Gregorig (1954) and Adamek (1983), ignored the condensation rates in the concave drainage channels (the fin root region), so they did not predict the condensation rates very well in a rectangular cross-section drainage channel. Adamek and Webb (1990b) presented a model to calculate the condensation in the concave channel region of vertical finned plates or tubes.

$$q = q_c + q_s + q_b \tag{10}$$

where the formulas for various q are not reported here due to the space limitation. The model predicted the experimental data of Kedzierski and Webb (1987) within ±12%. The predicted results showed that the condensation rate in the drainage channel accounted for 6-32% of the total on the Gregorig (1954) convex fin profile, as tested by Kedzierski and Webb (1987).

Rose (1994a) proposed a simple semi-empirical equation for condensation on horizontal integral-fin tubes with rectangular fins (see Fig. 1g). Further Rose (1994b) extended this model to trapezoidal fins (see Fig. 1e). The models combined the Nusselt (1916) approach for gravity-drained condensation on vertical plates and horizontal tubes and included the effect of surface tension. The proposed models had a good agreement with experimental data for various fluids from 19 investigations on condensation on copper tubes. Since conduction in the fins in their models were neglected (the entire fin was assumed to be at the fin root temperature), the model would be expected to become less accurate to predict condensation heat transfer rate with increasing values of he^2/tk_w. Briggs and Rose (1994) amended these models and included the fin efficiency effect into the model in an approximate way. They proposed the following model to calculate condensation heat transfer rate on horizontal integral-fin tubes.

$$q = q_f + q_{fl} + q_{int} \tag{11}$$

Refer to Briggs and Rose (1994) for the formulas for the right-hand terms.

Sreepathi et al. (1996) developed a generalized correlation to determine condensation heat transfer performance of horizontal integral-fin tubes (see Fig. 6). They divided the fin tube surface into two parts: flooded and unflooded. Then the fin surface was divided into three regions: (a) the fin tip and the upper corner of the fin, where the condensate flow was assumed to be surface tension driven, (b) the fin flank, where the gravity driven condensate flow was assumed, and finally (c) the valley region, for which an empirical relation was used to find the heat transfer rate. The fraction of the tube that was flooded due to condensate retention was calculated using the unflooded angle expression given by Honda et al. (1983).

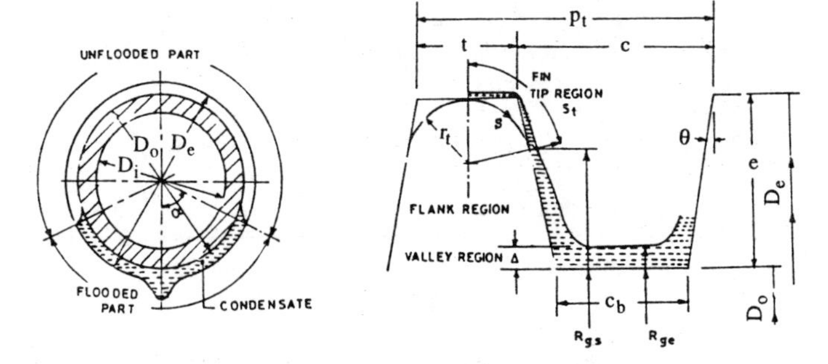

Fig. 6 Different regions of a typical fin tube analyzed by Sreepathi et al. (1996)

They assumed there was no heat transfer from the flooded region. Separate heat transfer coefficient correlations were used to calculate the heat transfer rates for each region. The total heat transfer rate from the tube per fin was given by

$$q = q_t + q_v + q_b = \overline{h}_p \pi D_e p_t \Delta T \tag{12}$$

where $\overline{h}_p$ was the predicted average heat transfer coefficient of the tube based on the fin tip diameter. After combining the heat transfer correlations in the fin tip, fin flank and unflooded valley regions, then simplifying with some assumptions, they obtained following correlation,

$$Nu = Ph^{n_1}\left(\frac{D_e}{p_t}\right)\left\{C_t(2-c_b')C_{ft}^{0.8}Su^{n_2}\psi^{1/2} + C_v(1-c_b')C_{fv}^{0.8}Ga^{n_3}\right\} + C_b(1-c_b')\left(\frac{D_o c_b}{p_t \rho}\right)^{n_4} \tag{13}$$

where, n_1, n_2, n_3, n_4, C_t, C_v and C_b are floating constants. Then they selected the 583 data points covering the fluids R-123, R-11, water and ethylene glycol of various fin geometries and arrived at the following values of the constants.

$$n_1 = 0.18, \quad n_2 = 0.25, \quad n_3 = 0.21, \quad n_4 = 0.51$$
$$C_t = 1.14, \quad C_v = 2.40, \quad C_b = 0.97 \tag{14}$$

Comparing with the available experimental data, they found that the above correlation predicted the heat transfer performance of refrigerants, like R-123 and R-11, with a deviation of less than ± 10%.

Briggs and Rose (1996) demonstrated that the circumferential variation of tube surface temperature had a stronger effect on the mean heat transfer coefficient for a low-finned tube than that for a smooth tube. They incorporated the interphase mass transfer into the model which led to condensation heat transfer coefficient dependent upon the saturation pressure, and found the model predicted steam data within ±10%.

From above analyses and models of quiescent vapor condensation outside integral-fin tubes, it is obvious that surface tension force is very important for condensation on integral-fin tubes except for the gravity force. Moreover, in some regions, surface tension force becomes a controlling force. The model which didn't include surface tension force effect, like the model of Beatty and Katz (1948), results in a large error or even totally fail to predict condensation heat transfer rates on some integral-fin tubes. To simplify and include the effect of surface tension force on condensation, most models incorporate Eq. (1) and usually assume having a constant pressure gradient over fin arc of complicated finned tube surfaces. The motivation to create fin curvature change is to develop a film liquid pressure difference over fin arc to effectively pull the film from fin tip and flank to the fin base. Therefore, fin geometry optimization, such as fin shape, fin density and relative parameters, are very important to design of high performance integral-fin tubes. Commercially available horizontal integral-fin tubes can be divided into the following classes: a two-dimensional (rectangular) constant thickness annular (radial) fin, Fig. 1g, and a three-dimensional fin with circumferential variation of the fin cross section (Fig. 1j). Mori et al. (1981) studied the effect of the fin shape on the condensation heat transfer on a vertical longitudinal finned tube. Three types of two-dimensional fins were analyzed in their study. They were the triangular fin with a small leading edge radii (see Fig. 2b), the wavy fin with rather large radii at the leading edge (see Fig. 2c), and flat bottom groove fin with a sharp leading edge and groove bottom (see Fig. 2d). The highest condensation coefficient was identified for the flat bottom groove fin. Wen et al. proposed an integral-fin tube with radiused fin root fillets having the radii equal to half the fin spacing (see Fig. 1i) to minimize the retention of liquid on the finned tubes. Briggs and Rose (1995) tested three fin geometries as shown in Fig. 1d, e and f. They found the trapezoidal fin tube having the best performance. Zhu and Honda (1993) employed the numerical method to analyze the condensation heat transfer performance of three rectangular fins [Figs. 1g, and 4 (two geometries)] which were studied by Kedzierski and Webb (1990), Mori et al. (1981), Honda et al. (1989), and Adamek and Webb (1990a). They identified that the highest heat transfer performance fin shape was that proposed by Kedzierski and Webb (1990) (see Fig. 4).

Condensation in High-Velocity Vapor

All of the results discussed above have been obtained for low velocity vapor, under gravity controlled or surface tension controlled conditions. However, the effect of vapor shear force on the condensation heat transfer on integral-fin tubes may be important in actual used condensers with high vapor velocity. Many investigators, such as Memory and Rose (1986), Honda et al. (1986), and Cavallini et al. (1986), have conducted experiments on the role of vapor shear force on condensation on a smooth single tube or a tube bundle.

Only a few investigators have presented the information on the effect of vapor velocity on the condensation on the integral-fin tubes. Gogonin and Dorokhov (1981) reported data for R-21 condensing on two different finned tubes with 800 fin/m fin density over a range of the vapor velocity (up to 8 m/s). They showed the effect of the vapor velocity on finned tubes was very small compared to the effect on smooth tubes. After calculating the range of vapor velocities in condensers and finding that the highest vapor velocity was 0.6 m/s, Webb (1984) pointed out that the shell-side vapor shear should not significantly affect the heat transfer coefficient of finned tubes in refrigeration condensers. However, for steam condensers, Yau et al. (1986) pointed out that almost the same influence of the vapor velocity on their finned tube data as their smooth tube data with three vapor velocities of 0.5, 0.7, and 1.1 m/s.

Bella et al. (1993) experimentally investigated the effects of vapor shear stress during condensation of a pure refrigerant, R-11 and R-113, on a single horizontal trapezoidal integral-fin tube (see Fig. 1c) with a vapor velocity range from 2 to 30 m/s. For their experimental data at the highest vapor velocity, there was a 50% increase in the heat transfer coefficient compared to the values measured during condensation of stagnant vapor; and the enhancement began when the vapor Reynolds number was greater than 10^5. No correlation of vapor shear effect was presented.

Cavallini et al. (1995) experimentally investigated the effect of vapor shear on condensation normal to the horizontal finned tubes. Two different extended tube geometries (fins and tubes made from copper) were tested: the first one was a 2,000 fin/m integral-fin tube with an outside diameter of 15.8 mm, fin height around 0.6 mm, and fin thickness around 0.2 mm; the second one was a commercial extended tube, the Thermoexcel by Hitachi with an outside diameter of 18.7 mm. Refrigerants R-11 and R-113 were the test fluids with their vapor pressure ranging from 111 to 190 kPa, average condensation temperature difference varying from 3.4 to 18.5°C and the maximum vapor velocity from 0.8 to 37.9 m/s. They concluded that on the 2000 fins/m integral-fin tube, the vapor velocity corresponding to $Re_v > 7 \times 10^5$ would give heat transfer enhancement around 80%.

It is clear form the above investigations that experimental results are available for some finned tube geometries to take into account the vapor shear effect; but no correlations are reported in the literature.

CONDENSATION INSIDE MICROFIN TUBES

While outside enhancement via integral-fin tubes was introduced in 1940s, an inside microfin tubes was first developed by Fujie et al. (1977) of Hitachi Cable, Ltd. Microfin tubes are commonly used in residential air conditioners. Many improvements have been made by Hitachi Cable, Ltd.; enhancement levels as much as 300% are obtained for both evaporators and condensers compared to smooth tubes.

As shown in Fig. 3, commercially available microfin tubes have short height triangular, rectangular, or trapezoidal fins. In each figure, the fin tips have round corners, because the tubes are usually mechanically expanded from inside to secure the tight contact between the outer surface of the tube and flat (plate) fins. The outer diameter D_o, the root diameter D_i, the fin height e and the half tip angle θ are major fin geometrical parameters as shown in Fig. 3. The other characteristic dimensions of microfin tubes are the spiral (helix) angle β of individual fins with respect to the tube axis (microfins may or may not be straight along tube length; spiral microfins along the tube length are used in some applications with a maximum spiral angle of 60°), the number of fins n and the ratio of the surface area of the microfin tube to that of a smooth tube if the microfins were removed. The tubes with fins having N = 50 – 70, e = 0.12 – 0.25 mm, β = 10 – 35° are referred to as micro-fin tubes, while the tubes with fewer fins (N < 30) and taller fins (e > 0.4 mm) are referred to as internally finned tubes.

Compared to the importance of surface tension force on condensation outside integral-fin tubes, only a few investigators emphasized and explained the importance of surface tension force on condensation in microfin tubes. Most investigators just reported experimental data of condensation heat transfer without any explanation of why surface tension was important and how surface tension was beneficial to condensation heat transfer in microfin tubes.

Experimental Results

Many data have been published on round microfin tubes for condensation. Shinohara et al. (1987) showed that since the condensate accumulated in the bottom of the plain horizontal tubes, there was very little heat transfer on the bottom surface of the tubes. However, for microfin tubes the condensate could be spread axially by surface tension force in micro-grooves through the entire circumference of the tube. As there was little condensate that accumulates in the bottom of a horizontal tube, the heat transfer rate in microfin tubes was much higher than that in plain tubes.

Torikoshi and Kawabata (1992) provided heat transfer coefficients and pressure drops for R-134/PAG lubricant oil mixtures in a microfin tube. The microfin tube had 60 fins, an 18° spiral angle, and outer diameter of 9.52 mm. Over the 50 – 200 kg/m^2s mass flux range tested, the heat transfer coefficients were increased by about 150% over the smooth tube heat transfer coefficients. Condensation pressure drops in the microfin tube were higher by about 100% over those for a smooth tube.

Cui et al. (1992) proposed an empirical correlation of R-502 evaporation data on nine microfin geometries. The correlation used the parameters of the plain tube correlation of Pierre (1964) with an additional parameter for the microfin geometry. In their experiment, they found that the microgrooves provided significant enhancement for condensation.

Eckels et al. (1994) reported condensation of R-134a with mixtures of a penta erythritol (SUS) ester mixed-acid lubricant in a smooth tube and a microfin tube. The 169-SUS and 369-SUS ester lubricants were tested at concentrations ranging from 0.5% to 5.0%. The mass fluxes of the refrigerant-lubricant mixtures ranged from 125-375 kg/m^2s in the tubes with outer diameter of 9.25 mm. Their experimental data showed that the pure R-134a heat transfer coefficients in the microfin tube were 100% to 200% higher than those in the smooth tube at the same mass fluxes. The pressure drops during condensation in the microfin tube were 20% to 50% higher than those in the smooth tube at the same mass fluxes. Also, the addition of the lubricant did not have a significant effect on the performance difference between the microfin tube and the smooth tube.

Chamra et al. (1996) developed an internal geometry that yields higher condensation coefficients than existing single-groove (single-helix) microfin designs. They presented experimentally R-22 condensation data for a new cross-groove microfin geometry applied to the inner surface of 15.88 mm outside diameter tubes. The single-groove geometry had 74-80 internal fins, 0.35 mm fin height, and 30° helix angle. The cross-groove geometries were formed by applying a second set of grooves at the same helix angle, but in the opposite angular direction to the first set. Data were provided for varying depths of the second set of groove without changing the depth of first set of grooves. The data showed that the heat transfer coefficient increased with the helix angle up to a 27° helix angle. The cross grooved tubes provided higher condensation coefficients than the single-helix geometries. The best cross-grooved tube with helix angle 27° provided 27% higher condensation coefficient than that for the single-helix tube. The pressure drop was 6% higher than that in the single-helix tube.

Yang and Webb (1996) provided experimental heat transfer data for R-12 condensation and subcooled single-phase convection in small hydraulic diameter, flat extruded aluminum tubes with and without microfins. Figure 3b shows such a tube with microfins. The flat tube outside dimensions were 16 mm (width) $\times$ 3 mm (height) $\times$ 0.5 mm (wall thickness). The tubes contained three internal membranes (webbs) which are primarily used to contain high internal fluid pressure. Two internal geometry were tested: one had a plain inner surface and the other had microfins, 0.2 mm high. Data were presented for the following range of variables: vapor qualities (12-97%), mass velocity (400-1400 kg/m^2s), and heat flux (4-12 kW/m^2). The overall heat transfer coefficient was measured for water-to-refrigerant heat transfer, and the modified Wilson plot method was used to determine the heat transfer coefficient for water-side flow in the annulus. The tube-side condensation coefficient was then extracted from the measured UA values. The authors proposed that surface tension force enhanced the condensation coefficients for vapor qualities greater than 50% but no modeling was presented for this effect.

Theoretical Models

Although a significant number of experimental investigations have been reported in the literature, only a few models are available and they provide a very little explanation for the physics of enhancement mechanisms for condensation in microfin tubes, and for the effect of surface tension force on the condensation in microfin tubes.

Kaushik and Azer (1989) were probably first to propose an analytical model to predict the condensation heat transfer coefficient inside longitudinally finned tube with fin height 1.6 mm, fin width 0.9 mm and fin number 6. Figure 7 shows a schematic view of the fin and condensate film for the development of their model for longitudinal fin tubes. They assumed that the flow of the liquid film from the microfins toward the fin base (on surface AB in Fig. 7) was laminar; the temperature of the fin was constant and equaled the tube wall temperature; only viscous and surface tension forces were significant. In Fig. 7, only the fin surface AB and the tube surface CD contributed to heat transfer, and heat transfer at the fin tip was negligible. They developed the average heat transfer coefficient over the fin surface AB as

$$\bar{h}_f = \frac{4k_l}{3\left[4C(e-\delta)^2\right]^{1/4}} \tag{15}$$

where
$$C = \frac{k_l \Delta T}{\frac{2\rho_l \sigma}{\mu_l}\left[\frac{1}{c_b}+\frac{1}{t}\right]\left[i_{lv}+\frac{3}{8}C_p \Delta T\right]} \tag{16}$$

They assumed that the liquid formed an annular film on the surface CD, as in smooth tubes, and they used the correlation of Soliman et al. (1968) to calculate the fin base heat transfer coefficient, h_b. Combining $\bar{h}_f$ and h_b, they average heat transfer coefficient over a given length of the tube.

$$\frac{1}{\bar{h}} = \frac{1}{x_{in}-x_{out}} \int_{x_{out}}^{x_{in}} \frac{\pi D_i dz}{h_b\left[\pi D_i - Nt - 2N\left\{4C(e-\delta)^2\right\}^{1/4}\right] + 2N\eta_f \bar{h}_f (e-\delta)} \tag{17}$$

The comparisons were made between the prediction of Eq. (17) and the experimental data of Royal (1975) for steam, and of Said (1982) for R-113, respectively. The results showed that 87% of the data of Royal and 47% of the data of Said were predicted to within $\pm 30\%$. The reason for such large discrepancies in the prediction was explained by authors as water has surface tension 3.8 times that for liquid R-113. Since this model is based solely on surface tension for in the draining condensate film, the better agreement only occurs with the fluid (steam) having higher surface tension. Hence, Kaushik and Azer (1989) model is good when surface tension force is important and vapor shear force is negligible.

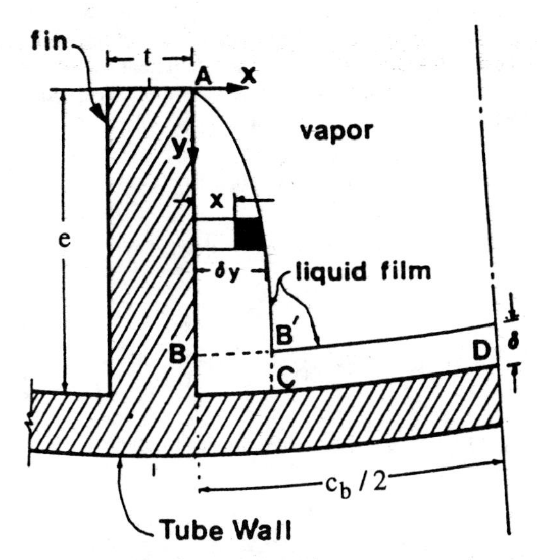

Fig. 7 Schematic view of the condensate flow model Ana-lyzed by Kaushik and Azer (1989)

In order to include both surface tension and vapor shear effects on condensation heat transfer in microfin tubes, recently Yang (1994) and Yang and Webb (1996) proposed a theoretical model to calculate the condensation heat transfer coefficient inside a horizontal microfin tube (see Fig. 3b). They divided a microfin tube into two regions: vapor shear controlled region, and both surface tension and vapor shear controlled region. For the vapor shear controlled region, the void fraction α should be less than A_e/A_c (where A_e is the flow core area and A_c is the tube cross sectional flow area), i.e. all of microfins were flooded by the condensate but the cross-section of microfin tube was not filled up by condensate. Therefore, the curvature of the condensate-vapor interface was nearly constant and no significant surface tension effect existed at this flow condition, only vapor shear force was important. In the vapor shear controlled region, they proposed that the condensation heat transfer coefficient can be calculated by a vapor shear controlled correlation. For both surface tension and vapor shear controlled region, the void fraction α should be greater than A_e/A_c, i.e. all of microfins were not completely flooded by the condensate. Therefore both surface tension and vapor shear forces were important in this region. They divided heat transfer surface area of an entire microfin tube into the flooded area (vapor shear controlled region), A_{fl}, and unflooded area (both vapor shear and surface tension controlled region), A_{uf}. Because of the condensate film liquid-vapor interface curvature change in the unflooded area, surface tension force drains the condensate from the convex surface into the concave region (the fin base). A high condensation rate occurs on the convex portion of the fins (fin tip and fin flank surface) due to condensate film drainage by the surface tension force, therefore both vapor shear and surface tension forces are important. They proposed the following formula to calculate the resulting heat transfer coefficients averaged over the total surface area.

$$h = h_{uf}\frac{A_{uf}}{A} + h_{fl}\frac{A_{fl}}{A} \tag{18}$$

The heat transfer coefficient h_{fl} in the flooded region can be calculated by various vapor shear controlled correlations for smooth tubes. They extended Akers et al. (1959) method to predict condensation heat transfer coefficient in unflooded region h_{uf}. They derived the condensation heat transfer coefficient in the unflooded region in terms of contributions due to vapor shear and surface tension forces.

$$h_{uf}^2 = h_{sh}^2 + h_{st}^2 \qquad (19)$$

They used Aker's (1959) correlation to calculate vapor shear term, h_{sh},

$$h_{sh} = h_{Akers} = 0.0265 \frac{k_l}{D_h} Re_l^{0.8} Pr_l^{1/3} \left[(1-x) + x \left(\frac{\rho_l}{\rho_v} \right)^{1/2} \right]^{0.8} \qquad (20)$$

They applied Eq. (1) and the Weber number to obtain surface tension term, h_{st},

$$h_{st} = C \frac{\tau_z - \tau_i}{dp/dz} k_l \frac{d(1/r)}{ds} \frac{Re Pr_l^{1/3}}{We} \qquad (21)$$

where, the terms τ_z, τ_i and dp/dz were obtained from measured single-phase and two-phase flow pressure drop data. The empirical constant C was obtained by fitting the experimental data. Their model greatly improved the prediction of condensation heat transfer rates in microfin tubes. They reported that their model agreed within an error band of $\pm 20\%$, or 95% of the data were within $\pm 16\%$ for R – 12 and R – 134a experimental data.

In general, condensation heat transfer research in microfin tubes is in early stage for three reasons. First, although Yang (1994) and Yang and Webb (1996) were successful to model condensation heat transfer in microfin tubes, a more general condensation heat transfer model is still not available since their model was only based on experimental data of two refrigerants, and didn't include the effects of fin efficiency, wall temperature variation and fin geometries. Second, the optimization of fin geometry of microfin tube is not attempted, although several fin geometries are suggested without any comparison and optimization. Third, the optimization of the inlet vapor velocity of a microfin tube may be very important to take advantage of both surface tension force and vapor shear force to thin condensate film resulting in improved thermal performance of a microfin tube.

DISCUSSION

Based on the preceding literature review, it is clear that several forces may significantly affect condensation heat transfer. These forces are: gravity force, viscous shear force between the liquid film and wall, surface tension force, and vapor shear force at the liquid-vapor interface. In a tube or tube bundle, the surface tension force is important only in the region between the points where condensation starts and where all of the extended surfaces are inundated with condensate liquid. This is because the pressure gradient caused by surface tension force exist only in the region where the curvature of the liquid-vapor interface is changing. Furthermore, it is clear that this is effective for all types of condensation surfaces, such as integral-fin tube, longitudinal fin tube, and inside microfin tube.

The vapor shear force is important inside microfin tubes as the vapor velocity is usually high enough to have a strong effect on condensation heat transfer, especially in the sections where condensation has just begun. However, the vapor shear force may not always be important for condensation heat transfer on outside integral-fin tubes. The vapor shear force has a less influence on condensation on outside integral-fin tubes than that of smooth tubes. The reason may be that the additional forces, such as the surface tension force at the liquid-vapor interface and the viscous force at the wall prevent vapor shear force from draining the liquid film on the integral-fin tubes.

For horizontal integral-fin tubes, significant progress has been made in the design of the fin shape, especially the fin profile along the fin tip to the fin bottom, in order to make the surface tension drainage more effective. It has been observed analytically and experimentally that a fin shape with a small fin tip radius can provide lager pressure gradient caused by surface tension force and therefore a thinner condensate film on the fin surface.

Optimized fin spacing of integral-fin tube also has been addressed by many investigators, such as Marto et al. (1986). Addition of fins results in an increase in heat transfer surface and possibly a reduction of effective heat transfer coefficient due to surface tension force for some fin spacings and associated complete flooding between fins. Condensation drainage devices (such as small strips at the bottom of the fin tips) provide an additional surface tension force to pull off the retained liquid in the bottom region of horizontal integral-fin tubes. Although these devices are very helpful in draining the condensate liquid, their manufacturing and installation may be difficult. It is a well known fact that the condensate liquid will retain in the bottom region of the finned tubes; however, almost no information is provided on how to design a special finned tube shape and geometry to minimize this detrimental effect. Such an investigation, when performed, should be very helpful in enhancing condensation heat transfer, especially in the bottom region of a tube bundle where a great deal of retained condensate liquid causes inundation. Although several very promising theoretical models exist for the prediction of condensation heat transfer coefficients for horizontal integral-fin tubes, no simple accurate and generalized theoretical, semi-empirical or empirical models are as yet available.

Vapor shear force may or may not be important for condensation outside integral-fin tubes, although, it is dependent upon the vapor velocity and the fluid properties. Further investigations are needed to discover critical dimensionless numbers that would indicate the existence of regions where the vapor shear force is important.

In condensation on longitudinal fin tube surface, the surface tension force always plays the positive role of pulling the condensate liquid from the fin tip (convex) region into the fin root (concave) region. Subsequently, the gravity force drains the condensate liquid through the fin root gutter, and as a result the condensation heat transfer is enhanced. The fin shape which has a good characteristic of drainage due to surface tension force in the upper portion of the fins in horizontal tube designs should also have a good condensation heat transfer characteristic in the vertical longitudinal fin tube or the inside microfin tube. The vapor shear force also improves the condensation heat transfer rate of longitudinal fin tubes.

Condensation heat transfer inside microfin tube is dependent upon surface tension and vapor shear forces; while only the surface tension may be more important for condensation outside integral-fin tubes. Before entering superheated vapor reaches the saturation temperature, the existing models or correlations for single-phase flows can be employed to predict the heat transfer rate. In this region of single-phase heat transfer, the dominating forces are inertia and viscous forces. A similar heat transfer and fluid flow mechanism can be found in the region where all of extended surfaces are completely flooded with the condensate. Once the vapor temperature reaches saturated temperature and condensation starts taking place, the surface tension force and vapor shear force usually become impor-

Table 2 Flow parameters in compact condensers

Condenser Type	Application	G (kg/m^2s)	Re_l	Re_v	We_l	We_v	Bo	D_h (mm)
Plate–fin	Cryogenic main condensers (nitrogen)	15–50	0–1000	0–12,000	0–0.75	0–25	1.0–10.0	1.5–3.0
Flat tube and corrugated fin	Automotive A/C condensers (R–134a refrigerant)	10–120	0–1000	0–12,000	0–1.0	0–50	0.6–20.0	1.0–3.0
Plate–and –frame	Energy conversion devices /chemical process systems	2–40	0–1200	0–30,000	0–0.60	0–450	3.0–12.0	2.0–8.0
Printed circuit	Chemical process systems	2–20	0–600	0–15,000	0–0.30	0–200	3.0–12.0	1.2–1.5
Shell–and–tube	Power plant (steam–water)	20–500	0–20,000	0–50,000	0–100	0–5000	3.0–85.0	12.7–25.4

tant. The condensate will be pulled by surface tension force from the fin tip regions and into every groove around the tube circumference for a microfin tube. In the case of smooth tubes, due to the absence of such surface tension effects, the condensate may form a stratified layer in the bottom of a horizontal tube at the same vapor velocity as in a microfin tube. Although some experimental data have been reported and several analytical and semi-empirical models have been proposed, the fundamental and applied research on film condensation in microfin tubes is still in the early stage. Significant efforts are needed to understand flooding in the static and dynamic conditions to help the development of more analytical or empirical models. A more complete analytical and empirical model should consider fin efficiency, helix angle effect, lubricant effect, and other microfin geometry effects. Newer, more sophisticated, fin geometries, helix angle and fin number must be proposed in order to achieve ultra high performance of microfin tubes.

In condensation heat transfer, there are several important forces, such as gravity and inertia in the condensate film, viscous force between liquid film and the wall, and surface tension and vapor shear at the liquid-vapor interface. The mechanism of condensation heat transfer becomes very complicated due to the influence of these different forces in the complex passages of heat transfer surfaces. Therefore, it is very helpful to introduce dimensional analysis in order to reduce the number and complexity of experimental variables that affect the condensation phenomenon. Several dimensionless numbers have been proposed and widely used in the literature. They are: liquid and vapor Reynolds numbers, Bond number [a ratio of gravity to surface tension forces, $Bo = (\rho_l - \rho_v)gD_h^2/s$], and Weber number [a ratio of inertia to surface tension forces, $We_l = G^2(1-x)^2D_h/(\sigma\rho_l)$] on which the Nusselt number is dependent. Although Srinivasan and Shah (1997) gave ranges of the Reynolds number, Weber number, and Bond number for several typical industrial compact and no-so-compact condensers (see Table 2), no specific information is available in the literature to identify the critical values of Bond and Weber numbers at which the surface tension starts to become quite important. Of course, one can identify the effect of surface tension as equally important or unimportant compared to the gravity force when the Bond number is of the order of unity. In general, the dimensionless condensation heat transfer coefficient, Nusselt number, should be a function of Re, We, Bo numbers and the effects of vapor shear force, the finned tube geometry and orientation, and fluids employed, that is,

$$Nu = Nu(Re, We, Bo, \text{Vapor Shear, Fin Geometry and Tube Orientation, Fluid Type}) \qquad (22)$$

No such generalized correlation is availabe today to predict condensation heat transfer coefficients even for a narrow range of fin shape, fin geometries, tube orientations, vapor velocity and fluids investigated so far.

AREAS OF FUTURE RESEARCH

Based on the assessment of the literature, the following three areas of future investigation are outlined.

(1) *Optimization of fin shape and fin spacing.* It is a well known fact that the condensate liquid will retain in the bottom region of integral-fin tubes. More information is needed to provide on how to design a special finned tube shape and geometry to minimize this detrimental effect. Such an investigation, when performed, should be very helpful in enhancing condensation heat transfer, especially in the bottom region of a tube bundle where a great deal of retained condensate liquid causes inundation. The optimization of microfin tube geometry is at early stage, more sophisticated fin geometries, helix angle and fin number must be proposed in order to obtain ultra high performance of microfin tubes.

(2) *Generalized models/correlations.* Although several very promising theoretical models exist for the prediction of condensation heat transfer coefficients for horizontal integral-fin tubes, no simple, accurate and generalized theoretical, semi-empirical or empirical models are as yet available. Although Yang (1994) and Yang and Webb (1996) were successful to model condensation heat transfer in microfin tubes, a more general condensation heat transfer model is still not available since their model was only based on experimental data of two refrigerants, and didn't include the effects of fin efficiency, wall temperature variation and fin geometries.

(3) *Nondimensional analysis.* The mechanism of condensation heat transfer is very complicated due to the influence of gravity, inertia, viscous, vapor shear, and surface tension forces in the complex passages of heat transfer surfaces. Therefore, it should be very helpful to introduce dimensional analysis in order to reduce the number and complexity of experimental variables that affect the condensation phenomenon. The resultant important nondimensional groups for the problem are liquid and vapor Reynolds numbers, Bond number and Weber number on which the Nusselt number will be dependent. Table 2 provides the typical ranges of these dimensionless groups for some industrial applications. No specific information is available in the literature to identify the critical values of Bond and Weber numbers for the surface tension to become important.

CONCLUDING REMARKS

A state-of-the-art review is made of condensation heat transfer over integral-fin tubes and longitudinal fin tubes, and inside microfin tubes; available correlations/models are summarized in Table 1 for condensation heat transfer coefficients. Emphasis is placed on the influence of surface tension for condensation heat transfer enhancement and drainage of condensate film to allow more unflooded fin surface available for condensation. Two types of flow induced by surface tension are identified: flow caused by the curvature change of liquid-gas interface and flow caused by local liquid temperature difference (which affects the surface tension coefficient σ). The first type of flow is more important for many condensation industrial applications. The second type of flow is referred as Marangoni flow which may be important when the gravity effect is negligible as in the outer space applications and orbital vehicles such as space station. It is shown that the surface tension effect is more important in the region where condensation has started taking place (i.e. the desuperheating of vapor has been completed locally), and the heat transfer surface is not yet completely flooded with the condensate.

REFERENCES

Adamek, T., 1981, "Bestimmung der Kondensationgrossen auf feingewellten Öberflachen zur Ausle-gun aptimaler Wandprofile," Wärme- und Stöffubertragung, Vol. 15, pp. 255-270.

Adamek, T., 1983, "Filmkondensation an gewellten und an eng berippten Rohroberflachen," Chenie-Ingr-Tech. 55, No. 9.

Adamek, T., and Webb, R. L., 1990a, "Prediction of Film Condensation on Horizontal Integral fin Tubes," Int. J. Heat Mass Transfer, Vol. 33, pp. 1721-1735.

Adamek, T., and Webb, R. L., 1990b, "Prediction of Film Condensation on Vertical Finned Plates and Tubes: A Model for the Drainage Channel," Int. J. Heat Mass Transfer, Vol. 33, pp. 1737-1749.

Akers, W. W., Deans, H. A., and Crosser, O. K., 1959, "Condensation Heat Transfer within Horizontal Tubes, " Chem. Eng. Prog. Symp. Series 29, Vol. 55, pp. 171-176.

Armstrong, R. M., 1945, "Heat Transfer and Pressure Loss in Small Commercial Shell-and-Finned-Tubes Heat Exchangers," Trans. ASME, Vol. 67, pp. 675-681.

Beatty, K. O., and Katz, D. L., 1948, "Condensation of Vapors on Outside of Finned Tubes," Chem. Eng. Prog., Vol. 44, pp. 55-70.

Bella, B., Cavallini, A., Longo, G. A., and Rossetto, L., 1993, "Pure Vapour Condensation of Refrigerants 11 and 113 on a Horizontal Integral Finned Tube at High Vapour Velocity," J. Enhanced Heat Transfer, Vol. 1, pp. 77-86.

Bergles, A. E., 1978, "Enhancement of Heat Transfer," Proc. 6th Int. Heat Transfer Conference, Heat Transfer 1978, Vol. 6, pp. 89-108.

Briggs, A., and Rose J. W., 1994, "Effect of Fin Efficiency on a Model for Condensation Heat Transfer on a Horizontal, Low-Fin Tube", Int. J. Heat Mass Transfer, Vol. 37, pp. 457-463.

Briggs, A., and Rose J. W., 1994, "Effect of Fin Efficiency on a Model for Condensation Heat Transfer on a Horizontal, Low-Fin Tube", Int. J. Heat Mass Transfer, Vol. 37, pp. 457-463.

Briggs, A., and Rose J. W., 1995, "Condensation Performance of Some Commercial Integral Fin Tubes with Steam and CFC 113," Experimental Heat Transfer, No. 8, pp. 131-143.

Briggs, A., and Rose, J., 1996, "Condensation on Low-Finned Tubes: Effect of Non-Uniform Wall Temperature and Interphase Matter Transfer," Process, Enhanced, and Multiphase Heat Transfer, A Festschrift for A. E. Bergles, Edited by R. M. Manglik, and A. D. Kraus, Begell House, Inc., New York, pp. 455-460.

Carnavos, T. C., 1965, "Thin-Film Distillation," Proceedings of First International Symposium in Water Desalination, SWD-17 Washington.

Carpenter, F. G., and Colburn, A. P., 1951, "The Effect of Vapor Velocity on Condensation Inside Tubes," Proceedings of the General Discussion of Heat Transfer, The Institute of Mechanical Engineers and ASME, July 1951, pp. 20-26.

Cavallini, A., Bella, B., Longo, G. A., and Rossetto, L., 1995, "Experimental Heat Transfer Coefficients during Condensation of Halogenated Refrigerants on Enhanced Tubes," J. Enhanced Heat Transfer, Vol. 2, pp.115-125.

Cavallini, A, Frizzerin, S., and Rossetto, L., 1986, "Condensation of R-11 Vapor Flowing Downward Outside a Horizontal Tube Bundle," Proc. 8th Int. Heat Transfer Conference, Heat Transfer 1986, Hemisphere, Washington, DC, Vol. 4, pp. 1707-1712.

Chamra, L. M., Webb, R. L., and Randlett, M. R., 1996, "Advanced Micro-Fin Tubes for Condensation," Int. J. Heat Mass Transfer, Vol. 39, pp. 1839-1846.

Cui, S., Tan, V., and Lu, Y., 1992, "Heat Transfer and Flow Resistance of R-502 Flow Boiling inside Horizontal ISF Tubes," in Multiphase Flow and Heat Transfer, Second International Symposium, X. J. Chen, T. N. Veziroglu, and C. L. Tien, eds., Hemisphere, New York, Vol. 1, pp. 662-670.

Dhir, V., and Lienhard, J., 1971, "Laminar Film Condensation on Plane and Axisymmetric Bodies in Nonuniform Gravity," ASME J. Heat Transfer, Vol. 94, pp. 97-100.

Eckels, S. J., Doerr T. M., and Pate, M. B., 1994, "In-Tube Heat Transfer and Pressure Drop of R-134a and Ester Lubricant Mixtures in a Smooth Tube and a Micro-Fin Tube: Part II - Condensation," ASHRAE Transactions: Research, part 2, pp. 283-294.

Fujie, K., Itoh, N., Innami, T., Kimura, H., Nakayama, N., and Yanugidi, T., 1977, "Heat Transfer Pipe," U. S. Patent 4,044,797, assigned to Hitachi Ltd..

Fujii, T., 1995, "Enhancement to Condensing Heat Transfer - New Developments", J. Enhanced Heat Transfer, Vol. 2, pp. 127-137.

Gogonin, I. I., and Dorokhov, A. R., 1981, "Enhancement of Heat Transfer in Horizontal Shell-and-Tube Condensers," Heat Transfer--Sov. Res., Vol. 3, No. 3, pp. 119-126.

Gregorig, R. 1954, "Film Condensation on Finely Rippled Surface with Consideration of Surface Tension," Z. Angew. Math. Phys., Vol. 5, pp. 36-49.

Hirasawa, S., Hijikata, K., Mori, Y., and Nakayama, W., 1980, "Effect of Surface Tension on Condensate Motion in Laminar Film Condensation (Study of Liquid Film in a Small Trough)," Int. J. Heat Mass Transfer, Vol. 23, pp. 1471-1478.

Honda, H., and Nozu, S., 1987, "A Prediction Method for Heat Transfer During Film Condensation on Horizontal Low Integral-fin Tubes," ASME J. Heat Transfer, Vol. 109, pp. 218-225.

Honda, H., Nozu, S., and Mitsumori, K., 1983, "Augmentation of Condensation on Horizontal Finned Tubes by Attaching a Porous Drainage Plate," Proc. 1st ASME-JSME Thermal Engineering Joint Conference, Y. Mori and W.-J. Yang, eds., Vol. 3, pp. 289-296.

Honda, H., Nozu, S., and Takeda, Y., 1989, "A theoretical Model of Film Condensation in a Bundle of Horizontal Low Finned Tubes," J. Heat Transfer, Vol. 111, pp. 525-532.

Honda, H., Nozu, S., Uchima, B., and Fujii, T., 1986, "Effect of Vapor Velocity on Film Condensation of R-113 on Horizontal Tubes in a Crossflow," Int. J. Heat Mass Transfer, Vol. 29, pp. 429-438.

Karkhu, V. A., and Borovkov, V. P., 1971, " Film Condensation of Vapor at Finely-Finned Horizontal Tubes," Heat Transfer-Sov. Res., Vol. 3, No. 2, pp. 183-191.

Katz, D. L., Hope, R. C., and Datsko, S. C., 1946, "Liquid Retention on Integral-Finned Tubes," Department of Engineering Research, University of Michigan, MI, Project No. M592.

Kaushik, N., and Azer, N. Z., 1989, "An Analytical Heat Transfer Prediction Model for Condensation Inside Longitudinally Finned Tubes," ASHRAE Transaction, part 2, pp. 516-523.

Kedzierski, M. A., and Webb, R. L., 1987, "Experimental Measurements of Condensation on Vertical Plate with Enhanced Fins," Boiling and Condensation in Heat Transfer Equipment, ASME HTD-Vol. 85, pp. 87-95.

Kedzierski, M. A., and Webb, R. L., 1990, "Practical Fin Shapes for Surface-Tensio-Drained Condensation, J. Heat Transfer, Vol. 112, pp. 479-485.

Lustenader, E. L., Richter, R., and Neugebauer, F. J., 1959, "The Use of Thin Films for Increasing Evaporation and Condensation Rates in Process Equipment," ASME J. Heat Transfer, Vol. 18, pp. 297-307.

Marto, P. J., 1988, "An Evaluation of Film Condensation on Horizontal Integral Fin Tubes," ASME J. Heat Transfer, Vol. 110, pp. 1287-1305.

Marto, P. J., Mitrou, E., Wanniarachchi, A. S., and Rose, J. W., 1986, "Film Condensation of Steam on Horizontal Finned Tubes: Effect o Fin Shape," Proc. 8th Int. Heat Transfer Conference, Heat Transfer 1986, Hemisphere, Washington, DC, Vol. 4, pp. 1695-1700.

Mashuda, H., and Rose, J. W., 1987, "Static Configuration of Liquid Films on Horizontal Tubes with Low Radial Fins: Implications for Condensation Heat Transfer," Proc. Roy. Soc. London, A410, pp. 125-139.

Memory, S. B., and Rose, J. W., 1986, " Film Condensation of Ethylene Glycol on a Horizontal Tube at High Vapor Velocity," Proc. 8th Int. Heat Transfer Conference, Heat Transfer 1986, Hemisphere, Washington, DC, Vol. 4, pp. 1607-1612.

Minckowycz, W. J., and Sparrow, E. M., 1966, "Condensation Heat Transfer in the Presence of Noncondensables, Interfacial Resistance, Superheating, Variable Properties, and Diffusion," Int. J. Heat Mass Transfer, Vol. 9, pp. 1125-1144.

Mori, Y., and Hijiikara, K., Hirasawa, S., and Nakayama, W., 1981, "Optimized Performance of Condensers with Outside Condensing Surfaces," ASME J. Heat Transfer, Vol. 103, pp. 96-102.

Nabavian, K., and Bromley, L. A., 1963, "Condensation Coefficient of Water," Chem. Eng. Sci., Vol. 18, pp. 651-660.

Nusselt, W., 1916, "The Surface Condensation of Water Vapor (in German)," Z. Ver. Dt. Ing., Vol. 60, pp. 541-546 and 569-575. (Translated into English by D. Fullarton, 1982, Chem. Eng. Fund., Vol. 1, no. 2, pp. 6-19).

Pierre, B., 1964, "Flow Resistance with Boiling Refrigerants," ASHRAE J., Vol. 6, No. 9, pp. 58-65; Vol. 6, No. 10, pp. 73-77.

Rose, J. W., 1994a, "Condensation on Low-Finned Tubes - An Equation for Vapor-Side Enhancement," Condensation and Condenser Design, St. Augustine, Florida, ASME, pp. 317-333.

Rose, J. W., 1994b, "An Approximate Equation for the Vapor-Side Heat Transfer Coefficient for Condensation on Low-Finned Tubes," Int. J. Heat Mass Transfer, Vol. 37, pp. 865-875.

Royal, J., 1975, "Augmentation of Horizontal In-Tube Condensation of Steam," Ph. D. Dissertation, Department of Mechanical Engineering, Iowa State University, Ames.

Rudy, T. M., and Webb, R. L., 1981, "Condensate Retention on Horizontal Integral-Fin Tubing," Advances in Heat Transfer, ASME HTD-Vol. 18, pp. 35-41.

Said, S., 1982, "Augmentation of Condensation Heat Transfer of R-113 by Internally Finned Tubes and Twisted Tubes Inserts," Ph. D. Dissertation, Dept. of Mechanical Engineering, Kansas State University, Manhattan, Kansas.

Shinohara, Y., Oizumi, K., Itoh, Y., and Hori, M., 1987, "Heat Transfer Tubes with Grooved Inner Surface," U. S. Patent 4,658,892, assigned to Hitachi Cable, Ltd..

Soliman, M., Schuster, J. R., and Berenson, P. T., 1968, "A General Heat Transfer Correlation for Annular Flow Condensation," ASME J. Heat Transfer, Vol. 90, pp. 267-276.

Srinivasan, V., and Shah, R. K., 1997, "Condensation in Compact Heat Exchangers," J. Enhanced Heat Transfer, Vol.4.

Sreepathi, L. K., Bapat, S. L., and Sukhatme, S. P., 1996, "Heat Transfer During Film Condensation of R-123 Vapour on Horizontal Integral-Fin Tubes," J. Enhanced Heat Transfer, Vol. 3, No. 2, pp. 147-164.

Sukhatme, S. P., 1990, "Condensation on Enhanced Surface Horizontal Tubes," Proc. 9th Int. Heat Transfer Conference, Heat Transfer 1990, Vol. 1, pp. 305-327.

Tatsumi, A., Oizumi, K., Hayashi, M., and Ito, M., 1982, "Application of Inner Groove Tubes to Air Conditioners," Hitachi Rev., Vol. 32, No. 1, pp. 55-60.

Tanasawa, I., 1994, "Recent Advances in Condensation Heat Transfer," Proc. 10th Int. Heat Transfer Conference, Heat Transfer 1994, Vol., pp. 291-312.

Torikoshi, K., and Kawabata, K., 1992, "Heat Transfer and Pressure Drops Characteristics of HFC-134a in a Horizontal Heat Transfer Tube," Proceedings of 1992 International Refrigeration Conference - Energy Efficiency and New Refrigerants, Vol. 1, pp. 167-176.

Traviss, D. P., Rohsenow, W. M., and Baron, A. B., 1973, "Forced-Convection Condensation Inside Tubes," ASHRAE Transaction, Vol. 79, Pt. 1, pp. 157-165.

Wanniarachchi, A. S., Marto, P. J., and Rose, J. W., 1986, "Film Condensation of Steam on Horizontal Finned Tubes: Effect of Fin Spacing," ASME J. Heat Transfer, Vol. 108, pp. 960-966.

Webb, R. L., 1984, "Shell-Side Condensation in Refrigerant Condensers," Trans. ASHRAE, Vol. 90, pt. 1, pp. 5-25.

Webb, R. L., 1994, "Advances in Modeling Enhanced Heat Transfer Surfaces," Proc. of the 10th Int. Heat Transfer Conference, Heat Transfer 1994, Vol. , pp. 445-459.

Webb, R. L., Rudy, T. M., and Kedzierski, M. A., 1985, "Prediction of the Condensation Coefficient on Horizontal Integral-Fin Tubes," ASME J. Heat Transfer, Vol. 107, pp. 369-376.

Wen, X. L., Briggs, A., and Rose, J. W., 1994, "Enhancement of Condensation Heat Transfer on Integral-Fin Tubes Using Radiused Fin-Root Fillets", J. Enhanced Heat Transfer, Vol. 1, No. 2, pp. 211-217.

Yang, C. Y., 1994, "A Theoretical and Experimental Study of Condensation in Flat Extruded Micro-Fin Tubes," Ph. D. Thesis, The Pennsylvania Sate University, State College, PA.

Yang, C. Y., and Webb, R. L., 1996, "Condensation of R-12 in Small Hydraulic Diameter Extruded Aluminum Tubes with and without Micro-Fins," Int. J. Heat Mass Transfer, Vol. 39, pp. 791-800.

Yau, K. K., Cooper, J. R., and Rose, J. W., 1986, "Horizontal Plain and Low-Finned Condenser Tubes - Effect of Fin Spacing and Drainage Strips on Heat Transfer and Condensate Retention," ASME J. Heat Transfer, Vol. 108, pp. 946-950.

Zhu, H. R., and Honda, H., 1993, "Optimization of Fin Geometry of a Horizontal Low-Finned Condenser Tube," Heat Transfer - Jap. Res., Vol. 22, No. 4, pp. 372-386.

HTD-Vol. 342, National Heat Transfer Conference
Volume 4
ASME 1997

INTERFACIAL MODELS FOR THE CRITICAL HEAT FLUX SUPERHEAT OF A BINARY MIXTURE

R. Reyes[1] **and P. C. Wayner, Jr.**[2]
The Isermann Department of Chemical Engineering
Rensselaer Polytechnic Institute
Troy, NY 12180-3590
Tel: (518) 276-6199, Fax: (518) 276-4030, wayner@rpi.edu

ABSTRACT

A constant partial vapor pressure model for surface shear in a two-component evaporating thin film is used to successfully predict previously published data on the effect of concentration and pressure on the superheat at the critical heat flux. Due to the dynamics of the high flux interfacial molecular processes, a small portion of the interface can be modeled using a constant partial vapor pressure boundary condition with a varying temperature and concentration in the liquid. This boundary condition demonstrates that, when a distilling fluid with a high initial concentration of the more volatile component flows towards a hotter contact line, the surface shear stress can change sign. Therefore, depending on the details of the distillation process in the microlayer-contact line region, enhanced flow can change to inhibited flow.

In addition, binary data on the effect of system pressure on the critical heat flux superheat of a binary mixture is succesfully predicted using a related adsorption model. We propose that the successes of these two interfacial models give an enhanced understanding of physical phenomena occurring in the microlayer-contact line region of a nucleation site at the critical heat flux.

[1]Current address : Universidad de las Américas-Puebla, Sta. Catarina, Cholula, Puebla,72820, Puebla, México

[2]Corresponding author.

NOMENCLATURE

$\bar{A}$	Hamaker constant
a	binary interaction parameters in Eq. (7)
A,B,C	coefficients in Antoine equation, Eq.(6)
c	constant stress field in the CLR
C	heat capacity
D	intermolecular distance for the calculation of work of adhesion
H	latent heat
P	pressure
q"	heat flux
T	temperature
w	liquid weight fraction
x	liquid mol fraction
y	vapor mol fraction
Δ	difference

Λ	temperature dependent interaction parameters, Eq.(7)
δ	liquid film thickness
η	dimensionless liquid film thickness
γ	activity coefficient
λ	interaction parameters in Eq.(7)
ρ	density
σ	surface tension

Subscript and Superscripts

bp	boiling point
C	Celsius
CHF	critical heat flux transition
CT	critical transition (CHF or Leidenfrost)
l	liquid
lim	limiting composition, Eq.(1)
lv	liquid-vapor
mix	mixture
M	molar property
o	adsorbed layer region
ref.	reference
ss	solid surface
SAT	equilibrium saturation
1	more volatile component, interaction parameters
2	less volatile component, interaction parameters

INTRODUCTION

Based on kinetic theory, the net interfacial mass flux in condensation or evaporation due to an interfacial temperature difference can be extremely large at the surface of a bulk liquid (see e. g., Plesset, 1952, Schrage, 1953). Recently, Carey (1992) discussed many of the limitations associated with this useful description. Using capillarity and Derjaguin's disjoining pressure concept, Potash and Wayner (1972) demonstrated that liquid flow and interfacial mass flux in an extended evaporating meniscus were functions of both the temperature and the stress (film thickness) profiles. They reasoned that, since the film thickness profile is a function of the intermolecular force profile in the region where the vapor-liquid-solid meet (the contact line region), the profiles can be used to describe the varying stress field and resulting transport processes in an extended meniscus which includes the contact line region, the interline and the adjacent intrinsic meniscus. The contact line (or interline) can be defined for an evaporating film as the junction of three regions: the vapor; the evaporating liquid in contact with a portion of the substrate; and a portion of the substrate with or without a non-evaporating adsorbed ultra thin film. This model gives the important boundary condition at the start of the evaporating portion of a thin liquid film and leads to a method to calculate the evaporative heat flow rate and liquid flow rate in a very thin liquid film.

For a pure fluid, the above concepts have been used to describe the intense evaporation process around a nucleation site in boiling (see e.g., Kenning and Torai, 1977 ,Stephan and Hammer, 1994, Unal et al., 1994, Yagov,1995, Lay and Dhir, 1995). A schematic drawing of the vapor stem at a nucleation site showing the relevant macrolayer, the microlayer, the contact line and the dry spot is presented in Fig. (1). In addition, Kelvin-Clapeyron spreading models for isothermal and heated plates have been presented by Wayner (1994) and Reyes and Wayner (1995). These spreading models were also used to predict the superheat at the critical heat flux.

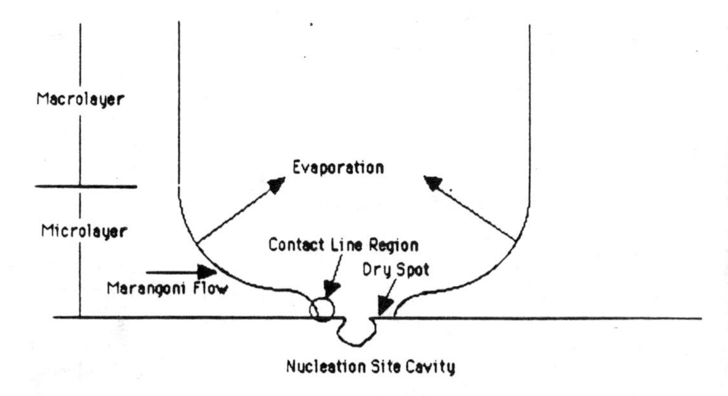

Figure 1 Depiction of a nucleation site showing the macrolayer, microlayer, contact line, and the dry spot.

For a liquid mixture, the net interfacial flux and fluid flow are also functions of the concentration. Specifically, the interfacial free energy and the liquid-vapor interfacial shear stress are functions of both the temperature and concentration. Liquid flow with evaporation in a thin film is then due to both the interfacial shear stress (Marangoni flows), and the varying vapor pressure field due to the thickness and temperature profiles. The resulting Marangoni flow fields in boiling and in evaporating thin films have been studied. The increase in the critical heat flux, q''_{CHF}, of mixtures is directly related to the increase in the superheat, ΔT_{CHF}, due to distillation of the more volatile component at the evaporating interface. In order to correlate the effect of mass transfer and surface tension gradients, various observations and hypotheses have been presented, such as: 1) the observation of a reduced bubble growth rate and departure diameter for positive mixtures in which the more volatile component has the lower surface tension (Van Stralen, 1979); 2) the presence of induced subcooling because of composition differences that makes the liquid bulk temperature lower than the vapor temperature (Reddy and Lienhard, 1989); 3) a description of the total driving force for positive mixtures which is a function of the surface shear and the equilibrium composition of the two phases at their maximum difference (Hovestreijdt, 1963); and 4) a predictive equation that is a function of the composition, the Helmholtz critical velocity, and the surface shear (Mc Gillis and Carey,1996).

In 1985, Wayner et al. optically demonstrated that small changes in the bulk composition significantly altered the characteristics of the transport processes occurring at relatively low fluxes in an evaporating extended meniscus. Using these experimental results and a constant vapor pressure boundary condition, Parks and Wayner (1985 and 1987) developed a model for the surface shear in a steady state two-component evaporating extended meniscus. They reasoned that, due to the dynamics of the high flux interfacial molecular processes, a small portion of the interface could be modeled using a constant partial pressure boundary condition with a varying temperature and concentration in the liquid. They found that surface shear was the most important contribution to flow in the film thickness range 10^{-6} m $< \delta < 10^{-5}$ m for the conditions they studied. They also found that, as a distilling fluid with a high initial concentration of the more volatile component flows towards the hotter contact line, the surface shear stress changes sign. This gave a physical reason for the experimentally observed hump in the film thickness profile.

Herein, we use the Parks and Wayner constant vapor pressure microlayer model to analyze the effect of concentration on the critical boiling heat flux superheat of a binary mixture reported by Van Stralen and Cole (1979), and Avedisian and Purdy (1993). In addition, binary data on the effect of system pressure on the critical heat flux superheat by Cichelli and Bonilla (1945) are analyzed using a related contact line adsorption model which was developed by Reyes and Wayner (1995). These interfacial models successfully predict the experimental observations. Therefore, we propose that they give an enhanced understanding of the physical phenomena occurring at the critical heat flux.

MARANGONI MODEL OF PARKS AND WAYNER (1987)

For a binary mixture of low molecular weight alkanes, the surface tension can be adequately represented by

$$\sigma_{mix} = x_1\sigma_1 + x_2\sigma_2 \tag{1}$$

Using Raoult's law, the chain rule and a constant partial vapor pressure boundary condition for the more volatile component, the following equation for the surface tension gradient was obtained.

$$\frac{d\sigma}{dT} = x_1 \frac{d\sigma_1}{dT} + (1-x_1)\frac{d\sigma_2}{dT} - x_1(\sigma_1 - \sigma_2)\frac{d \ln P_1^{sat}}{dT} \tag{2}$$

Equation (2) is presented in Fig. 2 for the pentane-heptane system which is discussed below.

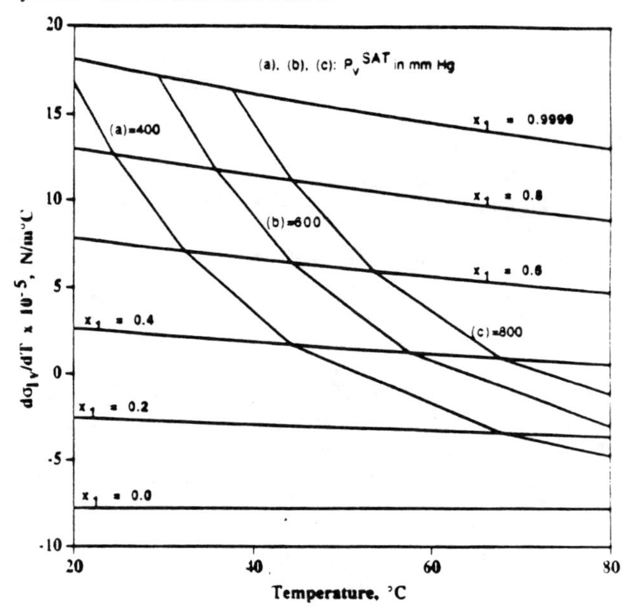

Figure 2 Surface Tension Derivative versus Temperature for Pentane-Propanol Mixtures.

We find that the interfacial shear stress changes sign at a particular value of the mol fraction of the more volatile component, $x_1 = x_{1\,lim}$:

$$x_{1\,lim} = \frac{d\sigma_2/dT}{(\sigma_1 - \sigma_2)(d \ln P_1^{sat}/dT) - d\sigma_1/dT + d\sigma_2/dT} \tag{3}$$

Since the Marangoni shear stress is the major cause of flow in a subregion of the microlayer, we would expect that a change in the sign of the surface shear from enhancing to inhibiting flow towards the contact line would show up as a maximum in a plot of superheat (or heat flux) versus concentration. To evaluate $x_{1\,lim}$, the equilibrium saturation pressure of a mixture is calculated from

$$P_{mix}^{sat} = x_1 P_1^{sat}\gamma_1 + (1 - x_1)P_2^{sat}\gamma_2 \tag{4}$$

which is the simplest realistic equation for VLE (Abbott and Van Ness,1989). For ideal liquid mixtures boiling at pressures close to atmospheric (e.g., mixtures of alkanes) $\gamma_i = 1.0$

Solving simultaneously Eqs.(3 and 4), we calculated the values of the saturation temperatures and $x_{1\,lim}$ at different saturation pressures for several mixtures. These predictions are compared with experimental data on the influence of the initial liquid mol fraction of binary mixtures on q''_{CHF} and ΔT_{CHF} in

Figs. 3 to 6. For this purpose, the surface tensions were evaluated using linear functions of temperature:

$$\sigma_{lv} = \sigma_{lv\ ref.} + \frac{d\sigma_{lv}}{dT}(T_c - T_{ref.}) \quad (5)$$

with $T_{ref.} = 0$ and T_C in degrees Celsius, and σ_{lv} in mN/m (Jasper, 1972). The uncertainty introduced by considering a linear variation of the surface tension with temperature is small in the range of experimental temperatures. The Antoine vapor pressure equation was used in the form:

$$\ln P_v^{sat} = A - B/(T_C + C) \quad (6)$$

with P_v^{SAT} in mm Hg and T_C in degrees Celsius (Reid et al., 1977). The uncertainty introduced by considering a linear variation of the surface tension with temperature is small in the range of experimental temperatures. The data used for these calculations are presented in Table (1).

Although the above simple analysis does not address the exact details of the complicated concentration and temperature fields present in the experiments, we find that the modeling results presented in Figs. 3-6 agree within reason with the experimental measurements. The results show that the Marangoni effect described by Eqs.3-4 predict the composition for the largest values of the heat flux and/or superheat at the CHF. Theoretically, the largest superheat and heat flux should occur at the same composition because the maximum subcooling generates the peak q''_{CHF}. However, this was not consistently observed experimentally. Therefore, we propose that this model appears to enhance our understanding of the evaporation process at the base of a bubble in boiling. We note that all the mixtures analyzed are "positive" : the less volatile component has the higher surface tension.

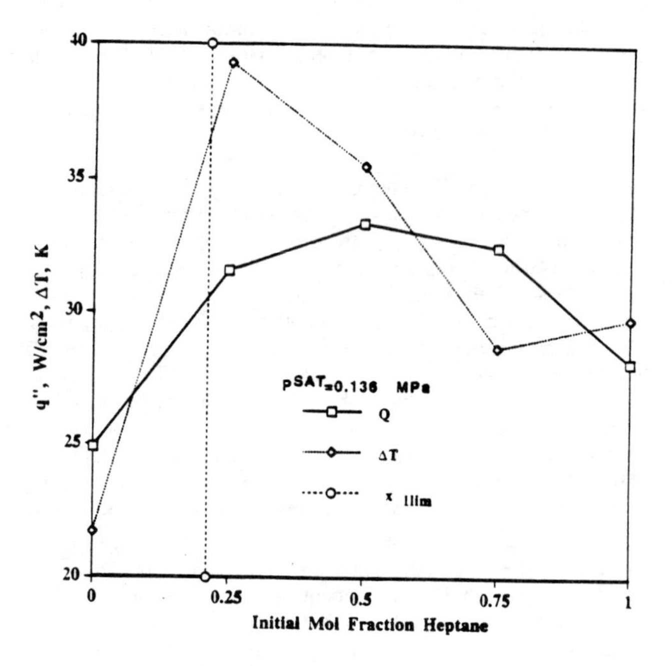

Figure 3 q''_{CHF}, ΔT_{CHF}, **and** $x_{1\,lim}$ **for mixtures of pentane-heptane at** $P_{mix}^{SAT} = 0.136$ **MPa (Avedisian and Purdy, 1993).**

Table 1: Experimental data for the calculation of the limiting composition for enhanced flow in the contact line region of binary mixtures.

	$\sigma_{lv\ ref.}$	$d\sigma_{lv}/dT$	A	B	C
Propane	9.22	-0.0874	15.726	1872.46	248.00
Pentane	18.25	-0.1102	15.833	2477.07	233.21
Heptane	22.10	-0.0980	15.874	2911.32	216.64
Propanol	25.16	-0.0777	17.544	3166.38	193.00
Decane	25.67	-0.0920	16.011	3456.80	194.48
Ethanol	24.05	-0.0832	18.912	3803.98	231.47
Water	75.83	-0.1477	18.304	3816.44	227.02

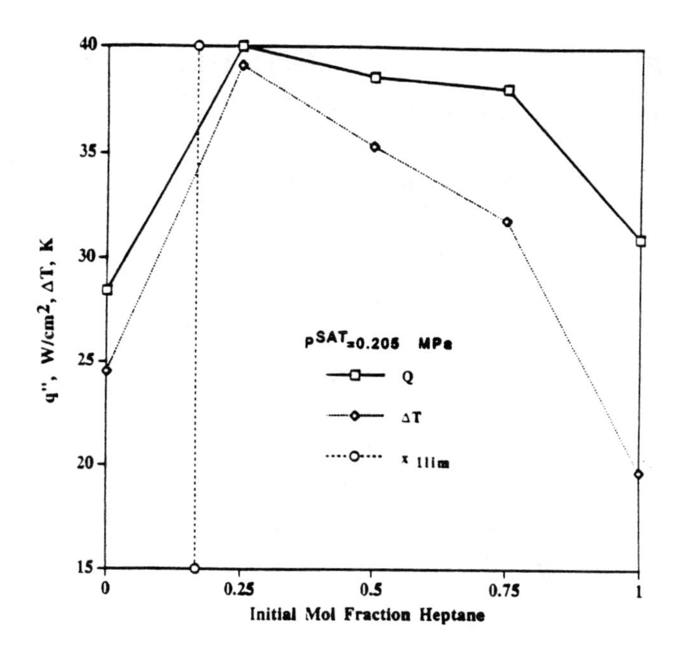

Figure 4 q''_{CHF}, ΔT_{CHF}, and $x_{1\,lim}$ for mixtures of Pentane-Heptane at P^{SAT}_{mix} = 0.205 MPa (Avedisian and Purdy, 1993).

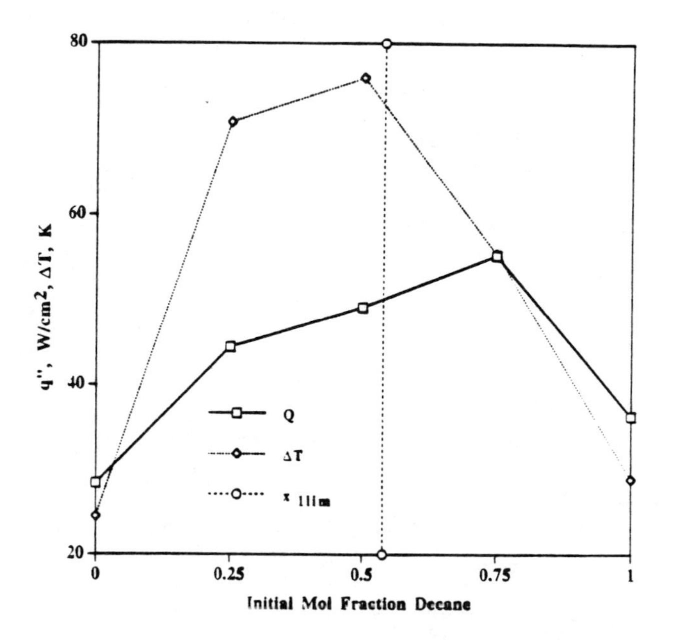

Figure 5 q''_{CHF}, ΔT_{CHF}, and $x_{1\,lim}$ for mixtures of Pentane-Decane at P^{SAT}_{mix} = 0.205 MPa (Avedisian and Purdy, 1993).

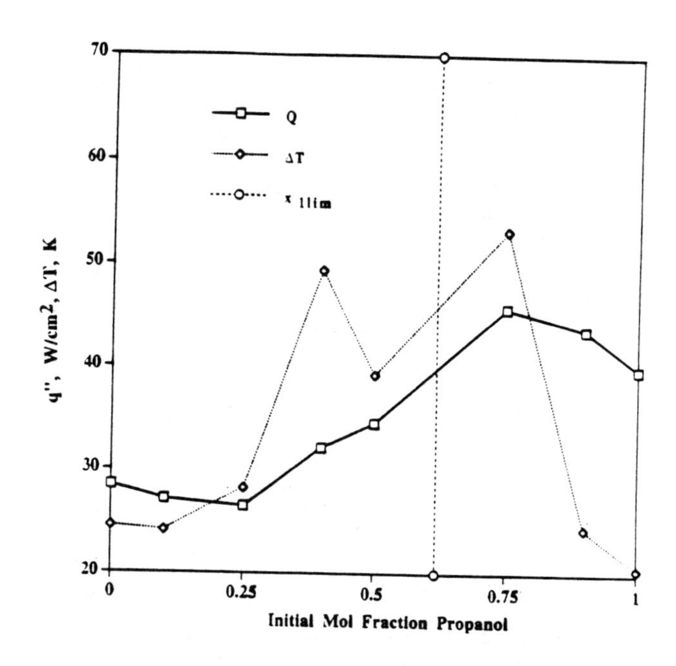

Figure 6 q''_{CHF}, ΔT_{CHF}, and $x_{1\,lim}$ for mixtures of Pentane-Propanol at P^{SAT}_{mix} = 0.205 MPa (Avedisian and Purdy, 1993).

For mixtures of polar liquids the calculation of thermodynamic properties has to account for their deviation from ideal models. For the calculation of x_{1lim} in a mixture of ethanol-water, γ was evaluated using the Wilson equation. The temperature dependent parameters in the Wilson equation allowed the calculation of γ_i at different saturation temperatures:

$$\Lambda_{12} = \frac{V_2^L}{V_1^L} \exp\left(-\frac{a_{12}}{RT}\right)$$

$$\Lambda_{21} = \frac{V_1^L}{V_2^L} \exp\left(-\frac{a_{21}}{RT}\right) \qquad (7)$$

$$\ln\gamma_1 = -\ln(x_1 + \Lambda_{12}x_2) + x_2\left(\frac{\Lambda_{12}}{x_1 + \Lambda_{12}x_2} - \frac{\Lambda_{21}}{\Lambda_{21}x_1 + x_2}\right)$$

$$\ln\gamma_2 = -\ln(x_2 + \Lambda_{21}x_1) + x_1\left(\frac{\Lambda_{12}}{x_1 + \Lambda_{12}x_2} - \frac{\Lambda_{21}}{\Lambda_{21}x_1 + x_2}\right) \quad (8)$$

From the experimental data in Gmehling et al. (1988), the parameters of the Wilson equation for the ethanol-water mixture in the range T_V^{SAT} of the calculations are $a_{12} = \lambda_{12}-\lambda_{11} = 427$, and $a_{21} = \lambda_{21}-\lambda_{22} = 1035$; $\lambda_{ij} = \lambda_{ji}$ are the parameters of binary interaction.

In Fig. 7, w_{1lim} compositions are compared with the composition of ethanol associated with the experimental peak values of q''_{CHF} measured by Van Stralen (1979) at different pressure levels.

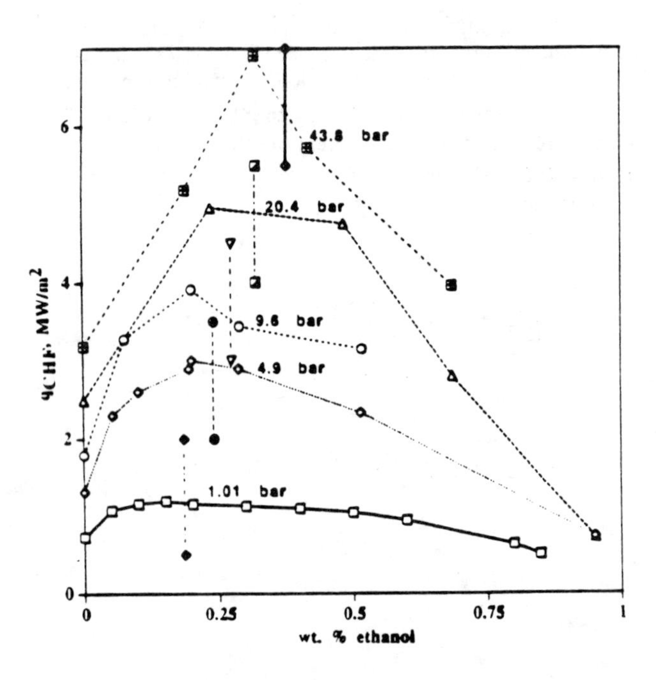

Figure 7 q''_{CHF} **and** w_{1lim} **for mixtures of ethanol-water (Van Stralen, 1979) at** P^{SAT} **in bars : (a) 1.01, (b) 4.90, (c) 9.60, (d) 20.4, and (e) 43.8.**

ADSORPTION MODEL FOR THE EFFECT OF SATURATION PRESSURE ON THE CRITICAL SUPERHEAT

We have also performed additional calculations to predict the effect of pressure on the superheat at the critical heat flux using Eq. (9).

$$\beta_{CT} = \left(\frac{c_{CT}^3}{-\bar{A}}\right)^{0.5} = \frac{\rho_l H_{lv}(T_{ss} - T_v)_{CHF}}{T_v \sigma_{lv}^{1.5}} \quad (9)$$

Reyes and Wayner (1995) successfully used Eq. (9) to accurately predict the effect of pressure on the critical heat flux superheats of pure fluids. The equation models the influence of the solid's surface properties on the superheat at the critical heat flux. Since the effect of the liquid-solid experimental system on adsorption at the contact line needs to be known in Eq.(9) to determine the dimensional constant, β, the experimental intermolecular force for a particular liquid-solid system in pool boiling is characterized using one experimental measurement of superheat. Using a set of measurements at one pressure to determine β, Eq. (9) is then used to calculate the value of ΔT_{CHF} at other pressures. The results are presented in Fig. 8. We note that at the CHF conditions, the contact line and the microlayer occupy an extremely small horizontal region.

In these calculations, simple mixing rules for the mixture properties and corresponding states calculation of superheat effects were used. The pseudo-critical properties are calculated from the initial concentration. A representative value of the surface free energy of the heater was chosen: σ_{ss} = 70 mJ/m2. Because of the proximity of the experimental data of the mixture with 33% mol propane to the critical temperature, it is impossible to use the correction factors for superheat in the evaluation of the thermophysical properties of the mixture. For this case, we used Eq.(9) with the ratios of thermophysical properties values at saturated conditions, at the specified total pressure.

The predictions of ΔT_{CHF} showed a good agreement with the experimental information (Cichelli and Bonilla, 1945) for a mixture of initial composition 67% mol propane, that is very close to x_{1lim}. Predictions of ΔT_{CHF} showed a bigger difference with the experimental data for the mixture of initial propane composition 33%. Seemingly, the composition of the evaporating interface is close to x_{1lim} only when the initial composition of the binary mixture is close to this value. This result indicates the lowest mass transfer resistance in the bulk liquid at this composition. The prediction of the composition of the evaporating interface for other initial compositions of the mixture is not possible under the premises proposed.

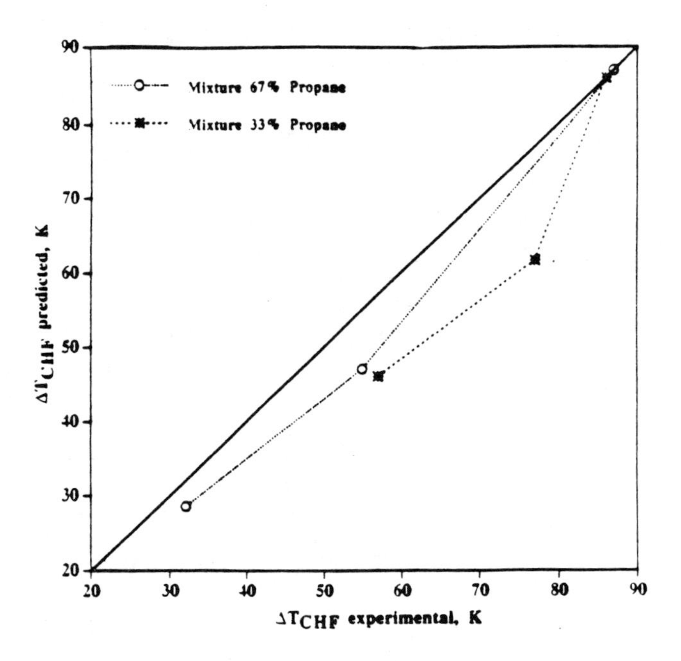

Figure 8 Comparison of predicted superheat at the critical heat flux (Eq.(9) with mixture properties evaluated at the initial concentration) and experimental values for mixtures of Propane-Pentane. The reference values for the calculation of β (ΔT_{CHF} = 87 and 86 K) are included.

CONCLUSIONS

1. A constant partial vapor pressure model for the limiting composition for enhanced Marangoni flow can predict the experimentally measured location of the maximum in a plot of superheat at the critical heat flux (or critical heat flux) versus concentration.

2. An adsorption model for the contact line can predict the effect of pressure on the critical heat flux superheat for a binary mixture.

3. These models appear to enhance our understanding of the evaporation process at the base of a bubble in the boiling of a binary mixture.

ACKNOWLEDGMENTS

This material is based on work partially supported by the National Science Foundation under grant # CTS-9123006 and by Fulbright/CONACYT Grant # 15922189. Any opinions, findings, and conclusions or recommendations expressed in this publication are those of the authors and do not necessarily reflect the view of the NSF or Fulbright/CONACYT.

REFERENCES

Abbott, M.M., and Van Ness, H.C., 1989, "Theory and Problems of Thermodynamics", Schaum's Outline Series, McGraw Hill, USA.

Avedisian, C.T., and Purdy, D.J., 1993, "Experimental Study of Pool Boiling Critical Heat Flux of Binary Fluid Mixtures on an Infinite Horizontal Surface," Proceedings of the 1993 ASME International Electronics Packaging Conference, Vol. 2, pp. 909-915.

Carey, V.P., 1992, Liquid-Vapor Phase Change Phenomena: An Introduction to the Thermophysics of Vaporization and Condensation Processes in Heat Transfer Equipment, Hemisphere Publishing Corp., Washington, DC.

Cichelli, M.T. and Bonilla, C.F., 1945, "Heat Transfer to Liquids Boiling Under Pressure, " Trans. AICHE, Vol.41, pp. 755-787.

Dhir, V.K., and Liaw, S.P., 1989, "Framework for a Unified Model for Nucleate and Transition Pool Boiling," J. Heat Transfer, Vol. 111, pp. 739-746.

Gmehling, J., Onken, U., Rarey-Nies, J.R., 1988, "Vapor-Liquid Equilibrium Data Collection", Dechema, Frankfurt am Main.

Hovestreijdt, J., 1963, ``The Influence of the Surface Tension Difference on the Boiling Mixtures," Chem Eng. Science, Vol. 18, pp. 631-639.

Jasper, J.J., 1972, "The Surface Tension of Pure Liquid Compounds," J. Phys. Chem. Ref. Data., Vol. 1, p. 841.

Katto, Y., 1995, "Critical Heat Flux Mechanisms," in Convective Flow Boiling, J. C. Chen, Editor, Taylor & Francis, Washington, DC, USA, pp 29-44.

Kenning, D. B. R. and Torai, A., 1977, "On the Assessment of Thermocapillary Effects in Nucleate Boiling of Pure Fluids," in Physiochemical Hydrodynamics II, Ed. D. Brian Spalding, pp 653-665, Advance Publications, London.

Lay, J.H., Dhir, V.K., 1995, "Shape of a Vapor Stem During Nucleate Boiling of Saturated Liquids," J. Heat Transfer, Vol. 117, pp. 394-401.

McGillis, W.R., and Carey, V.P., 1996, "On the Role of Marangoni Effects on the Critical Heat Flux for Pool Boiling of Binary Mixtures," J. Heat Transfer, Vol. 118, pp. 103-109.

Parks, C.J., and Wayner, P.C., Jr.,1987, "Surface Shear Near the Contact Line of a Binary Evaporating Curved Thin Film," AICHE Journal, Vol.33 (1) pp.1-10.

Parks, C.J., and Wayner, P.C., Jr., 1985, "Effect of Liquid Composition on Enhanced Flow Due to Surface Shear in the Contact Line Region: Constant Vapor Pressure Boundary Condition," Multiphase Flow and Heat Transfer, HTD-Vol.47, pp. 57-63.

Plesset, M.S., 1952, "Note on the Flow of Vapor Between Liquid Surfaces," J of Chem. Physics, Vol. 20 (5), pp 790-793.

Potash, Jr., and Wayner, P.C., Jr., 1972, "Evaporation from a Two-Dimensional Extended Meniscus," Int. J. Heat Mass Transfer, Vol. 15, pp 1851-1863.

Reddy, R.P., Lienhard, J.H., 1989, "The Peak Boiling Heat Flux in Saturated Ethanol-Water Mixtures," J. Heat Transfer, Vol. 111, pp 480-486.

Reid, R.C., Prausnitz, J.M., and Sherwood, T.K., 1977," The properties of Gases and Liquids," 3rd Ed., McGraw-Hill, New York, NY., Appendix A.

Reyes, R., and Wayner, P.C., Jr., 1995, "An Adsorption Model for the Superheat at the Critical Heat Flux," J. Heat Transfer, Vol. 117,pp. 779-782.

Sadasivan, P., Unal, C., Nelson, R., 1995, "Perspective: Issues in CHF Modeling - The Need for New Experiments," Vol. 117, pp 558-567.

Schrage, R. W., 1953, A Theoretical Study of Interphase Mass Transfer, Columbia University Press, New York.

Stephan, P., and Hammer, J., 1994, "A New Model for Nucleate Boiling Heat Transfer," Warme- und Stoffubertragung, Vol. 30, pp 119-125.

Unal, C., Sadasivan, P., Nelson, R.A., 1994, "Unifying the Controlling Mechanism for the Critical Heat Flux and Quenching: The Ability of the Liquid to Contact The Hot Surface," Proc. Tenth Int'l Heat Transf. Conf., Brighton, UK, pp .

Van Stralen, S., Cole, R., 1979, "Boiling Phenomena: Physicochemical and Engineering Fundamentals and Applications," Hemisphere Pub. Corp.,Washington, DC, Chap. 2.

Wayner, P.C., Jr., 1994, "Thermal and Mechanical Effects in the Spreading of a Liquid Due to a Change in the Apparent Finite Contact Angle," J. Heat Transf., 116, pp. 938-945.

Yagov, V. V., 1995, "The Principal Mechanisms for Boiling Contribution in Flow Boiling Heat Transfer," in Convective Flow Boiling, J. C. Chen, Editor, Taylor & Francis, Washington, DC, USA, pp 175-180.

AUTHOR INDEX

HTD-VOL. 342
1997 National Heat Transfer Conference
Volume 4

Book Number: H01090